Springer Biographies

The books published in the Springer Biographies tell of the life and work of scholars, innovators, and pioneers in all fields of learning and throughout the ages. Prominent scientists and philosophers will feature, but so too will lesser known personalities whose significant contributions deserve greater recognition and whose remarkable life stories will stir and motivate readers. Authored by historians and other academic writers, the volumes describe and analyse the main achievements of their subjects in manner accessible to nonspecialists, interweaving these with salient aspects of the protagonists' personal lives. Autobiographies and memoirs also fall into the scope of the series.

Richard de Grijs · Andrew Jacob

# William Dawes

## Scientist, Governor, Abolitionist: Caught Between Science and Religion

Springer

Richard de Grijs
School of Mathematical
and Physical Sciences
Macquarie University
Sydney, NSW, Australia

Andrew Jacob
Sydney Observatory
Museum of Applied Arts and Sciences
Sydney, NSW, Australia

ISSN 2365-0613 ISSN 2365-0621 (electronic)
Springer Biographies
ISBN 978-3-031-38776-0 ISBN 978-3-031-38774-6 (eBook)
https://doi.org/10.1007/978-3-031-38774-6



Front cover: Arrival of the *Supply*, on which Dawes arrived at Botany Bay on 18 January 1788. Artist: Charles Gore. (Mitchell Library, State Library of New South Wales; Ref. DG VIA/8; out of copyright). Back cover: Rendition of the 'Sydney Cove Map'. Photo: Richard de Grijs. And: A portable observatory tent. Engraving by Benard [after L. J. Goussier?]. Wellcome Collection. Public Domain Mark. Original publication: Wales and Bayly, 1777. Plate II.

This Springer imprint is published by the registered company Springer Nature Switzerland AG
The registered company address is: Gewerbestrasse 11, 6330 Cham, Switzerland

Paper in this product is recyclable.

# Preface

In mid-August 2020, we met in Andrew's office at Sydney Observatory. Sydney had just emerged from its first major COVID-19 lockdown, and in-person meetings were still rare.

During that first lockdown, Richard had redoubled his efforts to explore aspects of the history of the thorny "longitude problem" that had plagued generations of sailors during the "Age of Sail". The key issue at the time, roughly spanning the period from the mid-fifteenth to the mid-nineteenth centuries, was that once ships had ventured beyond coastal waters where landmarks could be used to determine progress, their exact position on the high seas was anyone's guess. Accurate timing devices were required to break the impasse, but in the absence of sufficiently reliable timepieces, at least until the end of the eighteenth century, astronomical observations and tedious calculations were the sailor's most promising tools of geographic position determination.

Having relocated from a previous senior appointment at the Kavli Institute for Astronomy and Astrophysics at Peking University in Beijing, China, to Macquarie University in Sydney in early 2018, Richard was keen to expand his research into aspects of the history of maritime navigation that involved early Australian colonial history. And so he had come across one William Dawes. However, "[t]o my surprise, there doesn't seem to be any book on the man (other than about his linguistic pursuits), and the astronomy information on the internet is not very detailed",[1] he told Andrew. Andrew's response—"Ah, the mysterious William Dawes …"—only added to the intrigue.

Our initial discussions focused on the location of Dawes' original observatory, which he had been tasked with establishing upon arrival in New South Wales. "And I can't even work out where his observatory stood!"[2] Andrew pointed out that the Australian prize-winning author, Catherine Elizabeth (Kate) Grenville, had published a fictional novel, *The Lieutenant* (The Text Publishing Company, 2008), which was

[1] de Grijs, R., 2020. Email to Andrew Jacob, dated 10 August 2020.

[2] Jacob, A., 2020. Email to Richard de Grijs, dated 11 August 2020.

loosely based on Dawes' efforts at trying to understand the local Aboriginal language and his developing friendship with a local Indigenous woman, Patyegarang.

In the book's accompanying "Notes for Readers",[3] Ms Grenville stated that an archaeologist had shown her the location of Dawes' original observatory in Sydney. In her response to Richard's questions (via her agent, Jane Novak), to her credit she quickly explained,

> I see that for the purposes of those readers' notes I took a shortcut, and I apologise. It's a little vague in my mind after more than fifteen years, but my memory is that I had a phone conversation with someone whose name I now can't remember ... and we discussed the possible location of the observatory. He was at pains to make sure that I realised that Dawes's observatory wasn't where the current observatory is, since that's a common misconception. So he would have directed me (but over the phone, I shouldn't have said he showed me) to the general area at the end of Dawes Point, under the Harbour Bridge, as being the location of the Dawes observatory. My memory … is that at that time there was a sign of some kind near the excavated rock wall there, saying that the observatory was thought to be somewhere in that vicinity. All that became compressed into something that I now see is misleading. I apologise for sending you on a false trail. I'll change that remark in the readers' notes and make it clear that the precise location remains unknown.[4]

Today, Ms Grenville's updated "Notes for Readers" indeed reflect that uncertainty better:

> Trying to understand Dawes, I looked for help in the landscape itself, in the place where the story happened. One of the first things I did was to walk from the bay where the first camp was established [Sydney Cove], out to the point where Dawes lived in his little observatory [now known as Dawes Point]. I could see that getting there in 1788 would have been a scramble up a steep rocky hillside, and he'd picked out the one place on the headland that was hidden from the settlement.
>
> Standing there, I felt I could work backwards from the place, to the kind of person who might choose it. I thought he might have been a man who enjoyed his own company, and had no fear of being alone in this foreign landscape.

Ms Grenville's memory of a plaque on the Harbour Bridge commemorating the site of Dawes' Observatory was indeed correct, as we discuss in Chap. 2 of this book. In fact, that plaque became an early focal point for our research into the actual location of Dawes' Observatory, and more broadly into Dawes' background. Those initial efforts led to two joint peer-reviewed, scholarly articles, published back-to-back in the March 2021 issue of the *Journal of Astronomical History and Heritage*. These articles form the basis for Chaps. 1 and 2 of this book.

A key conclusion of our initial investigation was that the location of Dawes' Observatory as indicated on that memorial plaque is, in fact, incorrect. This was based on extensive research in the archives and stacks of the State Library of New South Wales, the State Archives and Records of New South Wales, the National Library of Australia, the City of Sydney Archives, the Library of the Royal Australian Historical Society and numerous other repositories, both physical and virtual.

---

[3] https://kategrenville.com.au/books/the-lieutenant/.

[4] Grenville, K., 2020. Email to Richard de Grijs, dated 13 August 2020.

We are indeed indebted to the library staff at the Special Collections desk of the Mitchell Library, State Library of New South Wales, the archivists at the State Archives and Records of New South Wales and the curator of The Rocks Museum. In relation to Chaps. 1 and 2, we are particularly grateful for support received from (in alphabetical order) Ben Arnsfield, Archivist, Data and Information Management, City of Sydney; Donna Newton, Librarian, Royal Australian Historical Society; William (Bill) Phippen AOM; Matthew Stuckings, Reference Librarian, Pictures and Manuscripts Section, National Library of Australia; and Julie Sweeten, State Library of New South Wales, Special Collections Desk. We also thank Nikki Robertson of the Liverpool Plains Shire Council visitor information centre for providing background information about (and a personal guided tour of) the First Fleet memorial garden in Wallabadah, in the New England area of New South Wales. We further acknowledge discussions with Anthony (Tony) Webster (The University of Queensland) about the shape of the bedrock at Dawes Point.

Meanwhile, three additional articles based on the research conducted for this book were also published in the *Journal of Astronomical History and Heritage*. Edited versions of these articles form Chaps. 7 and 8 and the section on gravity measurements of Chap. 3. For Chap. 8, we are indebted to the library staff at the Special Collections desk of the Mitchell Library (at the time temporarily relocated to the Dixon Library) of the State Library of New South Wales. We also acknowledge the City of Sydney Archives for granting us physical access to one particular report. We additionally thank Conybeare Morrison International Pty. Ltd. for allowing us to reproduce some of their copyright materials, deposited at the State Library. Our analysis and discussion of Dawes' gravity measurements benefited, in particular, from the helpful insights provided by Case Bosloper, which we gratefully acknowledge. We took due account of all comments and suggestions made by the anonymous reviewers of all five peer-reviewed publications, and we believe that the resulting chapters have become more insightful as a result. The *Journal*'s editor, Wayne Orchiston, has been a constant source of encouragement throughout our own journey of discovery and exploration.

The remaining chapters in this book, *viz.* Chapters 4 through 6, paint a picture of Dawes as a brilliant but troubled yet high-performing individual. We will leave his legacy up to the reader's judgement. In the context of our research for Chap. 4, we acknowledge assistance received from Shelby Schellenger, Reference Coordinator, Kenneth Spencer Research Library, University of Kansas Libraries (USA), to source a hard-to-find document held at that institution.

It has been less than 2½ years since we started out on our journey of discovery into William Dawes' legacy. The journey has been fascinating, and we have been sharing our research with general audiences throughout this period. Our initial conclusions drew some press coverage,[5] and the story of Dawes and his struggles in the fledgling colony of New South Wales has clearly struck a chord with many. We hope that the comprehensive account of Dawes' life and legacy presented here will continue to be an inspiration moving forward.

## William Dawes' Family Tree

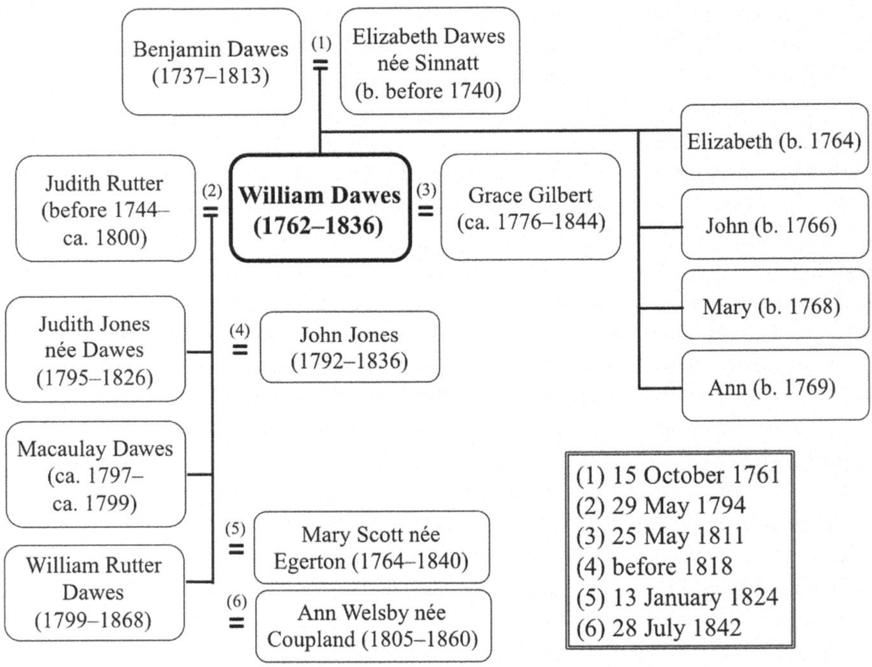


Sydney, Australia
January 2023

Richard de Grijs
Andrew Jacob

[5] Press release: https://lighthouse.mq.edu.au/article/march-2021/solved-the-site-of-australias-first-astronomical-observatory; press coverage included: https://www.miragenews.com/solved-site-of-australias-first-astronomical-521135/; https://www.nationaltribune.com.au/solved-site-of-australia-s-first-astronomical-observatory/.

# Contents

# Chapter 1
# Practical Astronomy on the 'First Fleet' from England to Australia

## 1.1 A Voyage to the Antipodes

In a move reeking of desperation, at daybreak on 13 May 1787 a convoy of 11 ships left from Spithead roads off Portsmouth, England, on an 8-month, 24,000 km undertaking billed as a "… voyage to Botany Bay …" in New South Wales.[1] Now known as the 'First Fleet',[2] it was a desperate attempt to rid the overcrowded English jails and prison hulks of the 'criminal classes'.

---

This chapter is based on de Grijs and Jacob (2021).

[1] At the time of the First Fleet's voyage, the British penal colony of New South Wales encompassed the area of present-day Australia east of the 135° East meridian, a North–South line bisecting present-day Australia about 20 km east of the South Australian town of Coober Pedy and ~115 km east of Alice Springs in the Northern Territory. Governor-designate Arthur Phillip's instructions of 25 April 1787 to proclaim the colony for Great Britain also included "… all the islands adjacent in the Pacific Ocean …" (Coltheart & The Museum of Australian Democracy, 2011; King, 1998) between latitude 10°37′ South—Cape York in far north Queensland—and 43°39′ South, the latitude of South East Cape, the southernmost point of the Tasmanian mainland. Incidentally, this area includes most of New Zealand.

[2] The term 'First Fleet' is a retrospective designation; at the time the fleet was made ready for departure, there were no plans for any subsequent fleets to be sent to the new colony (e.g., Atkinson, 1997: 59–61; Hughes, 1987).

 
R. de Grijs and A. Jacob, *William Dawes*, Springer Biographies,
https://doi.org/10.1007/978-3-031-38774-6_1

On 24 October 1786, H.M.S. *Sirius*, a sixth-rate[3] former merchantman equipped with 10 guns—four 6-pounder guns and six 18-pounder carronades—and crewed by a complement of 160 men, had been commissioned as the convoy's flagship. Under the command of Captain Arthur Phillip (1738–1814), Governor-designate of the new British colony, the *Sirius*, was accompanied by a faster vessel, H.M.A.T. *Supply*—an armed tender equipped with eight guns (four 3-pounders and four 12-pounder carronades) and a complement of 55 men. In addition, the convoy included the convict transports *Alexander*, *Charlotte*, *Friendship*, *Lady Penrhyn*, *Prince of Wales* and *Scarborough*, and the food and supply transports[4] *Borrowdale*, *Fishb[o]urn* and *Golden Grove* (for an in-depth account, see Pembroke, 2013).

Figures 1.1, 1.2 and 1.3 paint a vivid picture of the conditions in which the convicts lived in England. Figure 1.1 is a representative example of the numerous prison hulks the English Government operated in the late eighteenth century. Figure 1.2 shows a chain gang being moved to a new location in ca. 1782 (left) and a longboat carrying convicts to their transport (right), bound for Botany Bay in the early 1800s. Figure 1.3 shows the interior of the last Australian prison ship, the hulk *Success* (ca. 1926).

The total number of people carried by the First Fleet—convicts, Marines, seamen, civil officers and free people, male and female—will likely remain unknown forever. In fact, it appears from contemporary sources that the convoy's accounting for all of its crew was not done very carefully:

> On the 15th [May 1787] the signal was made for the transports to pass in succession within hail under the stern of the *Sirius*, when, on inquiry, it appeared, that the provost-marshal [the head of the military police] of the settlement (who was to have taken his passage on board the *Prince of Wales*) was left behind, together with the third mate of the *Charlotte* transport, and five men from the *Fishbourn* store-ship: the loss of these [latter] five persons was supplied by as many seamen from on board the [escort ship] *Hyæna* (Collins, 1798: Section I).

Only for the *Sirius* and *Supply*, neither of which transported convicts, do we know the names of all crew members. Gillen (1989) identified 1420 people who embarked on the First Fleet, with 1373 arriving in the colony after a voyage of 250–252 days, depending on their vessel. The voyage took 48 lives—45 of convicts or their

[3] In the British Royal Navy's rating system at the time of the First Fleet's departure, a sixth-rate warship carried 20–28 carriage-mounted '9-pounder' guns on a single deck. Sixth-rate ships carrying fewer than 28 guns were categorised as 'post ships', so that they could still sail under a full captain rather than a commander. The *Sirius* was an exception, in the sense that she only carried ten guns yet was still classified as a sixth-rate vessel.

[4] All ships, not only the food and supply store-ships, were stocked to the brim with provisions, agricultural and camp equipment, clothing for the convicts, baggage and numerous other items (Collins, 1798).

**Fig. 1.1** Representative example of an English prison hulk at Deptford (1826). Artist: Samuel Prout; engraved by George Cooke. (London, Longman & Co.; National Library of Australia; PIC Drawer 3841 #U3905 NK6402; out of copyright)

**Fig. 1.2** (*left*) "A fleet of convicts under convoy", ca. 1782. "Printed for & Sold by Carington Bowles, at his Map & Print Warehouse, No. 69 in St. Pauls Church Yard, London. Published as the Act directs." (Dixon Library, State Library of New South Wales; a128083/DL Pd 789; out of copyright); (*right*) Convicts embarking for Botany Bay, early 1800s. Artist: Thomas Rowlandson. (National Library of Australia; PIC Drawer 3842 #T2670 NK228; out of copyright)

**Fig. 1.3** Interior of the prison hulk *Success* (middle deck, looking forward), 1926. (State Library of Victoria; Ref. 3007866; out of copyright)

children—and 28 children were born in transit[5] (White, 1790). John White (1756–1832), the First Fleet's principal surgeon, painted a favourable picture of conditions on board:

> The newspapers were daily filled with alarming accounts of the fatality that prevailed among us; and the rumour became general, notwithstanding every step was taken to remove these fears, by assurances (which were strictly true) that the whole fleet was in as good a state of health, and as few in it would be found to be ill, at that cold season of the year, as even in the most healthy situation on shore. The clearest testimony that there was more malignity in the report than in the disease, may be deduced from the very inconsiderable number that have died since we left England; which I may safely venture to say is much less than ever was known in so long a voyage (the numbers being proportionate), even though not labouring under the disadvantages we were subject to, and the crowded state we were in (White, 1790: Section I).

However, the historical record remains inconclusive as to the total number of people on board. Other accounts report as few as 1044 arrivals, whereas Governor Phillip's first census of 1788, reported to the Home Secretary, Thomas Townshend

[5] A normal human pregnancy, from ovulation to natural childbirth, lasts 268 ± 9 days (standard deviation; Jukic et al., 2013: Table 1), which implies that all children born in transit were most likely conceived prior to the First Fleet's departure. Statistics of births in the colony are sparse for the first few years. In a letter of 12 February 1790 to Lord Sydney, Phillip (1790b: 144) states, "As near two years have now passed since we first landed in this country … [f]ifty-nine children have been born in the above time." Given the timeframe, this statement does not tell us whether any children were conceived during the voyage. Anecdotal evidence suggests that a number of births occurred in the first few months after the convoy's arrival, however (e.g., *Australian Town and Country Journal*, 1883; *Daily* Telegraph, 1888; National Library of Australia, n.d.; Nepean *Times*, 1883).

(1733–1800), First Viscount Sydney, returned a total white population of 1030 (Macquarie, 1988).[6] Even the number of convicts on board the ships is subject to confusion and debate. Gillen (1989) identified 775 convicts (582 men and 193 women) as having embarked, with 732 eventually landing at Sydney Cove. However, Judge Advocate and chronicler of the colony David Collins (1756–1810) counted 756 convicts across the ships—564 male and 192 female—as well as 13 of their children (Collins, 1798).

In addition to the convicts and crew, the First Fleet transported a total complement of 16 officers, 24 non-commissioned officers and 160 privates from the British Marine Corps (Laurie, 1988). The 11 ships also carried surgeon-general John White and his servant, on the *Charlotte*, as well as six assistant surgeons—two on the *Lady Penrhyn* and one each on the *Friendship*, *Scarborough*, *Sirius* and *Supply*. Two surgeon's mates were stationed on the *Sirius*, while the surveyor-general, August(us) Theodore Harman Alt (1731–1815), was on board the *Prince of Wales*. A chaplain, the Rev. Richard Johnson (ca. 1753–1827), his wife, and a servant and his wife sailed on the *Golden Grove*, whereas the commissary of stores, Andrew Miller (ca. 1759–1790), and his servant were on the *Sirius*. The *Sirius* also hosted three additional servants (to Captains Phillip and John Hunter, second captain of the *Sirius*; 1737–1821). Moreover, a peace officer had embarked on the *Lady Penrhyn* and a civilian settler on the *Alexander*. Finally, 28 soldiers' wives and their 14 children were also on board the convoy[7] (Collins, 1798).

## 1.2 An Astronomer on the First Fleet

The First Fleet's voyage carried it from England past Madeira, Tenerife (Canary Islands) and the Cape Verde Islands in the North Atlantic Ocean. Then, following the prevailing trade winds and ocean currents, they sailed to Rio de Janeiro and Table Bay at the Cape of Good Hope. Eventually, the voyage took the fleet through the southern Indian Ocean to Van Diemen's Land (present-day Tasmania) and the present-day New South Wales coast (see Fig. 1.4). Navigating long stretches of open ocean, low-lying islands and reefs in coastal areas, extended periods of calm in the tropics—the dreaded 'doldrums'—where the ships would be subject to a variety of ocean currents, and gale-force winds across the Indian Ocean's 'roaring forties' meant that accurate position determination was of the utmost importance at any time throughout the voyage.

In the late eighteenth century, the dominant sea-faring European nations continued to open up commercial shipping routes to ever more distant destinations, yet a reliable

[6] The small town of Wallabadah in the New England area of present-day New South Wales is home to the First Fleet memorial garden. The local visitors' centre maintains a list of those known to have sailed on the First Fleet, which as of 2021 contained 1186 names. However, the list does not fully tally with the names engraved on stone tablets scattered around the garden (Liverpool Plains Shire Council, 2020; N. Robertson, personal communication).

[7] See also https://history.cass.anu.edu.au/centres/ncb/first-fleet-ships-and-passengers.

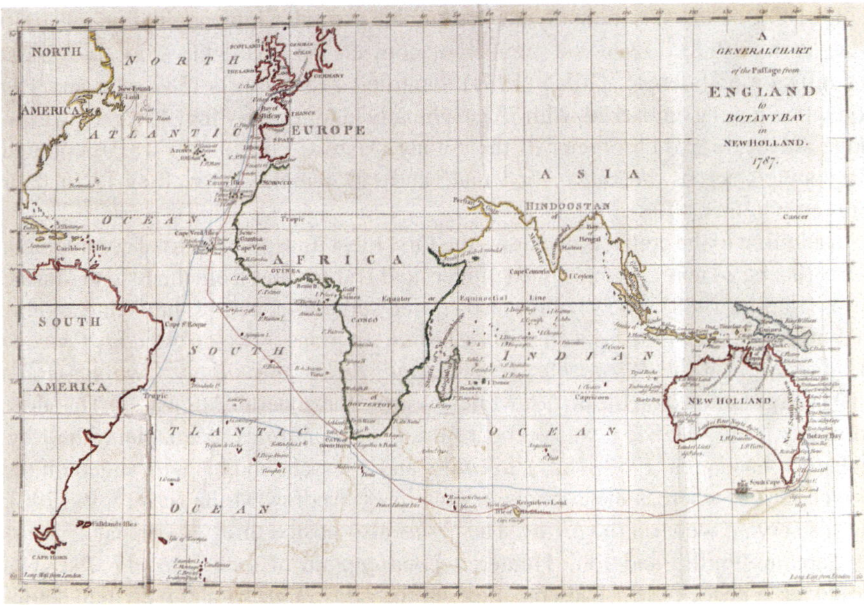


**Fig. 1.4** "A general chart of the passage from England to Botany Bay in New Holland." Artist: John Andrews (London, John Stockdale). The blue line represents the route taken by the First Fleet. (Mitchell Library, State Library of New South Wales; Ref. M2 118/1787/1; out of copyright)

means of geographic position determination at sea remained elusive. Determination of one's latitude was relatively straightforward: one simply needed to measure the height of the Sun or those of one or more bright stars at their meridian passage, corrected for seasonal variations, and a latitude measurement would follow naturally.

Determination of one's longitude at sea was significantly more complicated (for a recent review, see de Grijs, 2017; see also Andrewes, 1996). Longitude determination relies on knowing one's local time with respect to that at a reference location such as the Greenwich meridian. Solving the longitude problem occupied generations of scientist-scholars and navigators. It took until the second half of the eighteenth century before John Harrison (1693–1776) had perfected his maritime timepiece to work sufficiently accurately that he was awarded a significant fraction of the British Longitude Prize of 1714. Harrison's payment was delayed, however, until the British watchmaker Larcum Kendall (1719–1790) had managed to successfully copy Harrison's marine watch H4, in 1769. Kendall's copy is now referred to as his No. 1 Marine Timekeeper, 'K1'. James Cook (1728–1779) took K1 on his second and third voyages of discovery to the Pacific (in 1772–1775 and 1776–1779), calling it "… our trusty friend the Watch …" and "… our never failing guide the Watch …" (cited by Andrewes, 1996: 226, 252). Indeed, on Cook's second voyage, in 1775, after more than three years at sea, it was determined that the timepiece had gained, on average, just 13 s a day (e.g., Hawkins, 1979; Howse & Hutchinson, 1969).

Yet, although chronometers of sufficient operational accuracy to determine one's longitude at sea had become available by the time of Cook's second and third voyages, one had to remember to wind them up periodically and to re-calibrate them occasionally given that their rates would decay over time. In addition, chronometers were few and far between, and they were expensive to obtain. Alternative means of longitude determination had been developed, most involving angular distance measurements between the lunar limb and the positions of bright stars in the sky, an approach known as the 'lunar distance' method. In turn, these had to be reconciled with tabulated values for those same celestial distances as observed from Greenwich (for a recent review, see de Grijs, 2020). This same method was used to recalibrate the few operational chronometers that were available to contemporary navigators.

Starting in 1767, Dr. Nevil Maskelyne (1732–1811), Britain's fifth Astronomer Royal, had been instrumental in publishing his *Nautical Almanac*s. They included lengthy tables of lunar distances in time intervals of three hours, for Greenwich time. Shipboard lunar distance measurements, usually based on observations with a sextant, would allow accurate determination of the time in Greenwich. This could then be compared with the observer's local time to result in a longitude measurement. The calculations required to carry this process through to completion were highly complicated and tedious, however. For these reasons, it was common for voyages of discovery to include competent astronomers among their crews (e.g., Macleod & Rehbock, 1988).

These itinerant astronomers were often equipped with portable 'tent' observatories (see Fig. 1.5 and Chap. 7). In the seventeenth and eighteenth centuries, increasingly accurate astronomical observations from across the globe were urgently needed to improve shipping practices. In fact, Maskelyne complained in the 1788 *Nautical Almanac* that …

> It is indeed to be lamented that Persons, who visit distant countries, are not more diligent to multiply Observations of this kind; for want of which, the Observations made by Astronomers in established Observatories lose half their Use and the improvement of Geography is retarded (Maskelyne, 1787).

Maskelyne thus convinced Lord Sydney and the British Board of Longitude to support an astronomical component for the First Fleet, including the provision of a range of instruments and books on loan for the colony's new observatory (see Sect. 1.5). These would ostensibly be used for the determination of the settlement's precise position, to accurately determine the colony's local time, to provide an accurate time reference for the chronometers of visiting ships and to assist with surveying duties.

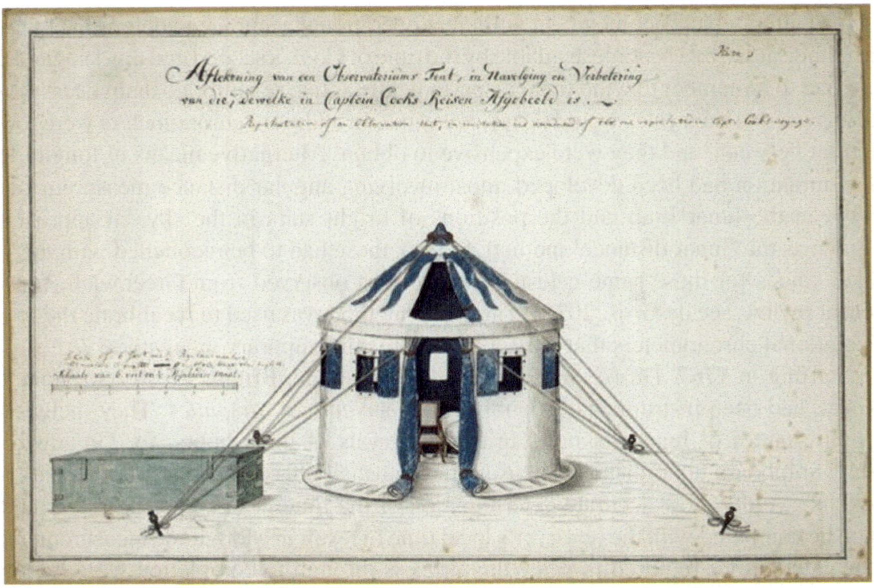

**Fig. 1.5** Typical example of the type of tent observatory Dawes would have pitched on Enchados island in the harbour of Rio de Janeiro (Artist unknown: *Aftekening van een observatoriums tent in navolging en verbetering van die, dewelke in Captain Cook's reisen afgebeeld is.* Plate 1, ca. 1780? Ref.: B-091-007. Alexander Turnbull Library, Wellington, New Zealand. Ref. /records/22320145; reproduced with permission)

However, from Maskelyne's personal letters we also get the distinct impression that he was personally invested in a positive outcome of the astronomical enterprise. Maskelyne had convinced the Board of Longitude to send Second Lieutenant William[8] Dawes (1762–1836; see Fig. 1.6) to New South Wales so as to test the Astronomer Royal's prediction of the return of the comet of 1532 and 1661 to the southern sky. The comet's earlier recorded appearances had led Edmond Halley (1656–1742), Britain's second Astronomer Royal, to predict that it would reappear in 1790. However, in a paper read to the Royal Society of London on 29 June 1786, Maskelyne suggested that the comet might reappear as early as 1788 or 1789, with a perihelion—its closest approach to the Sun—potentially as early as 1 January 1789:

> It will approach us from the southern parts of its orbit, and therefore will first appear with considerable south latitude and south declination; so that persons residing nearer to the equator than we do, or in south latitude, will have an opportunity of discovering it before us. It is to be wished that it first be seen by some astronomer in such a situation, and furnished with proper instruments for settling its place in the heavens, the earliest good observations being most valuable for determining its elliptic orbit … The Cape of Good Hope would be an excellent situation for this purpose (Maskelyne, 1786a).

---

[8] References to 'Nicolas' as William Dawes' middle name are found only on *Wikipedia*/*WikiTree*—https://en.wikipedia.org/wiki/William_Dawes_(British_Marines_officer), https://www.wikitree.com/wiki/Dawes-1222—but not anywhere else.

**Fig. 1.6** Second Lieutenant William Dawes, 1830s. Miniature oil painting, artist unknown. (Tasmanian Museum and Art Gallery, Hobart; ref. AG6048; out of copyright)

If Maskelyne's prediction were correct, this would be the second comet to be predicted correctly. The return of Halley's comet had been confirmed in 1759 following the astronomer's published prediction of 1705 (Halley, 1705), and so Maskelyne was particularly keen to make his mark alongside Halley. Therefore, he recommended that Dawes be appointed as the official astronomer on board the *Sirius*, with an expanded brief to 'recover' Maskelyne's comet and establish an observatory in the new colony.

In the remainder of this chapter, we will explore Dawes' background as well as his achievements on board the *Sirius* and, subsequently, the *Supply*. A high-level academic account of the scientific and political context in which the First Fleet's voyage was prepared and subsequently undertaken can be found in Saunders (1990: 24–74). Here, however, we focus in significantly greater detail on Dawes' contributions, while attempting to be as precise as possible in our descriptions and offering quantitative assessments wherever feasible.

## 1.3 Young William Dawes

Perhaps surprisingly given the leading intellectual role he was to be cast into eventually, little first-hand information about William Dawes survives:

> There is no man among the founders who ought to have given us so much information about himself and his views as Lieutenant Dawes, and there is no man among them who has given so little. He was the scholar of the expedition, man of letters and man of science, explorer, mapmaker, student of language, of anthropology, of astronomy, of botany, of surveying, and of engineering, teacher and philanthropist. The duty to posterity of such a man, in such singular circumstances, was that he should be always writing, and in fact he wrote nothing at all that can now be read. What we know about him we learn wholly from the writings of others; and the scantiness of our information is made the more exasperating by the fact that everything they write about him—apart from two differences of opinion with Governor Phillip, in respect to which a sensible historian will decline to arbitrate—is in tones of high praise both of character and ability (Wood, 1924: 1).

Currer-Jones (1930), Dawes' great-granddaughter, noted that the Dawes "… Family papers, many consisting of Dawes' letters, etc., … had been destroyed after the death of one of his grandsons …" and that

> I have heard from Antigua [where Dawes resided from 1813 until his death in 1836] that many of Dawes' papers were destroyed by the terrible hurricane[9] of [21 August] 1871. The utmost was done to decipher the remains of these, but it was found impossible.

Nevertheless, we have access to a sizable body of letters by his own hand. An important subset of this surviving correspondence is found in his letters to Maskelyne (this chapter and Chaps. 2–4). These are preserved among the papers of the Board of Longitude, which are held at the Cambridge University Library (UK). They are freely available online,[10] at high resolution. Moreover, in 1972 a set of 'language notebooks' from Dawes' time in Sydney Cove were (re-)discovered (Nathan et al., 2009). The latter contain 80 pages of interesting insights and will be discussed in Chap. 3. In addition, the Colonial Office papers and the Church Missionary Society's archive contain a large number of letters by and to Dawes (Chaps. 5 and 6), the latter covering the period 1820–1829, during his most active employment in Antigua. Most of the latter letters relate to routine management issues in his role as Superintendant of Sunday schools in the West Indies and as supporter of the Antigua Auxiliary Bible Society, the English Harbour Sunday School Society, the Distressed Females' Friend Society, the Female Refuge Society and the English Harbour Female Juvenile Association. The same archive contains cursory minutes (some barely or not legible; Thomas, 2011) covering Dawes' earlier life and achievements, from his first involvement with the Church Missionary Society in 1799.

William Dawes was the first-born son of Benjamin Dawes (1737–1813) and his wife Elizabeth (b. before 1740) of Portsmouth, England. William had four siblings, including his sisters Elizabeth (1764), Mary (1768) and Ann (1769), and his younger brother John (1766). The earliest available record of young William Dawes in the Church of England parish register of St. Thomas à Becket is that of his baptism on 17 March 1762 (Hampshire parish registers, 1653–1875). This suggests that he was in all likelihood born in early 1762, although his exact date of birth has not been recorded.[11] By the late eighteenth century, the time between birth and baptism had slipped from "… the Sunday, or other Holy day next after the child be borne, unless upon a great and reasonable cause declared to the Curate …" (Anglican Church Prayer Books, 1549 and 1552; cited by Basten, 2015), as commonly practised in the sixteenth and seventeenth centuries, to longer intervals. For instance, one study indicates that in the period 1771–1789, 75% of newborns were baptised within 38 days, while that time interval had increased to 64 days by 1791–1812 (Basten, 2015; see also Berry & Schofield, 1971).

---

[9] This was the Atlantic season's fourth hurricane, known as the 1871 Santa Juana hurricane (https://en.wikipedia.org/wiki/1871_Atlantic_hurricane_season).

[10] https://cudl.lib.cam.ac.uk/view/MS-RGO-00014-00048/509.

[11] English parish registers usually only recorded baptisms rather than births. In England, civil registration started in 1837.

Elizabeth Dawes née Sinnatt had been a widow until she remarried, to Benjamin Dawes, on 15 October 1761 in the same parish church (Blagg & Andrews, 1913: 128). Given that the couple married just five months prior to William's birth suggests a degree of social pressure. His father was employed as clerk-of-works (construction supervisor in charge of quality control) at the Ordnance Office of the Portsmouth Naval dockyard (Mander-Jones, 1966).

William would go far in life, despite his fairly humble beginnings. His intellectual pursuits and interests, in evidence during the early years of the new settlement in New South Wales (1788–1791), clearly showed that he had enjoyed an excellent and thorough education from an early age. If anything, as an adult Dawes developed into a true polymath, proficient in a number of European languages, including French and Latin.

Dawes' grasp of Latin grammar is demonstrated in his notebooks on the Sydney Aboriginal language which he would later compile at Sydney Cove (Clarke, 2015: 45–49; see Sect. 3.3). His Latin proficiency is evidenced by his almost casual use of verb paradigms and expressions such as 'imperative mood' and 'the ablative case' (Steele, 2005). Moreover, Dawes communicated in Latin with the Portuguese astronomers based at the observatory in Rio de Janeiro during the First Fleet's month-long respite there (Dawes, 1787m; see Sect. 1.6). In addition, shortly after the First Fleet's arrival in New South Wales, Dawes made the acquaintance of the French astronomer Joseph Lepaute Dagelet (1751–1788; see Chap. 2). The latter was a scientific member of the Lapérouse expedition, which was anchored in Botany Bay between 26 January and 10 March 1788. On 3 March, Dagelet sent Dawes a letter offering him advice on how to set up his observatory (Dagelet, 1788). The letter was written in French. At a later date, Dawes showed further proficiency in French in his notebooks, for instance by referencing "To make or do (faire in French)" (Dawes, 1791: *Notebook C*, 29).

Although it is unknown where Dawes received his early professional education, historians tend to agree that he most likely attended the Royal Naval Academy in Portsmouth (e.g., Steele, 2005; Crittenden, 2010; Clarke, 2015: 23). His skills in maritime navigation exhibited at a later stage of his career attest to a solid naval education. Nevertheless, Clarke (2015: 23) points out that he continued to perfect those skills as late as 1786 when, prior to the First Fleet's departure, he spent a few weeks under the tutelage of Maskelyne at Greenwich Observatory (Maskelyne, 1786b). In addition, William Bayly (1737–1810; Orchiston, 2016: 152–153), head-master of the Royal Naval Academy at Portsmouth and formerly Captain Cook's astronomer (Gibson, 2012), recommended the young Marine to Joseph Banks (1743–1820; Beaglehole, 1963; Gascoigne, 1994), then-President of the Royal Society of London, "… citing [Dawes'] knowledge of languages, botany, mineralogy and also astronomy …" (Howse, 2004). As regards the latter, he was said to be "… a tolerably good Astronomer & draws very well" (Bayly, 1786). Indeed, Dawes seems to have been an accomplished artist, although just four watercolour drawings of local plants attributed to him are known to have survived. One, '*Purple donkey orchid*', is described as

**Fig. 1.7** Royal Naval Academy, Portsmouth (1806). Artist: Hall, after I. T. Lee. (Wellcome Library no. 22276i; Creative Commons Attribution 4.0 International license)

> A Species of Orchi[d]s found growing in vast quantities on the steep downs between Pitt Water and the Sea by Lieu. William Dawes of the Marines $20^{th}$ September 1790 (Anemaat, 2014).

Crittenden (2010) verified that Dawes does not appear to have attended either the Naval College at Greenwich or that at Christ's Hospital, the other educational institutions where naval officers were 'taught the ropes'. The Portsmouth Naval Academy (see Fig. 1.7) had been established by Order-in-Council (legislation passed by King George II's Privy Council; reigned 1727–1760) in February 1729, to educate "… forty young [13–16 year-old] gentlemen, sons of noblemen and gentlemen …" for service in the British Navy. After all, as eventually proclaimed by King William IV (reigned 1830–1837), "… there was no place superior to the quarterdeck of a British man of war for the education of a gentleman" (Dickinson, 2007: 32).

However, that scenario had not developed as initially expected. The offspring of the city's better-off citizens disliked the institution, since it reminded them too much of school, with all its rules, responsibilities and restrictions. Upon graduation, the 'Academites' or 'College Volunteers' were equally disliked by the Navy captains required to accept them for sea service, since they lacked the rigorous attitudes of genuine naval officers trained on board as Captain's servants (e.g., Crittenden, 2010; Lloyd, 1966).

Moreover, in 1801 Admiral of the Fleet John Jervis (1735–1823), First Earl of St. Vincent, exclaimed that the Portsmouth Naval Academy was "… a sink of vice and abomination, which ought to be abolished" (Lloyd, 1952: 472). Given the Academy's bad reputation—"Are you so partial to that seminary as to hazard a son there?" (Jervis, 1801)—by 1773 only half of the approximately 40 places were filled by sons of the city's gentry and aristocrats. Therefore, the Admiralty extended entry to 15 sons of non-commissioned officers, aged 11–17, and waived their fees (Lloyd, 1966: 145).

Under those conditions, and probably aided by his father's position at the Naval dockyard, young William may have been admitted to the Royal Naval Academy in Portsmouth. Subjects taught at the time included French, drawing, fencing, use of the firelock (a type of musket) and Latin. If indeed the young Dawes was educated there, this would also account for his competence in mathematics, navigation and astronomy. Immersion in that latter subject would serve him well as the official astronomer on the First Fleet. It is highly likely that he received instruction in practical astronomy from William Bayly, who had made a name for himself as astronomer on Captain Cook's second voyage of discovery. Bayly became the Royal Naval Academy's headmaster by February 1785, two years before the First Fleet's departure. Therefore, he may have been involved in Dawes' advanced training in preparation for his duties during the voyage and in the new colony.

However, despite Dawes' likely education at the Royal Naval Academy, a career in the Navy from an entry-level position as midshipman[12] or Captain's servant was not open to him, given his common (non-aristocratic) background. Instead, he enrolled as a junior officer in the Marines. Crittenden (2010) suggests that, at one point, he may have considered a career in the Church, which would have been in line with his strong religious leanings and high morals (e.g., Clarke, 2015; Crittenden, 2010; James, 2012; Chaps. 5 and 6). For instance, following the First Fleet's arrival in New South Wales, Lieutenant Daniel Southwell (ca. 1764–1797), Master's mate on the *Sirius*, wrote to his mother,

> [Dawes] is a most amiable man, and though young, truly religious, without any appearance of formal sanctity. He is kind to everyone; but I am speaking of his many affabilities to myself, which are such that more could not be looked for from a relation. He has a great share of general knowledge, studious, yet ever cheerful, and the goodness of his disposition renders him esteemed and respected by all who know him (Southwell, 1790).

However, as in the Navy, clergy in the Church were also vying for positions and promotions. Thus, qualifications came second to the old boys' network (e.g., Crittenden, 2010), where the local nobility dominated. Crittenden (2010) muses that Dawes may have heard John Wesley (1703–1791; founder of the Anglican Church's Methodist movement) preach at one of his prayer meetings in the open air, while attempting to bring religion back to the masses. This would have appealed to young William. Instead, however, he opted for the advantage of the home court—after all, he grew up on the Navy dockyard—and joined the Marines.

Upon graduation from the Royal Naval Academy, on 2 September 1779, aged 17, Dawes was 'gazetted' (commissioned) as a second Lieutenant in the British Marine Corps.[13] He joined the 32$^{nd}$ Company of the Portsmouth Division (Steele, 2005; Wright, 1927). Great Britain was at war. The American War of Independence (1775–1783), also known as the American Revolutionary War, had been triggered by the British Government's plans to tax the colonies. This led to a general revolt, united

[12] Graduates from the Royal Naval Academy were often appointed as midshipman-by-order or 'midshipman ordinary' to make a clear distinction from the midshipmen who had been trained on board. The latter were paid higher wages (Lewis, 1939: 217).

[13] In the late eighteenth century, the British Marine Corps were soldiers associated with the Royal Navy, not with military forces. Wright (1927) reported that, on 27 September 1926, the Adjutant-General of the Royal Marines at the Admiralty in London responded to a request from the Royal Australian Historical Society about Dawes' career, confirming that "... [the] only particulars shown in official records are as follows: Granted Commission as 2nd Lieutenant, Royal Marines, 2nd September, 1779; ..." This is, in fact, an odd response from the official body, given that the Marines did not become *Royal* until 1802.

in their opposition—"... no taxation without representation ..." in Parliament. What had started as an uprising, with expedient attempts of suppression by the British Army and Navy, turned into a full-scale war by 1779. The French sailed in support of the American colonies, with Spain and the Dutch Republic following suit.

William Dawes was posted to H.M.S. *Resolution* on 5 April 1780 (Ship's muster, H.M.S. *Resolution*, The National Archives, Admiralty 36/8709), propelling him into the rough and tumble world of the warring nations. The *Resolution* was a third-rate[14] 'ship of the line' equipped with 74 guns. After Dawes had embarked, the company sailed promptly to the Americas. Almost two decades later, when being deposed in the House of Lords, Dawes had some trouble recalling the events of his youth:

> I think we arrived on the Coast of America in the Month of September 1780 ... It might have been August; I speak quite from Recollection. ... We lay chiefly in the New York [Hudson] River. ... I think the Fleet might have arrived off Sandy Hook [present-day New Jersey] in the Month of August, but she did not go into Port, if I recollect right, at that Time, but we were joined by Admiral [Mariot] Arbuthnot [1711–1794], and proceeded towards Rhode Island. I remained on the Coast of America till about the Middle of November 1780, when we sailed in Company of other Ships to the West Indies: We returned to America in the following Year, and sailed to the West Indies again near the Conclusion of 1781 (House of Lords Sessional Papers, 1714–1805: 171–172; HLSP).

Dawes saw action perhaps sooner than he had expected. The *Resolution* was part of the British fleet commanded by Rear Admiral Sir Thomas Graves (1725–1802). They engaged the French fleet under Rear Admiral François Joseph Paul (1722–1788), Comte de Grasse, in battle off Chesapeake Bay (Maryland) in early September 1781. Dawes was wounded during the main battle of 2 September, although probably not very seriously given that he continued to serve on the *Resolution* (Nicolas, 1845: 116–117).

Next, despite the *Resolution*'s new deployment as part of Admiral George Brydges Rodney's (1718–1792) British fleet, she saw renewed hostile engagement with de Grasse's fleet in the Battle of the Saintes (9–12 April 1782), off the coast of Dominica (West Indies). Whereas the Battle of Chesapeake had been a strategic victory for the French, the Battle of the Saintes—also known as the Battle of Dominica—was a clear British victory. It offered a significant morale boost to British troops at home and overseas.

His time in the West Indies also introduced Dawes to the realities of the slave trade, an experience that laid the foundations of his lifelong drive in support of the abolition of slavery. "I was on Two Plantations in Barbadoes [*sic*], besides walking over others in other Islands" (HLSP: 173). Asked about his feelings regarding the 'situation' in which he encountered the slaves, he responded, "I thought their Situation very unpleasant and hard. ... I mean that they must have felt themselves, from what I could discover, in a Situation of enduring much Hardship" (HLSP: 173). To get insights into how Dawes really felt about the slave trade at the time of his first encounter with the practice in Barbados, it is instructive to consider the following exchange from

[14] In the British Royal Navy's rating system, from the 1720s a third-rate warship carried 64 to 80 guns, typically on two gun decks.

18 June 1799 in the House of Lords between the House's legal counsel and Dawes, acting as witness in support of *An Act to prohibit the Trading for Slaves on the Coast of Africa, within certain Limits* (HLSP: 174):

> *Did you ever state this Sentiment of Hardship to your Messmates in the Ward Room of the* Resolution*?*
>
> I recollect no particular Instance of it; but I have often spoken my Sentiments respecting the Mode of Life which they lead; and, I think I might have done it occasionally to my Messmates.
>
> *Was the Slave Trade ever the Subject of Conversation in the Ward Room of the* Resolution*?*
>
> I think it may have been, and has been noticed by myself and others belonging to the Ship, how shocking a Situation it must be for the Slaves arriving in Ships, as they actually came into the Port, into Carlisle Bay [Barbados] from the Coast of Africa.
>
> *Do you, from your general Knowledge of the Navy and as a Sea Officer, conceive that the Officers of the Navy are prejudiced in favour of or against the Slave Trade?*
>
> Such as I have been more particularly in the Habit of conversing with have, to the best of my Recollection, expressed themselves averse to the Slave Trade. I could instance one very respectable Character. It may perhaps be reflected on me, that he is dead; but his Respectability, I presume, will not be called in question in this House; it was Lord Robert Manners, Commander of the *Resolution* at that Time.
>
> *Did Lord Robert Manners frequently express his Sentiments to that Effect or not?*
>
> I do not recollect that he has done so frequently; but on the Occasion of a Slave Ship arriving in Carlisle Bay it has been a Matter of general Concern to Officers who have been on Deck at the Time.
>
> *In the general Course of your Habits in the Ward Room of the* Resolution*, did the Generality of the Officers of the Navy, as near as you can now recollect, speak favourably or unfavourably of the Slave Trade?*
>
> Unfavourably.

Upon arrival back at their home port on 23 October 1782, Dawes left the *Resolution* (Ship's muster, H.M.S. *Resolution*, The National Archives, Admiralty 36/8713). He next enlisted in the 11$^{th}$ Company at Sheerness. This service period returned him to American waters—"I went One Voyage to Newfoundland …" (HLSP: 173)—while he was also engaged in patrolling the southern North Sea on the sloop H.M.S. *Merlin* (Ship's muster, H.M.S. *Merlin*, The National Archives, Admiralty 36/10463; Gillen, 1989: 101) between 1783 and 1784. On the *Merlin*, he was put in charge of a small Marine contingent, which continued until he left service on 15 May 1784. He was next assigned to the Marine barracks in Portsmouth. There, he pursued improvements of his surveying and engineering skills while also establishing himself as a competent astronomer.

## 1.4 Official 'First Fleet' Astronomer

Most accounts of Dawes' involvement in the First Fleet's voyage simply refer to him having volunteered for service as a Marine officer, without further consideration. This cavalier approach to referencing probably dates back as far as William Bradley's (1758–1833) journal entry of 5 December 1786 (Bradley, 1786–1792: 3; see Sect. 1.5) and fellow Marine Watkin Tench's (1758–1833) appeal to Earl Bathurst, Secretary of State for the Colonies of 1827. At the time, Tench requested that Dawes be awarded additional financial compensation[15] for his work in the new settlement: "… Your Lordship's Memorialist [Dawes] was among the Officers who volunteered their services to go to Botany Bay, in the Year 1786, and the only one from the Portsmouth Division" (Wright, 1926).

However, the sequence of events that eventually led to Dawes' appointment as Second Lieutenant of the Marines on the *Sirius* was complex; multiple obstacles had to be overcome. Dawes' good fortune seems to have hinged in large part on an early intervention by Bayly, already in his role as Master of the Portsmouth Royal Naval Academy. On 8 August 1786, Bayly wrote to Banks to recommend Dawes' "… great desire to go …" with the First Fleet, extolling the Marine's numerous useful technical and linguistic skills (Clarke, 2015: 26). He assured Banks that all of his lavish praise of Dawes was "… strictly true …".

Bayly soon developed into Dawes' main early patron. Since he had served as assistant to Maskelyne (Howse, n.d.), Bayly had a direct line to the nation's top scientist. He exploited his direct access to lobby the Astronomer Royal on Dawes' behalf.[16] In turn, Maskelyne wrote to Banks on 17 October 1786 that Dawes was "… well versed in most kinds of astronomical observations" (Mander-Jones, 1966; Maskelyne, 1786a), thus supporting the Board of Longitude's recommendation of Dawes' candidacy as astronomer on the voyage. He followed up on 8 November, further emphasising Dawes' suitability for the appointment:

> The bearer Mr. Dawes is the person whom I mentioned to you by a letter as desirous to make any observations that might be thought useful at Botany Bay, & whom you approved of in the answer you favored with. He has been a good deal here [at Greenwich Observatory] perfecting himself in the use of astronomical instruments & receiving instructions for discovering and observing the comet of 1532 & 1661 at its first approach to the sun & earth in the summer or autumn of 1788. He is very ready and will do the observing very well. He is desirous of paying his respect, to you, & I desire you will be pleased to allow this letter to be his introduction to you. He will bring you from me a copy of my paper on the return of this comet, which I desire your acceptance of. I saw the comet and got some good observations of it, on the 26$^{th}$ of last month. I have some faint hopes of being able to get a last observation

[15] Maskelyne had convinced Dawes that taking on the role of the voyage's official astronomer might lead to further career opportunities, and therefore Dawes did not seek financial compensation for the appointment (Clarke, 2015).

[16] Laurie (1988) suggests that Dawes may have approached Maskelyne directly, given his father's standing in the Portsmouth Ordnance Office and the elder Dawes' acquaintance with the Royal Naval Academy staff and with the family of Lord Samuel Hood (1724–1816), previously Commissioner at the dockyard and then-Commander-in-Chief at Portsmouth. This seems less likely than Bayly's approach, however.

> of it tomorrow or some other night soon if very fine before the moon rises. Having parted with [my assistant] Mr. [Joseph] Linley, I at present make all the observations myself, & therefore cannot attend the first meeting of the Royal Society (Maskelyne, 1786b).

Dawes received additional support from another local Portsmouth contact, Captain William Twiss (1745–1827), Royal Engineer in the British Army. Twiss was in charge of construction at the Naval dockyard and port. On 24 October 1786, he sent a letter to Sir Brook Watson (1735–1807), then a Member of Parliament and Deputy Governor of the Bank of England, recommending Dawes' services for the voyage to Botany Bay (Gillen, 1989). Watson forwarded Twiss's letter to Sir Evan Nepean (1752–1822), Under-Secretary of State for the Home Office with responsibility for naval and political intelligence (Twiss, 1786).

Although Dawes was clearly an intelligent young man, Maskelyne's choice of him as his emissary nevertheless seems odd, given his working-class background. As the nation's pre-eminent astronomer, Maskelyne was not short of well-qualified, university-educated volunteers, so Dawes must indeed have come with the highest praise. Dawes' successful appointment by the Home Secretary and the Board of Longitude is even more surprising given the enormous importance afforded to establishing the new colony. Other luminaries, including Banks and Nepean, also vied for patronage. Indeed, Maskelyne seems to have outwitted his nemesis Banks in securing Dawes' loyalty (e.g., Gibson, 2012).

The timeline now becomes confusing. We learn from a letter of 25 October 1786 from Sir Phillip Stephens (1723–1809), Secretary to the Admiralty, to Lieutenant-General William Sidney Smith (1764–1840), Commandant of the Marines, Portsmouth Division, that Dawes was granted leave to sort out his "… private affairs …" (Stephens, 1786). This included further instruction in and development of his technical skills as a practical astronomer at the Royal Greenwich Observatory (Maskelyne, 1786b; Clarke, 2015: 28). This timing of Dawes' leave for purposes of reviewing his technical astronomy skills appears odd, given that Maskelyne's proposal to the Board of Longitude to appoint Dawes as the First Fleet's astronomer was not discussed until 14 November that year. The Board of Longitude's minutes of 14 November 1786 record that Maskelyne, *ex officio* Board member,

> ... represented to the Board that Mr Dawes Lieutenant of Marines on board His Majesty's Ship the *Sirius* which is going with the Convicts to Botany Bay on the Coast of New Holland, is desirous of making useful Nautical & Astronomical observations in his passage thither, … (Board of Longitude, 1786)

From our current perspective, it may seem odd that sending an astronomer to the yet-to-be-established colony—even if he was also a Marine, a competent engineer, surveyor and explorer—was considered a high priority. Shortages of labour, materials and food would hamper the colony's early development and keep Governor Phillip occupied, yet Dawes was meanwhile commissioned to establish an astronomical observatory:

> … and during his stay there, if he could be allowed to use some Instruments belonging to the Commissioners of the Longitude, to enable him to do so; And the Astronomer Royal having at the same time laid before the Board a List of Instruments and Books proper to lend Mr. Dawes

> to enable him to make the said Observations, and at the same time [Maskelyne] informed the Board that Mr. Dawes was capable of making proper use of them. The Board took the same into Consideration and came to a Resolution to lend the Instruments and Books undermentioned for the purposes aforesaid, but being of the Opinion that if they were put into the Charge of Captain Philip [*sic*], the Commander of the *Sirius*, greater care would be taken of them, than could be done by Mr. Dawes, they directed that the said Instruments and Books should be accordingly delivered into the Charge of Captain Philip, and that he should be requested to give his Attention to the care and preservation of them (Board of Longitude, 1786).

## 1.5 Instrumentation for the New Observatory in New South Wales

Maskelyne thus managed to persuade the Board of Longitude to also provide Dawes with a set of astronomical and meteorological books and equipment for "... useful Nautical and Astronomical Observations ..." (Board of Longitude, 1786), which were selected by the Astronomer Royal to closely resemble the instrumentation carried on Cook's already legendary voyages of the late 1760s and 1770s (Howse, 1979; Howse & Hutchinson, 1969). Astronomy—whether a personal pursuit or of national strategic importance—thus became a foundational component of the new colony in New South Wales. Nevertheless, despite concerted efforts, Dawes did not find the comet. Astronomers in Europe did not find it either, thus suggesting that Maskelyne's predictions may have been wrong.

Dawes' initial request, submitted in early November 1786 and contained in the *List of Instruments proper for making astronomical Observations at Botany Bay* (Fig. 1.8, left; Dawes, 1786a), included a large number of instruments. However, the sponsoring organisations—the Board of Longitude, the Royal Observatory at Greenwich, the Royal Society of London and the British Admiralty—were unable to source all instruments requested prior to the First Fleet's departure, so that a significant number were crossed out. Eventually, the expedition was furnished with a set of instruments and books, in the care of Captain Phillip, contained in the much shorter *List of instruments to be lent by the Board of Longitude for making astronomical Observations at Botany Bay* (Fig. 1.8, right):

- A [12 in.; 30.5 cm] Astronomical Quadrant [by John Bird, for longitude determination on land]
- A 3½ Feet [~1 m] treble Object Glass Achromatic Telescope by Dollond with a micrometer with oblique wires, to be further provided with sliding tubes to steady the Telescope, and a common wire Micrometer for measuring Differences of Right Ascension and Declination [presumably a compact, portable 'comet sweeper', owned by Maskelyne]
- One Night Glass
- An Astronomical Clock [by John Shelton]

- A Journeyman Clock [probably also by Shelton]
- An Alarm Clock [maker unknown]
- A Pocket Watch, with a second hand & Ruby Pallets [possibly by Ellicott, No. 4659]
- An old [Hadley's] sextant by Ramsden
- A portable Barometer [by Burton]
- Two Thermometers
- A circular Protractor with a Glass having lines on it intersecting each other in the center [*sic*].

**Books**

- [William] *Gardiner's Logarithms* [*Tables of Logarithms*, 1742]
- Copy of [Henry] *Sherwin's Logarithms* [*Sherwin's Mathematical Tables … Table of Logarithms of Numbers from 1 to 101*].
- Copy of [Maskelyne's] *Requisite Tables*
- Copy [*sic*] of the *Nautical Almanacs* from 1787 to 1792.

An entry in William Bradley's journal of 5 December 1786 provides details about the suite of instrumentation that would be available to Dawes on the voyage (see also Laurie, 1988; Saunders, 1990: 41–44, 49–51):

> The Board of Longitude furnished the following instruments for the use of the voyage and the new colony, An astronomical Quadrant of one foot radius, A 3½ feet treble object glass achromatic telescope by Dollond, with a wire Micrometer for measuring diff.[ce] of right ascension & declination, also a Micrometer with oblique wires and a Quadrant fixed to it with a moveable short Telescope to take the distance of any object (nearly) from bright fixed stars. A night glass. An Astronomical clock. A Journeyman clock. An Alarum clock. An old sextant by Ramsden. A portable Barometer & two thermometers. These instruments Captain Phillip gave a receipt for [see Figure 1.9], promising to return them to the Board (the dangers of the sea and other unavoidable accidents excepted) at his return on a receipt from such Officer as may supercede [*sic*] him in the command. Lieut. Dawes of the Marines a volunteer for the Botany Bay Detachment having been introduced to Dr Maskelyne the Astronomer Royale [*sic*], he was acknowledged a proper person to make such observations on shore as might be judged of use (Bradley, 1786–1792: 3).

Among the instruments taken to the new colony (for a detailed discussion, see Laurie, 1988: 470–471), the clocks merit further discussion. The Board of Longitude had provided the expedition with both an astronomical pendulum clock and a journeyman (or assistant) clock. The latter was equipped with a gong that sounded every minute, while the seconds could be counted based on its ticking sound (James, 2012). The astronomical clock is thought to have been made by the specialist English regulator maker John Shelton (1712–1777). It was purchased in 1768 for £40 sterling and served as main reference of either sidereal or local time on Cook's second and

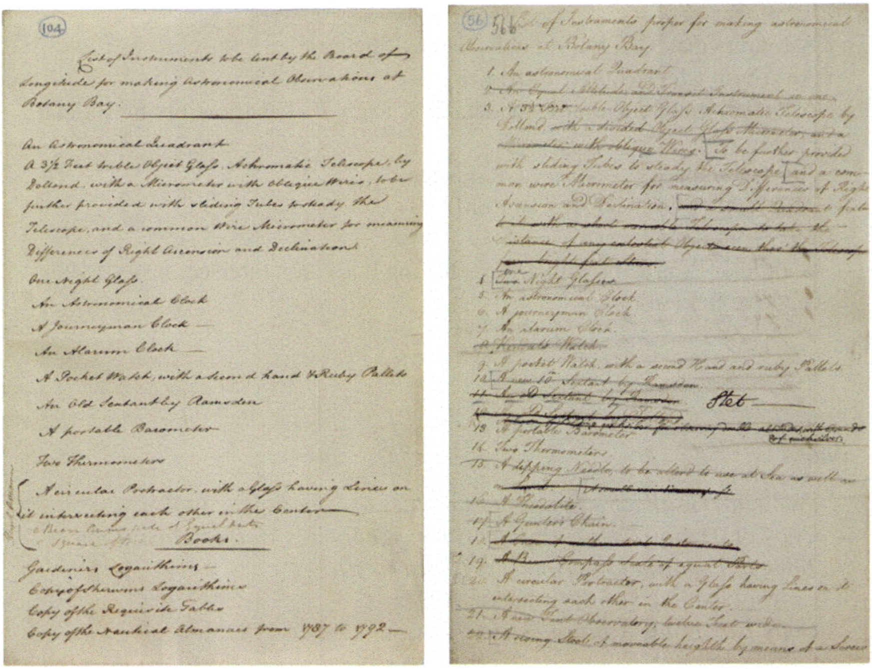

**Fig. 1.8** (*left*) Dawes' *list of instruments proper for making astronomical observations at Botany Bay*. (*right*) *List of Instruments to be lent by the Board of Longitude for making Astronomical Observations at Botany Bay*. (National Museum of Australia; out of copyright)

third voyages to the Pacific (see also Orchiston, 2016). This clock, most likely Royal Society Regulator No. 35 (Hawkins, 1979), showed hours, minutes and seconds to high accuracy.

Dawes only owned two instruments himself (Clarke, 2015: 27). Therefore, Maskelyne's intervention with Banks on the Marine's behalf (Maskelyne, 1786a) secured the success of the astronomical mission. In addition to the instruments lent by the Board of Longitude, Maskelyne was also instructed to supply Phillip with Kendall's 'K1' chronometer, which had served Captain Cook so well on his second and third voyages to the Pacific. During the First Fleet's voyage Dawes was given some additional instruments by Colonel Robert Jacob Gordon (1743–1795), commander of the Dutch troops in Cape Town, for use upon arrival at Botany Bay (James, 2012). There are no surviving records of any such equipment, however.

It is clear that the Board of Longitude valued the hierarchy on board the fleet. It was thus decided to lend the scientific instruments in the care of Captain Phillip, despite Maskelyne's assurance that Dawes was "… capable of making proper use of them". This decision was most likely driven by Phillip's more spacious and secure accommodation on board the *Sirius*. Although Dawes was initially pleased that the instruments were Phillip's responsibility, he expressed some concern about the inconvenience of the distance from his berth to Phillip's cabin:

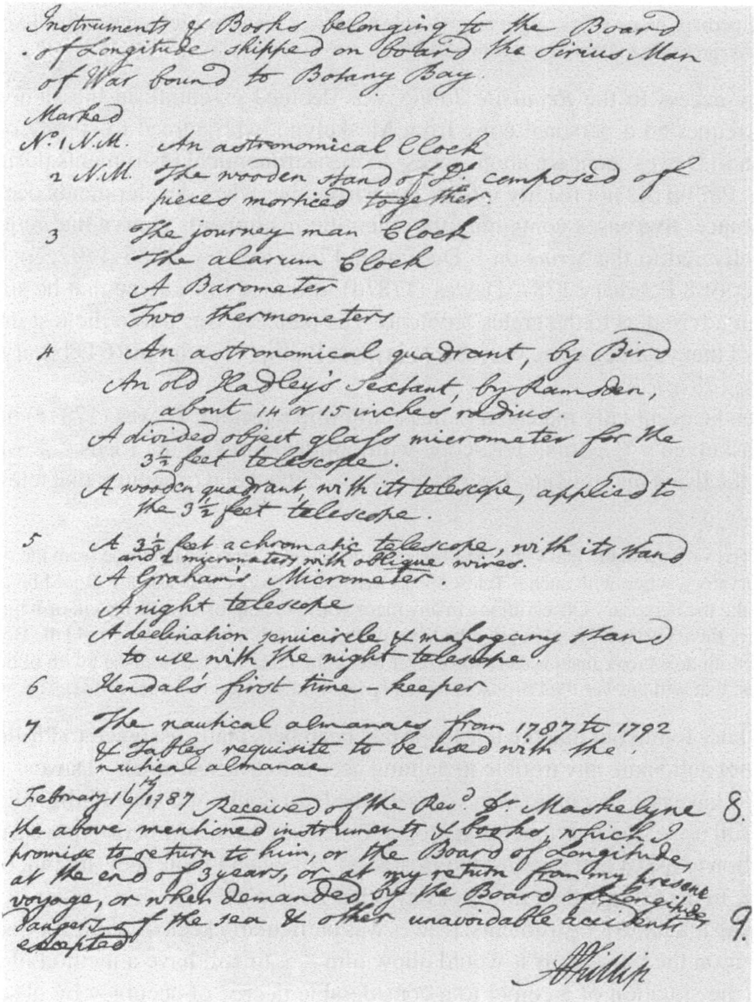

Instruments & Books belonging to the Board of Longitude shipped on Board the Sirius Man of War bound to Botany Bay.

Marked

No. 1 N.M. An astronomical Clock

2 N.M. The wooden stand of Do. composed of pieces morticed together.

3 The journey-man Clock

The alarum Clock

A Barometer

Two thermometers

4 An astronomical quadrant, by Bird

An old Hadley's Sextant, by Ramsden, about 14 or 15 inches radius.

A divided object glass micrometer for the 3½ feet telescope.

A wooden quadrant, with its telescope, applied to the 3½ feet telescope.

5 A 3½ feet achromatic telescope, with its stand and 2 micrometers with oblique wires.

A Graham's Micrometer.

A night telescope

A declination semicircle & mahogany stand to use with the night telescope.

6 Kendal's first time-keeper

7 The nautical almanacs from 1787 to 1792 & Tables requisite to be used with the nautical almanac

February 16/17 1787 Received of the Revd. Dr. Maskelyne the above mentioned instruments & books, which I promise to return to him, or the Board of Longitude at the end of 3 years, or at my return from my present voyage, or when demanded by the Board of Longitude, dangers of the sea & other unavoidable accidents excepted 8. 9.

A Phillip

**Fig. 1.9** Captain Arthur Phillip's signed receipt for the Board of longitude's instrumentation and books on loan. (Board of Longitude papers, Cambridge University; out of copyright)

In your letter of the 15 Inst[an]t you inform[ed] me that Capt$^{n}$. Phillip will give you a Receipt for all the Instruments and Books belonging to the Board, which I received from you, and will return me my Receipt, &c.; on the first perusal of the foregoing it did not occur to me that any Inconveniency could arise from such a Circumstance, but was rather pleased at the Idea of the whole Charge of valuable Books and Instruments being taken off my Shoulders; but on revolving it in my Mind, I cannot but imagine that unless the *Nautical Almanac*s and *Requisite Tables* be entirely at my own disposal and for my own peculiar Use, numberless Inconveniences and Delays will arise: my Cabin is at a considerable Distance from Capt$^{n}$. Phillip's and to be obliged on every Occasion to send to him for a *Nautical Almanac*, at which

> Time perhaps he may be using it and not able to spare it, must be exceedingly troublesome and disagreeable to both of us (Dawes, 1786b).

Since access to the *Requisite Tables* was deemed essential, in the same letter Dawes requested a personal copy from Maskelyne, who agreed to supply one. In addition to Dawes' concern about access to the astronomical instruments during the voyage, Phillip did not readily inform the astronomer when developments occurred. For instance, five cases containing the scientific instruments Dawes had requested were delivered to the *Sirius* on 5 December 1786 (Bradley, 1786–1792: 3), but in his letter of 8 February 1787, Dawes (1787d) informed Maskelyne that he still had not been advised as to the crates' contents. His patience was put to the test until 20 March of that year before he was able to inspect Phillip's receipt of 16 February 1787 (Dawes, 1787e).

When he eventually managed to inspect Phillip's receipt, Dawes (1787e) noticed that Maskelyne's "... small telescope with oblique wires in the focus ..." was not among the listed instruments. He was, however, rather keen to acquire that telescope, since

> ... it [is] very probable that I may be frequently detached some little Distance from the fix'd Observatory, when with such a Telescope as that, and a proper Stand to it I should be able to make the necessary Observations in any place with the help of a good Watch, of which I believe there will be several in [*sic*] the Ship, one of which I could borrow, should the Board of Longitude's prove insufficient; however if the night Telescope be furnished with oblique Wires; that will answer the Purpose effectually I should suppose (Dawes, 1787e).

Dawes later found out that the telescope had been personally assigned to Phillip, but he did not anticipate any trouble in gaining access to the instrument (Dawes, 1787f: 257v): "I have not, any more than yourself, the least doubt of Capt$^{n}$. Phillip allowing me the full use of the Instruments at proper Times; (he having hitherto shown a strong Inclination to assist and forward the Business) 'till when I think they are much more properly in his Charge than mine, or any other Person's".

Among his suite of instruments, Dawes was particularly keen to take a good sextant with him on the voyage, as it would allow him "... to still have a method of determining the situation of a comet to a considerable degree of accuracy by observing [its] distances from remarkable fixed stars ..." even if all other instruments failed (Dawes, 1787d). However, the instrument Bradley (1786–1792: 3) called the "... old sextant by Ramsden ..." and which Phillip referred to as an "... old Hadley's sextant, by Ramsden, about 14 or 15 inches radius ..." (McAfee, 1981: 3) developed a fault (a defective index glass; Dawes, 1787a; see also Saunders, 1990: 50), possibly because of an accidental drop, and so he requested a replacement instrument through Maskelyne's assistant,[17] George Gilpin (1754–1810). This expenditure was authorised by the Board of Longitude at its meeting of 3 February 1787. The new sextant, of the type made by Nairn and Blunt (1775; Dawes, 1787f) and supplied by the leading instrument maker Jesse Ramsden, cost £13 7*s*. 0*d*.; the cost of repairs to the other instruments came to £44 13*s*. 0*d*. (Board of Longitude, 1787). It arrived on 17 April

[17] Dawes was required to seek Maskelyne's approval for any adjustments to the instruments. At times, this caused strained relations between the men, since Dawes (based at Portsmouth) was left feeling as if he was being micromanaged from afar (Saunders, 1990: 46, 52).

1787, prompting an intense period of instrument testing, between 24 April and 30 May 1787 (Dawes, 1787f).

Meanwhile, Dawes' personal circumstances, where he had to satisfy multiple paymasters simultaneously, had the potential to derail the entire astronomical project. Indeed, despite the strong support from Maskelyne, Banks and the Board of Longitude, Dawes' anticipated roles as Marine, the First Fleet's astronomer-designate and the new colony's scientist in charge of establishing a permanent observatory were at risk. Whereas the home nation's scientific establishment might have found their man, the military hierarchy did not necessarily agree. Dawes' appointment was in some jeopardy. Although it is often suggested that he had joined the mission too late to be appointed as a member of the squadron's four shore-based companies of Marines (Eldershaw, 1972: 40; Tench, 1979: 77), Saunders (1990: 48) suggests that he was assigned to the Marine complement aboard the *Sirius* in good faith by the Admiralty, with the expectation that he "... might have leisure independent of other avocations ..." (Lord Howe, 1787) to pursue his scientific tasks. However, his shipboard allocation came with inherent disadvantages compared with membership of the general Marine contingent—in terms of his wages, opportunities and even career advancement.

Whereas members of the main, shore-based Marine contingent were offered a year's advance subsistence pay, the shipboard complement did not qualify for those benefits. Clearly unhappy, Dawes (1787a) appealed to Admiral Richard Howe (1726–1799), First Earl Howe, First Lord of the Admiralty and Chair of the Commissioners of the Longitude, arguing that he had been unfairly excluded. Lord Howe did not indulge the young Lieutenant. He responded tersely that the Marine complement had signed on for a fixed term in New South Wales, while the *Sirius* contingent might be redeployed at any time:

> The Officers of the Detachment are to remain on Garrison Service for the Term already communicated to them. – The party embarked in the *Sirius*, on the contrary, is liable to be ordered home with the Ship at the expiration of 12, 18 or any greater Number of Months as the Exigencies of the Service may require: and both is, & could only be considered in all cases, as similar in Circumstances with every other marine Detachment ordered for foreign Stations (Lord Howe, 1787).

This advice concerned Dawes greatly: as a ship's officer he was subject to the whims of his commanding officer and so he could be sent anywhere the ship would need to go, at any time. In the late eighteenth century, British Marines were soldiers enlisted as part of the Naval establishment, and so they were expected to live and remain on board the ship that served as their duty station. Moreover, while attached to the ship's contingent, Dawes was not automatically granted leave to go ashore to construct the observatory he had been tasked to establish. Both conditions could potentially jeopardise the fledgling astronomical endeavour in the new settlement. However, Lord Howe continued that he did not consider Dawes' orders to construct an observatory of paramount importance:

> The Astronomical Observations there might be opportunity to take are of very inferior moment and may be totally omitted, without prejudice to the intention of the Armament.

> Wherefore, if you are desirous of being relieved in the *Sirius*, and the time will admit of it with reasonable attention to the convenience of the Officer next in tour of Duty at the Division, I don't at present see any Objection to the concurrence of the Board in such request as you may see fit to make on that occasion.

Indeed, Lord Howe, not known as a pushover under any circumstances, called the young Marine's bluff.

Nevertheless, Dawes persisted in his efforts to enact a change in his circumstances, hoping for Maskelyne to exert his influence: "I think it will be allowed that my Situation is truely [*sic*] hard and distressing" (Dawes, 1787b). He argued that he was in danger of losing his seniority in his shipboard role, to which he forcefully objected in a letter to Maskelyne of 25 January 1787:

> I wrote to Capt$^{n}$. Phillip Yesterday informing him of what had passed relative to my appointment to the *Sirius* &c. & requesting he would endeavour to get me removed from her to the shore on the first Vacancy which may happen in the Detachment; this would not prevent my going out in the Ship. — I am very little solicitous about getting or saving Money, my greatest Fear is that my being on board the *Sirius* will prevent my promotion Case [illegible] Vacancies should happen above me, as I cannot possibly succeed to a superior Rank without being absolutely in the Detachment where the Vacancies (if any) will happen; I would not lose one Day's Rank for Tons of Gold & Diamonds & this may probably be the Case unless I get appointed to the Shore Detachment. I should think the loss of Rank the greatest Misfortune that could possibly befal[l] me; in short I would much sooner lose a Limb & would almost as soon lose my Life (Dawes, 1787c).

Therefore, Dawes promptly requested, although unsuccessfully, that he be reassigned to the shore-based contingent at Botany Bay. Dawes' insistence that he did not care much about being compensated financially does not fully ring true, however. This issue touches on the ultimate motivation for the men to volunteer for the Botany Bay expedition. Whereas courage, a sense of adventure, an inquisitive mind and an interest in exploration were undoubtedly important character attributes (e.g., Clarke, 2015: 10; Wood, 1924), their realistic alternative looked rather bleak, with limited opportunities for career progression and retention on half pay only (e.g., Clarke, 2015: 10; Wright, 1926). It is, therefore, not surprising that Dawes did not raise the issue of his shipboard assignment again in his subsequent correspondence with Maskelyne. He appears to have accepted his appointment on the *Sirius* without further complaints,[18] although he requested to be considered for the first available shore-based vacancy (Dawes, 1787b).

Dawes spent the remaining time prior to their departure from England on calibrating his instruments, preparing an observing plan for their sojourn at the Cape of Good Hope and determining the accuracy of the chronometer's clock rate. The latter aspect caused him great concern. In particular, he was worried that the change in the clock's rate between the start and the end of the voyage, and the accumulated errors resulting from that change, would preclude him from performing the gravity experiments that had been proposed by Maskelyne (Dawes, 1787e).

---

[18] Dawes' relatively quick acceptance of his assignment on the *Sirius* was likely driven, at least in part, by the fact that he was near the top of the Marines' rotation roster, which implies that he would have shortly been allocated to another vessel if he had remained in Portsmouth (Bayly, 1786).

In collaboration with Bayly, he carefully compared the chronometer's rate of decay with the clocks at the Portsmouth Royal Naval Academy and found worrying irregularities. These appeared to depend on the extent to which the timekeeper's spring was wound.

Shortly before their departure, Dawes told Maskelyne that Captain Phillip had arrived on board the *Sirius* and, more importantly, that he was favourably disposed of Dawes' astronomical endeavours (Dawes, 1787g). Bradley (1786–1792: 11) highlighted Phillip's apparent interest in the scientific aspects of the voyage at that time and their final preparations:

> May 7.$^{th}$ Captain Phillip arrived at Portsmouth. He brought with him a timekeeper made by Mr. Kendal[l] and a sextant, both furnished by the Board of Longitude for the use of the voyage. The timekeeper was sent by an officer to the Royal Academy at Portsmouth, and left in charge of Mr. Bayley [*sic*] the head master. ...
>
> The 11.$^{th}$ Waiting only for a wind to carry the ships to sea, an officer was sent to bring on board the timekeeper, the rate of its going determined by Mr. Bayley. The 3 days which he had it, it lost 1″38 per day.

It later transpired that this period of appeasement between Dawes and Phillip, which appears to have lasted until Dawes inadvertently allowed K1 to run down following their departure from the Cape of Good Hope (see Sect. 1.7),[19] was the only time that the men were on cordial terms.[20] Dawes had apparently been warned about Phillip's allegedly difficult personality:

> I am well convinced of the Propriety of paying every proper respect and Attention to Capt.$^{n}$ Phillip, and shall on all Occasions endeavour to make my self as serviceable to him as possible, and am well aware that in Case of difference of Opinions on any subject: if cool reasoning and fair Argument have not the desired Effect, a more harsh Method of proceeding whether with Superiors or Inferiors cannot by any Means do Good but may do much Harm (Dawes, 1787h).

## 1.6 *En Route* to Botany Bay

Upon the First Fleet's departure from English shores (see Fig. 1.10), Dawes' most important daily task consisted of maintaining the flagship's main chronometer, K1, given its importance for the convoy's navigation to Botany Bay. Phillip had issued a set of meticulous orders detailing the protocol that was to be adhered to. Bradley's notes (Bradley, 1786–1792: 11) specify that the 'timekeeper' had to be rewound every day at noon in the presence of Phillip with either Hunter or Dawes (it required two keys), as well as the guard on duty outside Phillip's cabin. Subsequently, the day's readings had to be confirmed by the officer of the watch:

---

[19] http://www.visitsydneyaustralia.com.au/william-dawes.html.

[20] Dawes' name and achievements are conspicuously absent from Phillip's (1790a) account of the voyage to New South Wales.

The precautions necessary to prevent the timekeeper from being let down were ordered by Captain Phillip who, with Captain Hunter or Mr. Dawes, were always to be present at the winding it at noon. And it was ordered to be the duty of the lieutenant who brought 12 o'clock to see it done and the officer who relieved him was not to take charge of the deck until he was informed that it was done. The sentinel at the cabin door was also ordered to plant himself inside the cabin, on hearing the bell ring at noon, and not to go out to be relieved until he was told, or saw, that the timekeeper was wound up by one of the officers (Bradley, 1786–1792: 11).

K1 was not the only portable timepiece on board the *Sirius*. Dawes had been lent an Ellicott watch[21] by Maskelyne (Board of Longitude, 1786; Laurie, 1988), whereas Hunter owned a "... reliable chronometer ..." made by the London watchmaker John Brockbank (1747–1806; see Sects. 1.7 and 1.8). We will discuss Hunter's use of his own chronometer in navigating the southern Indian Ocean in more detail below.

Dawes' other scientific duties involved recording daily observations of the weather, including the temperature and air pressure (e.g., Ashcroft, 2016; McAfee, 1981). He also routinely recorded the Earth's gravity on the basis of pendulum timings (e.g., Forbes, 1975: 172), compass deviations as the voyage progressed (e.g., Hunter, 1793: 23) and their position at sea using either the timepiece or lunar distance

**Fig. 1.10** Departure of the First Fleet, passing the Isle of Wight's 'Needles', 13 May 1787. Artist: William Bradley, *A Voyage to New South Wales* (ca. 1802). (Mitchell Library, State Library of New South Wales; Ref. 412997; out of copyright)

[21] This watch had been taken on Cook's second and third voyages of exploration to the Pacific in the care of his astronomer, William Wales (Howse and Hutchinson, 1969: 139–140).

measurements, or ideally both, to derive their actual longitude. Collins (1798: Section II) explains,

> The longitude, when calculated by either altitudes of the sun, for the time-piece, (of Kendal[l]'s construction, which was sent out by the Board of Longitude,) or by the means of several sets of lunar observations, which were taken by Captain Hunter, Lieutenant Bradley, and Lieutenant Dawes, was constantly shewn to the convoy, for which purpose the signal was made for the whole to pass under the stern of the *Sirius*, when a board was set up in some conspicuous part of the ship with the longitude marked on it to that day at noon.

Hunter further clarified that their longitude determination "... was [routinely] marked with chalk in large characters on a black painted board, and shewn over the stern to the convoy" (Hunter, 1793: 36).

As a matter of protocol, Dawes compared his results with those of Hunter and Bradley, and he sent his observations back to Maskelyne at every possible opportunity (e.g., Dawes, 1787i). The first such opportunity occurred on 20 May 1787 by means of the homeward-bound frigate H.M.S. *Hyæna* under Captain Michael de Courcy (Coursey; d. 1824). The *Hyæna* had escorted the convoy from Spithead roads (anchorage) to a point some 300 miles west of the island of Ushant (Ouessant, France), at "Latitude 47° 57½ N, Longitude 12° 14½ W by Timekeeper this Day at Noon" (Dawes, 1787h).

Hunter, clearly impressed by Dawes' careful work, referred to Dawes as "... a young gentleman very well qualified for such a business, and who promises fair, if he pursues his studies, to make a respectable figure in the science of astronomy" (Bach, 1968: 13). Hunter was a proficient astronomer and navigator in his own right, as we will see below. As a case in point, consider the following passage from his journal shortly after their departure:

> On the 15th [May 1787], at sun-set, the Start Point bore north-east half east by compass, distant seven or eight leagues: at noon on this day (which finishes the nautical and begins the astronomical day) the longitude, by account [i.e., based on dead reckoning], was 5°01′ west of the meridian of Greenwich, and by a time-piece made by Mr. Kendal[l], with which the Board of Longitude had supplied us, it was 4°59′ west; ... our lunar observations, which we never failed to make at every opportunity, constantly confirmed the truth of the watch (Hunter, 1793: 5, 15).

Having passed close by Madeira and the Savage Islands (Salvage Islands, Ilhas Selvagens), the convoy reached Santa Cruz bay at Tenerife in the Canary Islands on 3 June 1787 (see Fig. 1.11), "... after a Passage, for the most Part pleasant and agreeable ..." (Dawes, 1787i). Dawes continued, "... we made as much use of the Time Keeper & our Sextants as the Weather would permit us, we only miss'd getting the Longitude by the T.K.r [Time Keeper] two Days during the whole Passage". Although Dawes was keen to make observations from Santa Cruz during their week-long stay there, Phillip refused permission to unload the pendulum clock and the quadrant, since they were stored deep in the hold, below the convoy's stores of bread[22]:

---

[22] In good Admiralty tradition, scientific instruments were usually stored near a ship's supply of bread, which was kept in the driest place on the ship (King, 1779; Saunders, 1990: 67).

> Immediately on our Arrival at the Place I applied to Capt[n]. Phillip to get the Clock and Quadrant on Shore; but he informed me that it was absolutely impossible to make any satisfactory Use of them at this Place, as he is determined to get away on Thursday [7 June 1787] if possible ... besides which, the Motion of the Ship has caused the Bread to cover the Cases so entirely that they could not be got at without much difficulty & the Risk of getting a good deal of the Bread damaged; but on our Arrival at Rio [de] Janeiro, we shall not only have more Time, but the Bread will then be so much reduced as to obviate any difficulty in that Respect. ... I shall attend most assiduously to the T.K.[r] [Time Keeper] while we remain at this Place & hope with the Sextant to be able to attain to almost as much accuracy as if I had the Astronomical Quadrant on Shore (Dawes, 1787i).

The ships' sojourn at Tenerife was meant to restock their wine, water and other provisions for the next leg of the voyage down to the South Atlantic. Dawes wrote two letters to Maskelyne from the Santa Cruz roads, both dated 5 June 1787, which included a range of longitude and latitude measurements (Dawes, 1787i, 1787j).

**Fig. 1.11** "Santa Cruz on the SE side of Teneriffe; *Sirius* & Convoy in the Roads. June 1787. The Peak Shewing in a Gap between two other Mountains." Artist: William Bradley, *A Voyage to New South Wales* (ca. 1802). (Mitchell Library, State Library of New South Wales; Ref. 412997; out of copyright)

It would take another eight weeks following the convoy's departure from Tenerife on 10 June before they arrived at St. Sebastian (Rio de Janeiro) by 5 August 1787 (see Fig. 1.12). They had been forced to abandon their plans for a brief stop in the Cape Verde archipelago because of adverse wind conditions at the St. Jago (Santiago) roads. Since a number of officers on the First Fleet kept detailed accounts of the voyage in their personal journals, we will reproduce some of the most pertinent passages related to the convoy's navigation and the astronomical endeavours here:

> On Thursday the 2^d^ of August we had the coast of South America in sight; and the head-land, named Cape [Cabo] Frio, was distinctly seen before the evening closed in. Our timepiece had given us notice when to look out for it, and the land was made precisely to the hour in which it had taught us to expect it.[23] It was not, however, until the evening of the 4^th^ that we anchored within the islands at the entrance of the harbour of Rio de Janeiro.
>
> …
>
> On the morning after our arrival the intendant of the port, with the usual officers, repaired on board the *Sirius*, requiring the customary certificates to be given, as to what nation she belonged to, whither bound, the name of her commander, and his reason for coming into that port; to all which satisfactory answers were given; and at eleven o'clock the day following Captain Phillip, accompanied by the officers of the settlement, civil and military, waited upon Don Louis Vasconcellos, the viceroy of the Brazils, at his excellency's palace, who received them with much politeness, readily assenting to a tent being pitched on shore for the purpose of an observatory; … (Collins, 1798: Section II).

The day after their arrival at St. Sebastian (see Fig. 1.13), Dawes was indeed busy. He wrote an urgent letter to Maskelyne requesting that some of his spoilt books be replaced at the first opportunity. The letter was taken back to Europe by a passing Portuguese ship bound for Lisbon:

> My Cabbin [*sic*] being under Water (the greater part of it) I have frequent Seas coming into the Scuttle in consequence of which & the constant Use I have made of them, the *Requisite Tables* are much the Worse ... (Dawes, 1787j; see also Dawes, 1787l, for a repeat request)

On account of his earlier service in the Portuguese navy (e.g., Pembroke, 2013), Captain Phillip was received and treated with the highest regard, and so his officers were given unprecedented freedom to move around the city. This preferential treatment also included permission to establish a temporary tent observatory. Dawes (1787m: 269r) requested it to be pitched on the harbour island of Enchados (Ilha das Enxadas), close to the city:

> 9.^th^ [August 1787] Captain Phillip and Mr. Dawes went to look at a room which had been granted for the purpose of making the necessary observations for the timekeeper etc., but it not being thought proper for the purpose, Captain Phillip obtained leave from the Viceroy to make use of an Island (Enxadas) lying to the NE of the town, which spot being approved, a tent was erected on the 11^th^ and on the 12^th^, the instruments landed and fixed under the direction of Mr. Dawes (Bradley, 1786–1892: 36).

[23] However, Hunter (1793: 17) cast doubt on the party's longitude determinations: "It will appear hereafter that we had not the true rate of the watch, and consequently that the above longitude is not correct".

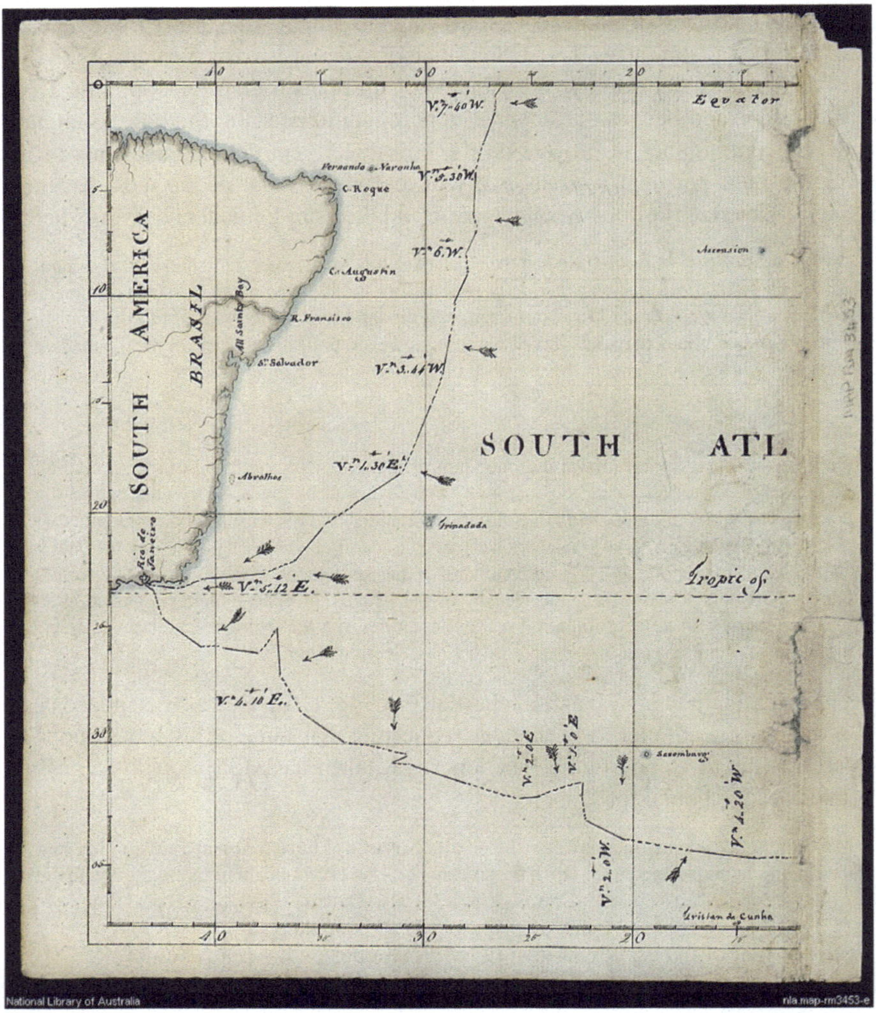


**Fig. 1.12** Chart of the South Atlantic showing the route of the *Sirius*, 1787–1788. (George Raper; National Library of Australia; Ref. nla.map-rm3453; out of copyright)

During our stay here, we were permitted to erect a tent on the island of Enchados, (a small island about a mile and a half farther up the harbour than where we lay with the ships,) for the purpose of landing a few of the astronomical instruments which were necessary for ascertaining the rate of the time-keeper; they were put under the charge and management of Lieutenant William Dawes of the marines … (Hunter, 1793: 20–21).

The weather was rather unfavourable, during the time the instruments were on shore for ascertaining the rate of the time-keeper, but as constant attention was paid, every opportunity that offered was made use of, and the watch [rate] was found to be 2″-27 which is near a second more than was its rate at Portsmouth (Hunter, 1793: 21).

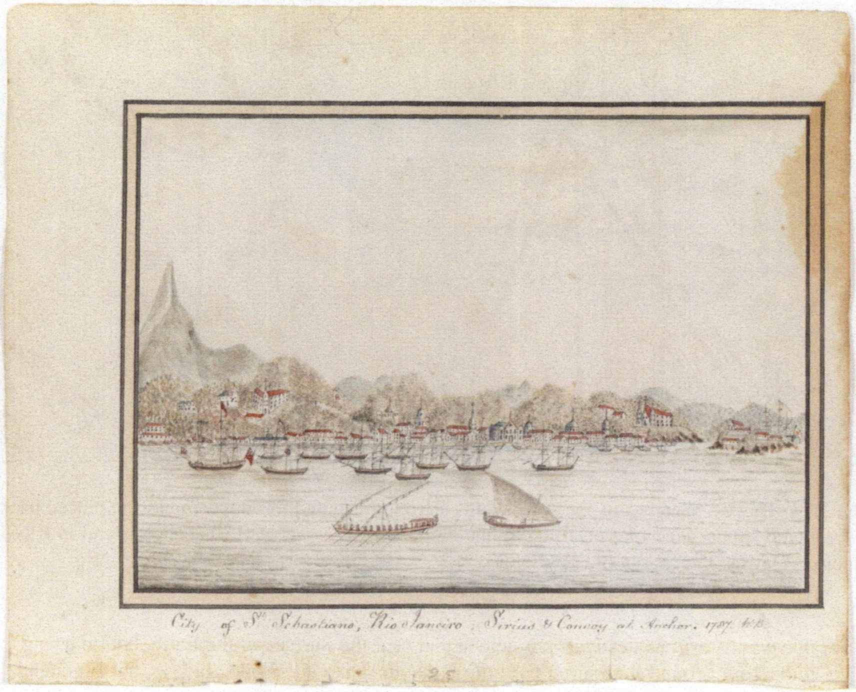

**Fig. 1.13** "City of St. Sebastian, Rio Janeiro: *Sirius* & Convoy at Anchor. 1787." Artist: William Bradley, *A Voyage to New South Wales* (ca. 1802). (Mitchell Library, State Library of New South Wales; Ref. 412997; out of copyright)

Despite the inclement weather (see also King, 1787–1790), for the next month Dawes, assisted by two young sailors from the *Sirius* (Collins, 1798: Section II), proceeded apace with his observational programme. His observations included measurements of the ephemerides—the positions on the sky—of Jupiter's satellites, and particularly of the eclipses of the planet's third largest moon, Ganymede (Dawes, 1787m: 271v). He made sure to express his appreciation of Phillip's encouragement and support, particularly of the fact that Phillip himself had covered the costs associated with establishing the temporary observatory. The Portuguese authorities provided guards to ensure that no one would disturb the observatory's business (Dawes, 1787m)—or perhaps to keep an eye on their progress: "The 13.$^{\text{th}}$ [August 1787] The Timekeeper was sent to the tent and all boats belonging to the transports strictly forbidden landing on that island" (Bradley, 1786–1792: 36).

Dawes had meanwhile also ascertained the rates of his clocks and felt confident of their performance. In his letter of 3 September 1787 (Dawes, 1787j), he sketched the set-up of his astronomical and journeyman clocks (see Fig. 1.14), placed back-to-back, explaining the need for his rather intricate support system:

> The Earth being very hard it was not in my Power to get the Clock Frame set up 'till Wednesday Morning, when the Earth was rammed exceedingly well all round it, so that the

**Fig. 1.14** Dawes' sketches of the setup of his astronomical and journeyman clocks, placed back-to-back and supported by large struts, in his temporary observatory at Enchados island in Rio de Janeiro. (Dawes, 1787g; out of copyright)

> Frame was as firm & steady as possible at and near the Surface of the Earth, but on trying it at the Top, I found it required but little Force to shake it considerably, so that I thought necessary to get four large Pickets & as many Shoars made … the Pickets being driven into the Earth very firmly at about three Feet distance from the Angles of the Frame & the Shoars placed as represented in the Plan sketched below …

Dawes appears to have been a careful and thorough observer (Bradley, 1786–1792; White, 1790; Tench, 1979: Introduction):

> If Circumstances permit I always take 6 at least Observations at one Time, whether Altitudes for the Longitude by Time Keeper or Distances for the Lon[gitude] by the Lunar Method. I divide them into two Setts [*sic*] of three each and work the Mean of each three separately & then take a mean of the Result of each Sett; by this Means I think Mistakes may almost always be detected. The Means of each Sett have in general agreed exceedingly well. The greatest Difference has only amounted to about 4″ of Time, but they have almost always agreed within 2″ of each other. – I shall not remit in the least from endeavouring by a great Number of Observations to get the true Rate of the Time Keeper as well as the Longitude of this or any other Place as nearly as possible by Distances, Occultations &c.[a] (Dawes, 1787j).

As a case in point, he determined the longitude of Cape Frio to within a third of a degree of its actual longitude (equivalent to approximately 19 nautical miles at the latitude of Rio de Janeiro; Laurie, 1988). Rio de Janeiro served as a useful calibration benchmark for the convoy's position determinations (see Fig. 1.15). To his credit, Dawes diligently compared the performance of his own instruments with those at the local observatory, and he would do the same upon their arrival at the Cape of Good Hope. On 3 September, he wrote to Maskelyne,

> ... on Saturday the 11.[th] [August 1787] ... I went on board to inform C.P. [Captain Phillip] that the instrument might be sent on shore that afternoon, which he immediately gave directions

for, and I accordingly took them on shore with me the same afternoon. On Monday Morning I set up the Quadrant, and in the evening, the clock being set up ...

... On the 17.$^{th}$ [August 1787] in the forenoon I was invited by two astronomers employed here by the Court of Portugal to settle the limits between the Portuguese & Spanish Settlements in this Country. They were brought to the Tent by Capt$^{n}$. Phillip & I immediately foresaw that I might possibly get some observations from them which would be acceptable to you, & therefore determined if possible to return their Visit before our quitting this Place: they admired the Quadrant & took notice how equal the Beats of the Clock were. – On the 1.$^{st}$ of Sept.$^{r}$ [1787] I went on shore to return their Visit and found one of them at home who very willingly allowed me to copy the Observations contained in the two Sheets accompanying this ... The Gentleman told me he had made many meteorological observations ...

The Portuguese astronomers Dawes referred to in his letter to Maskelyne were based at the city's observatory. Watkin Tench provides a little more detail about the key issue of contention in Brazil at the time:

Among other public buildings, I had almost forgot[ten] to mention an observatory, which stands near the middle of the town, and is tolerably well furnished with astronomical instruments. During our stay here, some Spanish and Portuguese mathematicians were endeavouring to determine the boundaries of the territories belonging to their respective Crowns. Unhappily, however, for the cause of science, these gentlemen have not hitherto been able to coincide in their accounts, so that very little information on this head, to be depended upon, could be gained. How far political motives may have caused this disagreement, I do not presume to decide; though it deserves notice, that the Portuguese accuse the Abbee de la Caille, who observed here by order of the King of France, of having laid down the longitude of this place forty-five miles too much to the eastward (Tench, 1979: Chapter V).

City of St. Sebastian.

| | Deg. | Min. | Sec. | |
|---|---|---|---|---|
| Latitude - - - - - - | 22 | 54 | 13 | ſouth. |
| Longitude, deduced from our time-keeper of the meridian of Greenwich, and which agrees with that laid down in the new requiſite tables, but which certainly are not correct. | 42 | 44 | 00 | weſt |
| Longitude, determined by two aſtronomers ſent from Portugal for that and other purpoſes | 43 | 18 | 45 | weſt. |
| Longitude, by an eclipſe of Jupiter's third ſatellite, taken by Lieutenant Dawes, on the iſland Enchados - - - - | 43 | 19 | 00 | weſt. |
| Longitude, by a mean of ſeveral diſtances of ☉ and ☽ taken by me at the outer anchorage | 43 | 11 | 15 | weſt. |
| Ditto, by Lieutenant Bradley - - | 43 | 33 | 00 | weſt. |

**Fig. 1.15** Comparison of the First Fleet's longitude determinations of the city of St. Sebastian (Rio de Janeiro) as recorded by Hunter (1793: 50). (out of copyright)

From both Rio de Janeiro and the Cape of Good Hope, Dawes encouraged Maskelyne to keep in touch with the local astronomers:

> ... he [Senhor Bento Sanchez d'Orta] would with pleasure communicate to you if you should think proper to enter into a correspondence with him ... I also informed him of the comet expected in 1789 [see below] which he promised to pay as much attention to as he could (Dawes, 1787m: 273v).

Almost a month after their arrival in Rio de Janeiro the convoy left, on 4 September 1787, on their way to the Cape of Good Hope. They arrived in Table Bay on 13 October (see Fig. 1.16):

> This run, from about lat[itude] 22° south, long[itude] 43° west of London, to lat[itude] 34° south, long[itude] 18° east of London, a distance of about four thousand miles, was performed in thirty-nine days: for having left Rio on the 4th of September, 1787, on the 13th of October the ships came to anchor in Table Bay (Phillip, 1790a: 43).

Dawes informed Maskelyne that "... the whole Fleet is in perfect good Order & every body exceedingly healthy; only abt [*sic*] 15 Deaths have happened since our departure ..." (Dawes, 1787n). Although Captain Phillip had initially planned to stay at the Cape for just a week, the convoy did not depart until a month had passed, on

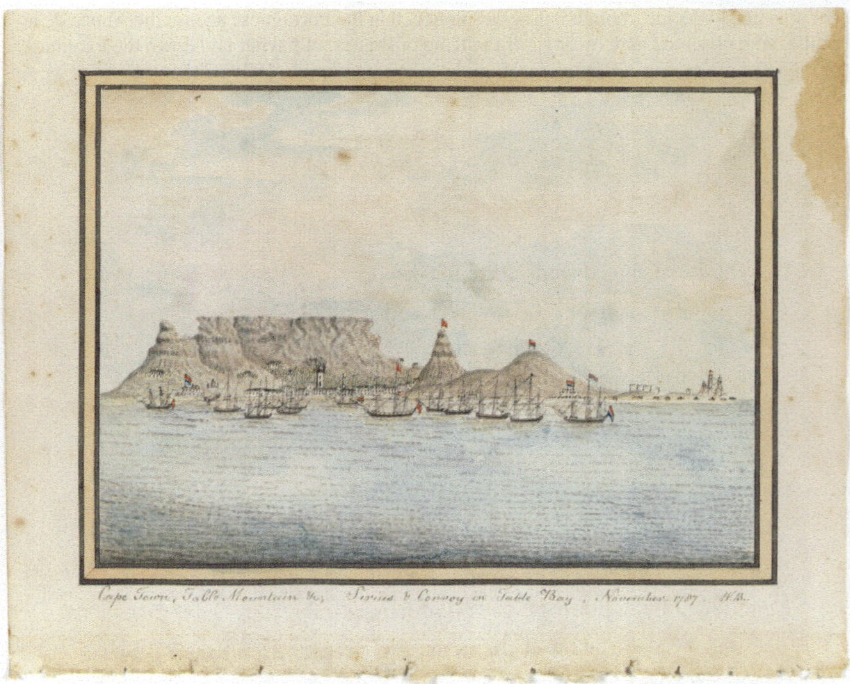

**Fig. 1.16** "Cape Town, Table Mountain &c.; *Sirius* & Convoy in Table Bay, November. 1787." Artist: William Bradley, *A Voyage to New South Wales* (ca. 1802). (Mitchell Library, State Library of New South Wales; Ref. 412997; out of copyright)

12 November 1787. Yet, and to Dawes' great disappointment, Phillip once again did not permit him to disembark his instruments, this time citing safety concerns and fearing that the instruments might sustain damage in the rush of loading cattle and provisions (e.g., Dawes, 1787n; Laurie, 1988; Clarke, 2015: 31–32). And so Dawes' carefully prepared observing programme for the Cape came to nothing.

Shortly after the convoy's arrival in Table Bay, Hunter questioned the accuracy of the timepiece:

> By altitude taken this morning [15 October 1787] for the time-keeper, it appear'd that we had not had sufficient time at Rio de Janeiro for ascertaining the true rate of the watch's going, having determined what we have allowed this passage, *viz.* 2″-33 from a very few observations, and those not to be relied on, the weather having been very unfavourable; for, by the difference of time between the meridian of Rio de Janeiro and the Cape, both which places are well determined, the watch has lost at the rate of 3″-17, which we shall hereafter allow to be the true rate; and as a proof of that having been really its rate all along, by allowing it from the time of our leaving Portsmouth, until our arrival at Rio de Janeiro, we shall have the longitude of that place 43°33′30″ west of the meridian of Greenwich, which is 45′45″ to the westward of that laid down in the new *Requisite Tables*, and which agrees very nearly with the observations made on the spot (Hunter, 1793: 30).

Meanwhile, as the ships were being stocked one final time with provisions, stores and livestock for the new colony, Dawes was introduced to Colonel Robert Jacob Gordon, commander of the Dutch troops at the Cape, and a great host to the visiting convoy:

> … a gentleman whose thirst for natural knowledge amply qualified him to be of service to us, not only in procuring a great variety of the best seeds and plants, but in pointing out the culture, the soil, and the proper time of introducing them into the ground (Collins, 1798: Section II).

Gordon owned a Ramsden quadrant, an Arnold watch "… in gold, price 60£ spared to him by Capt$^{n}$. Cummings in the East India Service …" (Dawes, 1787n), various telescopes, almanacs and astronomical tables. However, he remained a novice in terms of having done any observing other than taking latitudes (Laurie, 1988). Dawes convinced him to help with their efforts to observe the comet that Maskelyne had predicted would reoccur in the southern hemisphere in 1788, possibly as early as 1 January of that year (Maskelyne, 1786c; Saunders, 1990): "We have ever since [having been invited to stay with Gordon] been fully employed in putting things in a train for observing the comet which is expected about this time next year" (Dawes, 1787n). Enthusiastically, Gordon ordered a telescope from Dollond through Maskelyne, which Gordon arranged to be paid for by the Dutch consul in London. He also initiated a subscription to *Taylor's Logarithms*, and asked Maskelyne to provide him with the necessary instructions to embark on an observational programme aimed at finding Maskelyne's comet.

Dawes was clearly enamoured with the man and saw his potential as an ally in the southern hemisphere:

> This gentleman is rema[rk]able for his great love of science and Attention to persons e[m]ployed in promoting it ... he has some knowledge of astronomy and is more determined to employ the greatest part of his leisure time to the study of it ... it occurred to me,

that it would be doing you the greatest pleasure possible, to embrace so fair an opportunity of adding one to the small number of observers in the southern hemisphere ... (Dawes, 1787n)

Dawes' next letter to Maskelyne was sent from New South Wales.

## 1.7 Uncharted Waters

At this point, we should probably pause for a moment and consider what the next and final leg of the voyage meant to the settlers. The mixed emotions that may have been felt by a number of the men are perhaps best reflected by David Collins' eloquent reminiscence upon their departure from Table Bay:

> It was natural to indulge at this moment a melancholy reflection which obtruded itself upon the mind. The land behind us was the abode of a civilized people; that before us was the residence of savages. When, if ever, we might again enjoy the commerce of the world, was doubtful and uncertain. The refreshments and the pleasures of which we had so liberally partaken at the Cape, were to be exchanged for coarse fare and hard labour at New South Wales. All communication with families and friends now cut off, we were leaving the world behind us, to enter on a state unknown; and, as if it had been necessary to imprint this idea more strongly on our minds, and to render the sensation still more poignant, at the close of the evening we spoke a ship from London. The metropolis of our native country, its pleasures, its wealth, and its consequence, thus accidentally presented to the mind, failed not to afford a most striking contrast with the object now principally in our view (Collins, 1798: Section II).

The First Fleet left the Cape on 12 November 1787, a day later than anticipated because of adverse winds. Dawes excitedly told Maskelyne in his letter of 9 November 1787 (Dawes, 1787n) that Gordon had supplied him with "... several books & instruments which will be of service ... among which are several spare barometer tubes". Meanwhile, Hunter had determined the longitude of Cape Town on a number of occasions during their visit, which both Hunter and Collins recorded in their respective journals (although the actual values differ slightly):

> During the time we lay in this bay [Table Bay], I took a considerable number of lunar observations, by a mean of which I make Cape Town, in longitude 18°24′30″ east of the meridian of Greenwich: latitude observed in the bay, 33°55′ south, ... (Hunter, 1793: 32)

> Before we quitted the Cape Captain Hunter determined the longitude of the Cape-town in Table-bay to be, by the mean of several sets of lunar observations taken on board the *Sirius*, 18°23′55″ east from Greenwich (Collins, 1798: Section II).

Less than a fortnight after their departure, on 25 November "... being then only 80 leagues eastward of the Cape ..." (Phillip, 1790a: 49), Captain Phillip, Lieutenant Philip Gidley King (1758–1808) and Dawes[24] transferred from the *Sirius* to the faster *Supply*, aiming to reach Botany Bay sooner than the main convoy (see also Hunter, 1793: 32):

---

[24] Note that Phillip only mentions King as having transferred with him to the *Supply* (Phillip, 1790a: 50). He does not mention Dawes.

> For several days after we had sailed, the wind was unfavourable, and blowing fresh, with much sea, some time elapsed before we had reached to the eastward of the Cape of Good Hope. On the 16th, Captain Phillip signified his intention of proceeding forward in the *Supply*, with the view of arriving in New South Wales so long before the principal part of the fleet, as to be able to fix on a clear and proper place for the settlement. Lieutenant Shortland was at the same time informed, that he was to quit the fleet with the *Alexander*, taking on with him the *Scarborough* and *Friendship* transports. These three ships had on board the greater part of the male convicts, whom Captain Phillip had sanguine hopes of employing to much advantage, before the *Sirius*, with that part of the fleet which was to remain under Captain Hunter's direction, should arrive upon the coast. This separation, the first that had occurred, did not take place until the 25th [November 1787], on which day Captain Phillip went on board the *Supply*, taking with him, from the *Sirius*, Lieutenants [Philip Gidley] King and Dawes, with the [K1] time-keeper (Collins, 1798: Section II, xxxv).

In Hunter's account of the event, we learn that Phillip also took along with him "… several sawyers, carpenters, blacksmiths and other mechanics …", while King (1787–1790: 19 November 1787) wrote "Capt. Phillip, Lieut[t] Dawes of the Marines, and myself are to go in the *Supply* with 6 Artificers" (see also Tench, 1979: Chap. 7). In addition, he specifically addresses the fate of Kendall's timepiece:

> On the 25th [November 1787], being in latitude 38°40′ south, and longitude 25°05′ east, Captain Phillip embarked on board the *Supply*, in order to proceed singly in that vessel to the coast of New South Wales, where he made sure of arriving a fortnight or three weeks before us, as some of the convoy sailed very heavy; he took with him from the *Sirius*, Mr. Philip Gidley King, second lieutenant, and Lieutenant Dawes, of the marines, who had hitherto kept an account of the time-keeper, which he also took with him; … (Hunter, 1793: 32)

Phillip's transfer to the *Supply* and his decision to take the K1 chronometer with him on the faster vessel caused significant consternation among the officers remaining behind (e.g., Saunders, 1990: 62, 64). Arthur Bowes-Smyth (1750–1790), the convoy's Surgeon-General—known simply as 'Bowes' in the colony—referred to that decision as "… a mere abortion of the brain, a whim which struck him at the time as the sequel will evince" (Choat, 2020: 61). Meanwhile, King assessed the *Supply* "… much too small for so long a voyage, which, added to her not being able to carry any quantity of provisions, and her sailing very ill, renders her a very improper vessell [*sic*] for this service" (King, 1787–1790: 513), and thus he thought that Phillip "… flattered himself …" (King, 1787–1790: 531) in believing that he would be able to make up a fortnight on the remainder of the fleet. Captain James Campbell (d. 1795?) of the *Lady Penrhyn* was incensed and expressed his dismay that Phillip was "… indifferent about everything but his own safety …" (Campbell, 1788: 98) in deciding to proceed with "… the Don Quixote scheme of separating our little Fleet—leaving them to work their way through an immense sea but little known, and to which all were strangers" (Campbell, 1788: 96).

And alas, the transfer of K1 from the *Sirius* to the *Supply* indirectly led to accidental negligence when, on 18 December 1787, the timepiece was not rewound at noon and instead ran down as it was forgotten: "[On 25 January 1788,] We received the Timekeeper from the *Supply* where it had unfortunately been let down on the passage to this place" (Bradley, 1786–1792: 64). It took a significant number of lunar distance measurements (e.g., James, 2012) and lengthy calculations before Dawes was able to reset and recalibrate the clock mechanism (Bradley, 1786–1792: 6; King, 1787). Dawes later explained to Maskelyne the circumstances leading to the accidental stoppage:

> Whenever there is any thing to be got out of the hold, it is a very awkward rather dangerous thing to go from the quarter deck to the cabbin [*sic*] it was owing to this that on the 18th Dec.r [1787] the time-keeper was let down, as Captn. Phillip could not get down at noon to wind it up & it was not thought of afterwards by anyone till near six o'clock at which time it had been down above an hour; however the next day I got some exceeding[ly] good altitudes, from which the longitude was carried on, so that we were only liable to the inaccuracy of the log for two days (Dawes, 1788).

However, when K1 stopped, rewinding it did not fully return it to its former operation; rewinding it caused the timepiece's rate to change as well. Dawes' careful observations led him to conclude that the watch had been idle for a total of 1 h 9 min 44 s (Morrison & Barko, 2009; see also King, 1787–1790). It took until September 1788, well after their arrival in Sydney Cove, before Dawes was able to fully recalibrate K1 (Morrison & Barko, 2009).

Meanwhile, with K1 on the *Supply* and the rest of the convoy split into two smaller squadrons, the ship's captains were left to their own devices. Captain Hunter of the *Sirius* was the most qualified astronomer remaining with the main fleet. Both Hunter and Collins provide detailed accounts of their progress across the southern Indian Ocean.

> After the time-keeper was taken from the *Sirius*, I kept an account of the ship's way by my own watch, which I had found for a considerable time, to go very well with Kendal[l]'s; I knew it could be depended on sufficiently to carry on from one lunar observation to another, without any material error; for although its rate of going was not so regular as I could have wished, yet its variation would not in a week or ten days have amounted to any thing of consequence; it was made for me by Mr. John Brockbank, of Cornhill, London, upon an improved principle of his own. The lunar observation, which I never failed to take [at] every opportunity, and which Lieutenant Bradley also paid constant attention to, gave me reason to think, by their near agreement with the watch, that it continued to go well (Hunter, 1793: 33).

> For three successive days [from 1 December 1787] both Mr. Bradley and myself had a variety of distances, by which our account seemed to be very correct. I now determined (if I could avoid it) never to get to the northward of latitude 40°00′ south, and to keep between that parallel and 43° or 44° south. After the 3d [December 1787], I found, by altitudes taken for the watch, that we went farther to the eastward than the log gave us, and no opportunity offered for getting a lunar observation to compare with it until the 13th, when both Mr. Bradley and I got several good distances of the ☉ [Sun] and ☽, by which our longitude was 70°22′ east, by the watch 70°07′ east, and by account [dead reckoning] 67°37′ east. On the 14th, the weather being very clear, we had another set of distances, which gave our longitude 73°06′ east, by the watch 73°09′ east, and by account 70°34′ east. Again, on the 15th, I observed with two different instruments, one by Ramsden, and the other by Dollond, and the results agreed within ten miles of longitude; the mean was 75°18′ east, by the watch 75°16′ east, and by account 72°49′ east. Mr. Bradley's mean was also 75°18′ east: so that, as I have already observed, the ship seemed [to be] gaining on the account; but there was no reason to believe, that in the middle of this very extensive ocean we were ever subject to much current: I therefore attribute this set to the eastward, to the large following sea, which constantly attended us, since we had taken a more southerly parallel (Hunter, 1793: 33–34).

In early January 1788, the astronomers on the *Supply* and the *Sirius* both sighted New Holland. Dawes, on the *Supply*, determined the longitude of Tasmania's South Cape on 3 January. It compared favourably with Cook's surprisingly accurate 1777 measurement of 146°07′00″ East (i.e., within 10 km of the present-day value; Hunter, 1793: 39; Morrison & Barko, 2009), which had also been obtained by reference to Kendall's K1 chronometer.[25] This allowed Dawes to refine the recalibration of K1 by adjusting the timepiece (negatively) by 3 min 21.7 s (King, 1788).

Following a beautiful display of the aurora australis on 6 January 1788 (see also Bradley, 1786–1792: 51; Tench, 1979: Chap. 7),

> This night the aurora austreales [*sic*] were very bright, of a beautiful crimson colour, streaked with orange, yellow, and white, and these colours were constantly changing their places: the highest part was about 45° above the horizon, and it spread from south by east to south-south-west (Hunter, 1793: 37).

… we learn from Collins (see also Southwell, 1788) that by …

> … a lunar observation taken at ten o'clock of the forenoon of Monday the 7th, the fleet was then distant seventeen leagues from the South Cape of New Holland; and at five minutes past two in the afternoon the signal was made for seeing the land.
>
> Nothing could more strongly prove the excellence and utility of lunar observations, than the accuracy with which we made the land in this long voyage from the Cape of Good Hope, there not being a league difference between our expectation of seeing it, and the real appearance of it (Collins, 1798: Section III).

The *Supply* reached Botany Bay on 18 January 1788 (see Fig. 1.17), followed by the *Alexander*, *Friendship* and *Scarborough* on the 19th, while the remainder of the convoy, including the *Sirius*, arrived at their destination on 20 January 1788 (see Fig. 1.18). Dawes was formally transferred back to the *Sirius* on 24 January.

**Fig. 1.17** Arrival of the *Supply* at Botany Bay, 1788. Artist: Charles Gore. (Mitchell Library, State Library of New South Wales; Ref. DG VIA/8; out of copyright)

[25] Hunter (1793: 38–39) noted that "The latitudes and longitudes of the different points or capes, seem to have been very correctly determined by Captains Cook and [Tobias] Furneaux, when they were here; … Such observations as we had an opportunity of making near this coast, agree very well …".

**Fig. 1.18** "Botany Bay. *Sirius* & Convoy going in: *Supply* & Agents Division in the Bay, 21 Jan$^{ry}$ 1788." Artist: William Bradley, *A Voyage to New South Wales* (ca. 1802). (Mitchell Library, State Library of New South Wales; Ref. 412997; out of copyright)

## 1.8 Measurement Quality

During the 250–252-day voyage of the First Fleet, latitudes and longitudes were determined as often as possible, subject to the prevailing weather. Latitudes were most easily measured, sometimes even in slightly overcast conditions. Determination of the Sun's altitude at its meridian passage would often suffice. Longitude determination based on reading off the time on either K1 or Hunter's personal timepiece was intrinsically more complicated. It required altitude determinations of celestial objects—the Sun, the Moon or certain bright stars—combined with their predicted locations as tabulated in the relevant *Nautical Almanac* and the *Requisite Tables*.

As far as we are aware, a full and comprehensive account of the First Fleet's positions as the voyage progressed is not generally available (but see Gergis et al., 2010). Dawes included a table of the *Sirius*' longitudes and latitudes between 14 May and 2 June 1787 in his letter to Maskelyne of 9 June 1787 (Dawes, 1787k: 268r). Meanwhile, Hunter (1793: 26–28) included a table of their geographic locations, as well as the prevailing weather conditions, on the leg from Tenerife to Rio de Janeiro. Since most journals of First Fleet officers are now readily accessible online (Project Gutenberg Australia, 2019), we consulted the journals of John Hunter

(1793), William Bradley (1786–1792), David Collins (1798), Arthur Phillip (1790a, 1790b), Arthur Bowes-Smyth (Choat, 2020), John White (1790) and Daniel Southwell (Bladen, 1893). We also perused William Dawes' correspondence contained in the Board of Longitude papers. We thus collected a total of 287 unique geographic location references.[26]

Bradley included a number of charts of the *Sirius*' detailed itinerary in his journal based on longitude and latitude measurements. Here, we have reproduced his charts of the first leg of the voyage from England across the Equator in the eastern Atlantic Ocean (see Fig. 1.19) and an overview chart of the southern hemisphere section, from a location just north of the Equator to Rio de Janeiro, Table Bay and New South Wales (see Fig. 1.20). Combined with Fig. 1.12, this selection of charts offers detailed insights into the movements of the *Sirius* during her long voyage to the new settlement (for a modern representation, see Gergis et al., 2010; their Fig. 1.3). Note, in particular, the short-term deviations from the expected direct routes, which were due to adverse winds (indicated by wind arrows on the charts), or the absence of wind in the doldrums. In addition, some deviations were owing to currents, mostly in the Atlantic Ocean where the convoy was subject to strong West→East currents south of the Equator (e.g., Hunter, 1793).

It is interesting to note the discrepancies among the different observers, particularly as regards their crossing of the Equator and the Tropic of Cancer (in the northern hemisphere). While Collins, Tench, Hunter and White note that they crossed the Equator on 14 July 1787, Bradley and King refer to the 15$^{th}$ and Phillip cites 5 July 1787 (this may be a transcription error). One may think that this discrepancy could be related to the use of the astronomical (or nautical) versus the civil day, with the astronomical day starting at 12 noon rather than at 12 midnight. However, that explanation seems unlikely. Among those who cited their crossing of the Equator, Hunter and Bradley were the most experienced observers, and it is likely that the others followed their lead.[27]

Yet, Hunter indicates that the Equator was crossed during the "evening" of 14 July, while Bradley and King recorded specific times of, respectively, 5 p.m. and 8 p.m., but on the 15$^{th}$. These timings are well matched, and so perhaps the discrepancy is owing to the officers having completed their journal entries from memory at a later date. Similarly, both Collins and Bradley note that the Tropic of Cancer was crossed on 15 June 1787, but White's record refers to 10 June. We suspect that this latter

[26] See http://astro-expat.info/Data/First_Fleet_positions.html. This webpage has been saved for posterity at http://web.archive.org/web/20201116085541/http://astro-expat.info/Data/First_Fleet_positions.html by the *Wayback Machine*.

[27] As noted in our online database (note 24), White—who was based on the *Charlotte* without access to a chronometer or astronomical equipment—indicates time and time again that his geographic locations were obtained from measurements displayed by the *Sirius*.

**Fig. 1.19** "*Sirius* track from England to the Equator, & of the *Waakzaamheydt* with the crew of the *Sirius* from the Equator to England." The *Sirius*' itinerary is indicated by the dotted line. Chart from William Bradley's journal, *A Voyage to New South Wales*, ca. 1802. (Mitchell Library, State Library of New South Wales; Ref. 404927; out of copyright)

record is a mistake given the accuracy of Bradley's astronomical observations and of Collins' careful chronicling of the convoy's itinerary.

Our complete data set of 287 entries includes a total of 37 longitude and latitude determinations of well-defined geographic locations from First Fleet journals. To these, we added two of Cook's position determinations of features on the Tasmanian coast (Bradley, 1786–1792), given that they were cited by the First Fleet's chroniclers. Here, we will make an attempt at comparing the accuracy and precision of the contemporary measurements with modern values. Our complete data set, as well as a direct comparison with modern (2020) *Google Maps*® locations, is provided below; 'Time-keeper' refers to K1 measurements.

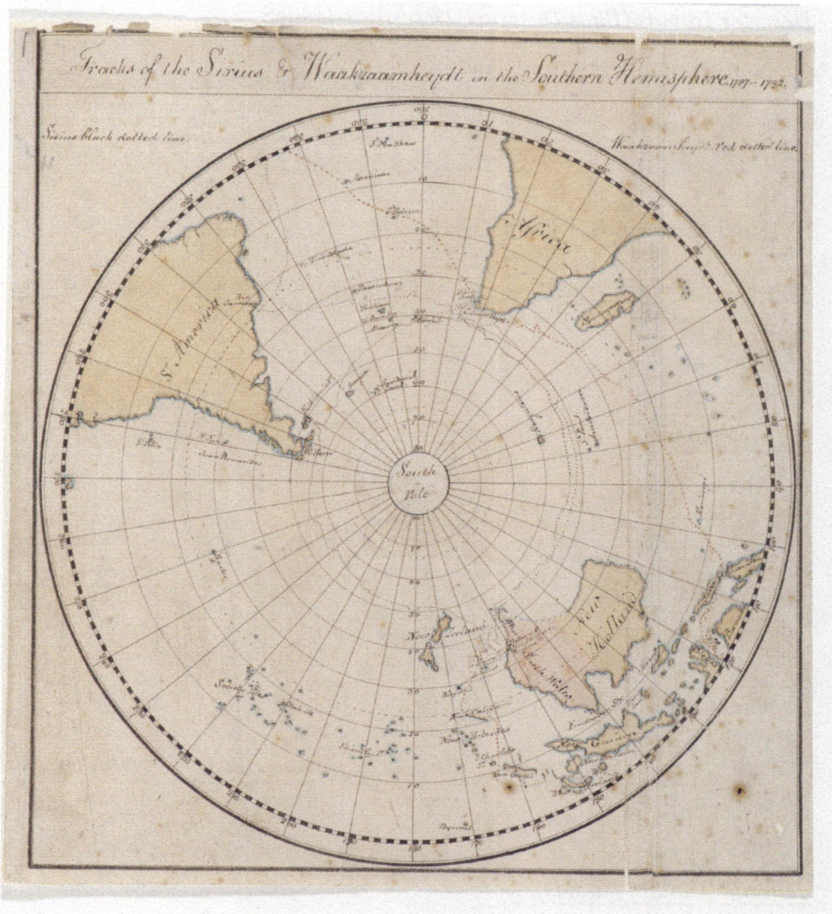


**Fig. 1.20** "Tracks of the *Sirius* & *Waakzaamheydt* in the Southern Hemisphere, 1787–1792." The *Sirius*' itinerary is indicated by the thin dotted line. Note the distinction between 'New Holland' and 'New South Wales' (including the size of the latter compared with today's state by the same name) on the map of Australia. Chart from William Bradley's journal, *A Voyage to New South Wales*, ca. 1802. (Mitchell Library, State Library of New South Wales; Ref. 404927; out of copyright)

1. *Deserter Islands (Ilhas Desertas, Desertas Islands)*:

a. Northernmost Deserter Island

| *Latitude* | *Longitude* | *Means* | *Observer* | *Modern Latitude* | *Modern Longitude* |
|---|---|---|---|---|---|
| 32°27′N | 16°35′W | Time-keeper | Bradley | 32°35′N | 16°33′W |

b. Southeasternmost Deserter Island

| *Latitude* | *Longitude* | *Means* | *Observer* | *Modern Latitude* | *Modern Longitude* |
|---|---|---|---|---|---|
| 32°29′N | 16°38′W | Time-keeper | Hunter | 32°25′N | 16°28′W |
| 32°28′N | | | Tench | | |

2. *Salvage Islands*:

c. Mean

| *Latitude* | *Longitude* | *Means* | *Observer* | *Modern Latitude* | *Modern Longitude* |
|---|---|---|---|---|---|
| 30°12′N | 15°56′W | Time-keeper | Bradley | 30°08′N | 15°52′W |
| 30°13′N | 15°56′W | Time-keeper | King | | |
| 30°12′N | 15°53′W | Time-keeper | Hunter | | |

d. Great Salvage, eastern side

| *Latitude* | *Longitude* | *Means* | *Observer* | *Modern Latitude* | *Modern Longitude* |
|---|---|---|---|---|---|
| 30°12′N | 15°39′W | | Tench | 30°08′N | 15°52′W |

3. *Tenerife (Canary Islands)*:

a. Santa Cruz bay

| *Latitude* | *Longitude* | *Means* | *Observer* | *Modern latitude* | *Modern longitude* |
|---|---|---|---|---|---|
| 28°30′N | 16°16′W | Time-keeper | Bradley | 28°29′N | 16°14′W |
| 28°29′N | 16°18′W | Time-keeper | Hunter | | |
| | 16°18′W | Time-keeper | King | | |
| | 16°18′W | | Dawes | | |

b. Santa Cruz (old town)

| *Latitude* | *Longitude* | *Means* | *Observer* | *Modern latitude* | *Modern longitude* |
|---|---|---|---|---|---|
| 28°28′N | 16°18′W | | Tench | 28°28′N | 16°15′W |

c. Teide volcano (peak)

| *Latitude* | *Longitude* | *Means* | *Observer* | *Modern latitude* | *Modern longitude* |
|---|---|---|---|---|---|
| 28°18′N | 16°31′W | | Bradley | 28°16′N | 16°39′W |

4. *Cape Verde Islands*:

a. Sal, northern end

| *Latitude* | *Longitude* | *Means* | *Observer* | *Modern latitude* | *Modern longitude* |
|---|---|---|---|---|---|
| 16°48′N | 23°03′W | Time-keeper | Hunter | 16°51′N | 22°55′W |
| 16°50′N | 23°02′W | Time-keeper | Bradley | | |

b. Sal, southern end

| *Latitude* | *Longitude* | *Means* | *Observer* | *Modern latitude* | *Modern longitude* |
|---|---|---|---|---|---|
| 16°40′N | 23°02′W | Time-keeper | Bradley | 16°35′N | 22°55′W |
| 16°40′N | 23°05′W | | Tench | | |
| 16°15′N | 22°51′W | Time-keeper | King | | |

c. Bona Vista (Boa Vista), northern end

| *Latitude* | *Longitude* | *Means* | *Observer* | *Modern latitude* | *Modern longitude* |
|---|---|---|---|---|---|
| 16°13′N | 22°51′W | Time-keeper | Hunter | 16°14′N | 22°47′W |

d. Bona Vista (Boa Vista), southern end

| *Latitude* | *Longitude* | *Means* | *Observer* | *Modern latitude* | *Modern longitude* |
|---|---|---|---|---|---|
| 15°59′N | 23°02′W | Time-keeper | Bradley | 15°58′N | 22°48′W |
| 15°52′N | 23°08′W | | Tench | | |
| 16°00′N | 22°51′W | Time-keeper | Hunter | | |

e. Isle of May (Maio), southern end

| *Latitude* | *Longitude* | *Means* | *Observer* | *Modern Latitude* | *Modern Longitude* |
|---|---|---|---|---|---|
| 15°11′N | 23°26′W | | Tench | 15°07′N | 23°10′W |

f. St. Jago, Port Praya fort

| *Latitude* | *Longitude* | *Means* | *Observer* | *Modern latitude* | *Modern longitude* |
|---|---|---|---|---|---|
| | 23°37′W | | Tench | 14°44′N | 23°31′W |

5. *Brazil*:

## a. Cape Frio (Cabo Frio)

| *Latitude* | *Longitude* | *Means* | *Observer* | *Modern latitude* | *Modern longitude* |
|---|---|---|---|---|---|
| 23°05′S | 41°40′W | | Tench, Dawes | 22°47′S | 41°56′W |
| 23°00′S | 41°44′W | Time-keeper | Bradley | | |
| 22°58′S | 41°40′W | Time-keeper | Hunter (*) | | |

(*) "It will appear hereafter that we had not the true rate of the watch, and consequently that [this] longitude is not correct" (Hunter, 1793: 17)

## b. Rio de Janeiro

| *Latitude* | *Longitude* | *Means* | *Observer* | *Modern latitude* | *Modern longitude* |
|---|---|---|---|---|---|
| 22°54′S | 42°40′W | | Bowes-Smyth | 22°54′S | 43°11′W |
| 22°54′S | 42°44′W | Time-keeper | Hunter (*) | | |
| 22°54′S | 43°19′W | | White | | |
| 22°54′S | 43°11′W | Lunar distances | Hunter | | |
| 22°54′S | 43°33′W | Lunar distances | Bradley | | |
| 22°54′S | 43°19′W | Jupiter's 3rd moon | Dawes, Portuguese astronomers | | |

(*) See the note to Table 5a

## c. Enchados (Enxadas) Island, Rio de Janeiro

| *Latitude* | *Longitude* | *Means* | *Observer* | *Modern latitude* | *Modern longitude* |
|---|---|---|---|---|---|
| | 43°21′W | Time-keeper | King | 22°53′S | 43°11′W |

6. *Final leg: from Table Bay to Botany Bay*:

## a. Table Bay

| *Latitude* | *Longitude* | *Means* | *Observer* | *Modern latitude* | *Modern longitude* |
|---|---|---|---|---|---|
| 34°22′S | 18°45′E | | Bowes-Smyth | 33°53′S | 18°27′E |

## b. Cape Town (old town)

| *Latitude* | *Longitude* | *Means* | *Observer* | *Modern latitude* | *Modern longitude* |
|---|---|---|---|---|---|
| 33°55′S | 18°24′30″E | Lunar distances | Hunter | 33°54′S | 18°25′E |
| | 18°23′ 55″E | Lunar distances | Collins | | |

c. Mewstone

| *Latitude* | *Longitude* | *Means* | *Observer* | *Modern latitude* | *Modern longitude* |
|---|---|---|---|---|---|
| 43°48′S | 146°25′E | | Tench, Cook | 43°44′S | 146°22′E |

d. Eddystone and Swilly Rock[28]

| *Latitude* | *Longitude* | *Feature* | *Observer* | *Modern latitude* | *Modern longitude* |
|---|---|---|---|---|---|
| 43°54′S | 147°09′E | Eddystone | Tench | 43°47′S | 147°01′E |
| 43°53′S | 147°09′E | Eddystone | Cook | | |
| 43°54′S | 147°03′E | Swilly Rock | Tench | 43°47′S | 147°01′E |

It is interesting to compare the accuracy of the contemporary longitude and latitude determinations with current-best estimates. On the whole, it transpires that the First Fleet's measurements were systematically offset by 2.4′ towards larger longitude differences with respect to the Greenwich meridian and by 3.6′ in latitude (offset towards northern latitudes), respectively, but with large standard deviations of 11.5′ and 7.0′, respectively. The systematic offset in longitude is reminiscent of a similar (but larger) systematic offset *in the same sense* recorded by Dawes, Hunter and Bradley for the geographic location of Dawes' observatory in the new colony compared with modern measurements (see Chap. 2). A similar discrepancy was found for the location of the semi-permanent observatory on the shores of Botany Bay established by the Lapérouse expedition. In Chap. 2 we will show that these systematic offsets are most likely owing to problems related to the accuracy of the contemporary almanac tables.

We can directly convert the systematic angular differences just cited to time differences. In this context, we are particularly interested in the longitude difference: the systematic offset of 2.4′ corresponds to a time discrepancy of 9.6 s. The longitude offset for Dawes' observatory amounted to almost 29 s of time (see Chap. 2). Therefore, at the latitude of Sydney, 33°52′08″ South, the corresponding average offset in longitude achieved during the First Fleet's voyage would be less than 4 km. Consideration of this accuracy, achieved during the First Fleet's voyage *at sea and on shore*, then begs the question as to why the longitude determination of Dawes' observatory *on shore only* was off by as much as 11 km.

Whereas some contemporary position determinations rivalled the accuracy and precision of modern measurements, particularly those obtained in well-known locations such as Santa Cruz de Tenerife, Rio de Janeiro and Cape Town, shipboard measurements of distant landmarks often incurred significant uncertainties. The most significant differences are found for some of the Cape Verde islands, for Cape

[28] The modern coordinates provided for 'Swilly Rock' are only approximate. Matthew Flinders (1774–1814) identified Swilly Rock as the westernmost of a pair of islets: "They had, a little before, passed two cliffy islets lying to seaward; of which the westernmost (*Swilly* of Furneaux) is like *Pedra Blanca* near the coast of China; the easternmost (*Eddystone* of Cook) resembles an awkward tower, and is about sixteen miles from the mainland." (Flinders, 1814: Section IV, part I).

Frio on the Brazilian coast and, surprisingly, for the geographic location of Table Bay. In fact, the Table Bay measurement recorded by Bowes-Smyth is one of the most discrepant determinations in our database: "13 October 1787 … Anchored at Cape of Good Hope. Lat[itude] 34°22′S, long[itude] 18°45′E" (Choat, 2020: 53). This location corresponds to the southeastern entrance to False Bay, south of Cape Town and just off the coast of present-day Pringle Bay village in the Western Cape. It is clear from Bradley's maps, however, that the convoy anchored in Table Bay (Bradley, 1786–1792). Bowes-Smyth's contemporary record may have been incorrectly copied, given that he was assigned to the *Lady Penrhyn* and thus without a direct means of geographic position determination. We suspect that some of the offshore measurements of island features may reflect the ships' locations rather than the island features themselves, but this cannot be verified based on the contemporary record.

Among the more experienced observers from whom we have sufficient numbers of measurements (Hunter: 9; Bradley: 9; and Tench: 10), Hunter's longitude determinations were most accurate (with a mean difference compared with modern measurements closest to zero). Hunter's nine longitude measurements deviate, on average, by −1.4′, compared with longitude differences of +4.2′ and +4.1′ resulting from, respectively, Bradley's and Tench's observations. The corresponding standard deviations for all three observers span a narrow range from 9.7′ to 11.4′. Latitude determinations have always been easier to obtain and were usually more accurate and precise (i.e., corresponding to smaller standard deviations) than longitude determinations. On average, with respect to modern measurements, our three most prolific observers determined their latitudes to within 1.9′ (Hunter), 2.0′ (Bradley) and 2.7′ (Tench), with all observers reporting latitudes offset from modern values to the north. Hunter's precision exceeded that of his companions, with a standard deviation of 3.9′ for his set of latitude determinations, compared with 5.1′ and 6.8′ for Bradley and Tench, respectively.

Whereas Dawes was known to be a careful astronomer, his measurements were good but not better than those of Hunter. It is perhaps surprising that Dawes only recorded a small number of longitude determinations. Nevertheless, it is clear from our careful perusal of his letters to Maskelyne that he obtained numerous lunar distance measurements. However, it appears that he did not regularly convert those observations to the convoy's longitude at the time of observation. This would have required lengthy calculations. With Dawes occupied on the First Fleet's vanguard ship, *Supply*, the remainder of the convoy under Hunter's command was clearly in safe hands.

From our modern perspective, it is challenging to adequately visualise Captain Phillip's remarkable leadership. Given the navigational means available at the time, it was indeed a major achievement to direct an entire convoy of 11 ships on a 24,000 km, 8-month voyage half-way around the world without the loss of a single ship, and to ensure that all ships arrived safely within a few days of one another with minimum loss of life despite the cramped living quarters. They could not have done it without a firm grasp of the principles of practical astronomy.

# References


Andrewes, W. J. H. (Ed.). (1996). *The quest for longitude*. Collection of Historical Scientific Instruments, Harvard University.

Anemaat, L. (2014). *Natural curiousity. Unseen art of the First Fleet*. University of New South Wales Press. Cited by Clarke (2015: 26, note 46).

Ashcroft, L. (2016). Australia's meteorological anniversary. Retrieved November 2, 2020, from https://lindenashcroft.com/2016/01/26/australias-meteorological-anniversary/

Atkinson, A. (1997). *The Europeans in Australia. A history* (Vol. 1). Oxford University Press.

*Australian Town and Country Journal*. (10 February 1883). *The Late Mrs Oakes* (Vol. XXVII, p. 17).

Bach, J. (ed.). (1968). *John Hunter: An Historical Journal of Events at Sydney and at Sea 1787–1792*. Angus and Robertson/Royal Australian Historical Society.

Barton, G. B. (1889). *History of New South Wales from the Records. I. Governor Phillip 1783–1789*. Charles Potter.

Basten, S. (2015). *Birth–Baptism Intervals for Family Historians*. Retrieved November 5, 2020, from https://www.familysearch.org/wiki/en/Birth-Baptism_Intervals_for_Family_Historians

Bayly, W. (1786). Letter to Sir Joseph Banks, dated 8 August 1786. Banks Papers, JBK/1/4, folio 237. Library Archives, Royal Botanic Gardens, Kew.

Beaglehole, J. C. (ed.). (1963). *The Endeavour Journal of Joseph Banks (Second Edition)*. Two volumes. Angus and Robertson.

Berry, B. M., & Schofield, R. S. (1971). Age at baptism in pre-industrial England. *Population Studies, 25*, 453–463.

Bladen, F. M. (ed.). (1893). Journal and letters of Daniel Southwell (The 'Southwell Papers'). In *Historical Records of New South Wales*. II (Grose and Patterson): Appendix D (pp. 661–732).

Blagg, T. M., & Andrews, S. (1913). *Hampshire Parish registers. Marriages*. XV. Phillimore & Co. Retrieved November 5, 2020, from https://archive.org/stream/hampshireparishr15phil#page/128/mode/2up

Board of Longitude (1786). Confirmed minutes, 14 November 1786. RGO 14/6, folios 101–102.

Board of Longitude (1787). Accounts and minutes, 21 April 1787. RGO 14/18, folio 257r and RGO 14/6 (p. 109).

Bradley, W. (1786–1792). In *A Voyage to New South Wales: The Journal of William Bradley RN of HMS Sirius, 1786–1792*. William Dixon Foundation Publication, 11. Ure Smith (1969).

Campbell, J. (1788). Letter to Baron Dulcie, dated 12 July. Unpublished. Cited by Mackaness (1937).

Choat, C. (ed.). (2020). *Arthur Bowes Smyth's Journal of a Voyage from Portsmouth to New South Wales and China*. Retrieved November 13, 2020, from http://gutenberg.net.au/ebooks20/2000051h.html

Clarke, R. M. (2015). *Vanguards of empire: The lives of William Dawes, Watkin Tench and George Worgan*. Ph.D. thesis. The Australian National University.

Collins, D. (1798). *An Account of the English Colony in New South Wales: with remarks on the dispositions, customs, manners, &c. of the native inhabitants of that country to which are added, some particulars of New Zealand, compiled, by permission, from the Mss. from Lieutenant-Governor King*. Vol. 1. London, T. Cadell Jun. and W. Davies. Retrieved September 30, 2020, from http://adc.library.usyd.edu.au/data-2/colacc1.pdf

Coltheart, L., & The Museum of Australian Democracy. (2011). Governor Phillip's Instructions 25 April 1787 (GB). In *Australia's Story: Documenting a Democracy*. Retrieved October 17, 2020, from https://www.foundingdocs.gov.au/item-sdid-68.html

Crittenden, V. (2010). William Dawes: The unknown man of the First Fleet. In *MARGIN: Life & letters in early Australia*. Mulini Press.

Currer-Jones, A. (1930). *William Dawes, R.M., 1762–1836*. Torquay, Smith.

Dagelet, J. L. (1788). Letter to William Dawes, dated 3 March 1788. Mitchell Library, State Library of New South Wales, Sydney. Microfilm CY 1414, frames 31–37.

*Daily Telegraph*, (23 January 1888). Australian Patriarchs. Centennial Supplement. P. 6.
Dawes, W. (1786a). *List of instruments proper for making astronomical observations at Botany Bay*. Royal Society of London. Miscellaneous MS vol. 7, no. 56A (ca. 1 November 1786); Board of Longitude, *Minutes*, 14 November 1786. RGO 14 (Vol. 6, pp. 100–104). Cambridge University Library.
Dawes, W. (1786b). Letter to Nevil Maskelyne, dated 28 December 1786. RGO 14/48, folio 239v.
Dawes, W. (1787a). Letter to Nevil Maskelyne, dated 16 January 1787. RGO 14/48, folio 244r.
Dawes, W. (1787b). Letter to Nevil Maskelyne, dated 19 January 1787. RGO 14/48, folios 245r–v.
Dawes, W. (1787c). Letter to Nevil Maskelyne, dated 25 January 1787. RGO 14/48, folio 250r.
Dawes, W. (1787d). Letter to Nevil Maskelyne, dated 8 February 1787. RGO 14/48, folio 251r.
Dawes, W. (1787e). Letter to Nevil Maskelyne, dated 20 March 1787. RGO 14/48, folios 255r–256v.
Dawes, W. (1787f). Letter to Nevil Maskelyne, dated 1 May 1787, RGO 14/48, folio 257r–v.
Dawes, W. (1787g). Letter to Nevil Maskelyne, dated 8 May 1787, RGO 14/48, folio 259v.
Dawes, W. (1787h). Letter to Nevil Maskelyne, dated 20 May 1787, RGO 14/48, folio 261r.
Dawes, W. (1787i). Letter to Nevil Maskelyne, dated 5 June 1787, RGO 14/48, folios 262–263.
Dawes, W. (1787j). Letter to Nevil Maskelyne, dated 5 June 1787, RGO 14/48, folio 264r.
Dawes, W. (1787k). Letter to Nevil Maskelyne, dated 9 June 1787, RGO 14/48, folios 265r–v.
Dawes, W. (1787l). Letter to Nevil Maskelyne, dated 6 August 1787, RGO 14/48, folio 268r.
Dawes, W. (1787m). Letter to Nevil Maskelyne, dated 3 September 1787, RGO 14/48, folios 269–274.
Dawes, W. (1787n). Letter to Nevil Maskelyne, dated 9 November 1787, RGO 14/48, folios 277–279.
Dawes, W. (1788). Letter to Nevil Maskelyne, dated 10 July 1788, RGO 14/48, folios 283–284.
Dawes, W. (1791). *The notebooks of William Dawes on the Aboriginal language of Sydney*. Vol. 3. London, School of Oriental and Asian Studies. Retrieved September 30, 2020, from https://www.williamdawes.org/
de Grijs, R. (2017). Time and Time again: Determination of Longitude at Sea in the 17th Century. Bristol UK. Institute of Physics Publishing.
de Grijs, R. (2020). A (not so) brief history of lunar distances: Lunar longitude determination at sea before the chronometer. *Journal of Astronomical History and Heritage, 23*, 495–522.
de Grijs, R., & Jacob, A. (2021). William Dawes: Practical astronomy on the "First Fleet" from England to Australia. *Journal of Astronomical History and Heritage, 24*, 7–40. Chapter 2.
Dickinson, H. W. (2007). *Educating the Royal Navy*. Routledge.
Eldershaw, M. B. (1972). *Phillip of Australia*. Discovery Press.
Flinders, M. (1814). *A Voyage to Terra Australis* (Vol. 1). W. Bulmer and Co. Retrieved January 6, 2021, from http://gutenberg.net.au/ebooks/e00049.html
Forbes, E. G. (1975). *Greenwich Observatory: The story of Britain's oldest scientific institution, the Royal Greenwich Observatory at Greenwich and Herstmonceux, 1675–1975*. Taylor and Francis.
Gascoigne, J. (1994). *Joseph Banks and the English Enlightenment*. Cambridge University Press.
Gergis, J., Brohan, P., & Allan, R. (2010). The weather of the First Fleet voyage to Botany Bay, 1787–1788. *Weather, 65*, 315–319.
Gibson, R. (2012). *26 Views of the Starburst World: William Dawes at Sydney Cove 1788–91*. University of Western Australia Press.
Gillen, M. (1989). *The founders of Australia: A biographical dictionary of the First Fleet*. Library of Australian History.
Halley, E., 1705. *Astronomicae cometicae synopsis* (pamphlet). Oxford; English translation: Halley, E., 1705. A Synopsis of the Astronomy of Comets. London, John Senex. https://archive.org/stream/synopsisofastron00hall#page/n3/mode/2up Accessed 4 November 2020.
Hampshire parish registers. (1653–1875). *Parish Church of Portsmouth (Hampshire)*. Church of England.
Hawkins, J. B. (1979). Observatories in Australia 1788–1830. *The Australasian Antique Collector, 19*, 98–104.

House of Lords Sessional Papers (HLSP). (1714–1805). Evidence of William Dawes. Minutes of Evidence taken on the Second Reading of the Bill. In *An act to Prohibit the Trading for Slaves on the Coast of Africa, within Certain Limits* (Vol. 3).

Howe, R. (Lord Howe). (1787). Letter to William Dawes, dated 18 January 1787. RGO 14/48, copied by Dawes. Folios 245v–246r.

Howse, D. (n.d.). *Bayly, William (bap. 1738, d. 1810).* Oxford Dictionary of National Biography. Retrieved November 9, 2020, from https://doi.org/10.1093/ref:odnb/1769

Howse, D. (1979). The principal scientific instruments taken on Captain Cook's voyages of exploration, 1768–80. *Mariner's Mirror, 65*, 119–135.

Howse, D. (2004). Dawes, William (1762–1836). In *Oxford Dictionary of National Biography.* Oxford University Press. https://doi.org/10.1093/ref:odnb/50961

Howse, D., & Hutchinson, B. (1969). *The clocks and watches of Captain James Cook, 1769–1969.* Antiquarian Horological Society.

Hughes, R. (1987). *The Fatal Shore. A history of the transportation of Convicts to Australia, 1787–1868.* Collins Harvill.

Hunter, J. (1793). *An historical journal of the transactions at Port Jackson and Norfolk Island, with the discoveries which have been made in New South Wales and in the Southern Ocean, since the publication of Phillip's voyage, compiled from the official papers; including the journals of Governors Phillip and King, and of Lieut. Ball; and the voyages from the first sailing of the Sirius in 1787, to the return of that ship's company to England in 1792.* John Stockdale.

James, A. (2012). The William Dawes observatory. In *Southern astronomers and Australian astronomy.* Retrieved November 2, 2020, from http://www.southastrodel.com/Page031b.htm

Jervis, J. (Lord St. Vincent). (1801). The Letters of Lord St. Vincent, I. *Publications of the Navy Records Society*, LV (1921–1922). Accessed November 6, 2020, from https://archive.org/stream/navyrecords55navyuoft/navyrecords55navyuoft_djvu.txt

Jukic, A. M., Baird, D. D., Weinberg, C. R., McConnaughey, D. R., & Wilcox, A. J. (2013). Length of human pregnancy and contributors to its natural variation. *Human Reproduction, 28*, 2848–2855. https://doi.org/10.1093/humrep/det297

King, P. G. (1787). *Private Journal*, ca. 19 December 1787. Cited by Morrison and Barko (2009).

King, P. G. (1787–1790). *The Journal of Philip Gidley King, Lieutenant, R.N.*; Australian Documents Library. (1980). Retrieved November 12, 2020, from http://purl.library.usyd.edu.au/setis/id/kinjour

King, P. G. (1788). *Private Journal*, 3 January 1788. Cited by Morrison and Barko (2009).

King, J. (1779). Letter to the Board of Longitude, dated 10 June. *Banks Papers*. Published in Barton (1889: 422).

King, R. J. (1998). Terra Australis, New Holland and New South Wales: The Treaty of Tordesillas and Australia. *The Globe, 47*, 35–55.

Laurie, P. S. (1988). William Dawes, and Australia's first observatory. *Quarterly Journal of the Royal Astronomical Society, 29*, 469–482.

Lewis, M. (1939). *England's sea-officers.* Allen and Unwin.

Liverpool Plains Shire Council. (2020). *First Fleet Memorial Gardens, Wallabadah, NSW*. Retrieved November 5, 2020, from https://www.visitliverpoolplains.com.au/attractions-in-wallabadah, https://static1.squarespace.com/static/5b28b2d1372b966c17b905fe/t/5dfac4f69a89ef5096bdde2b/1576715524601/First-Fleet-Brochure-Names-list.pdf

Lloyd, C. (ed.). (1952). *Naval Miscellany*. Naval Records Society.

Lloyd, C. (1966). The Royal Naval Colleges at Portsmouth and Greenwich. *The Mariner's Mirror, 52*, 145–156.

Mackaness, G. (1937). *Admiral Arthur Phillip: Founder of New South Wales.* Angus and Robertson.

Macleod, R., & Rehbock, P. F. (Eds.). (1988). *Nature in Its Greatest Extent: Western Science in the Pacific.* University of Hawaii Press.

Macquarie, L. (1988). Convictism and Colonization, 1788 to 1828. *Journal of the Australian Population Association, 5*(Supplement 1), 31–43.

Mander-Jones, P. (1966). Dawes, William (1762–1836). In *Australian Dictionary of Biography* (Vol. 1). Retrieved September 28, 2020, from http://adb.anu.edu.au/biography/dawes-william-1968

Maskelyne, N. (1786a). Letter to Sir Joseph Banks, dated 17 October 1786. RGO 35/63 (photocopies of material held in other archives). National Library of Australia.

Maskelyne, N. (1786b). Letter to Sir Joseph Banks, dated 8 November 1786. RGO 35/64 (photocopies of material held in other archives). National Library of Australia.

Maskelyne, N. (1786c). Advertisement of the expected return of the Comet of 1532 and 1661 in the Year 1788. *Philosophical Transactions of the Royal Society of London, 76*, 426–431.

Maskelyne, N. (1787). *The Nautical Almanac and Astronomical Ephemeris for the Year 1767*. W. Richardson and S. Clark.

McAfee, R. J. (1981). Dawes's meteorological journal. Bureau of Meteorology, *Historical Note* No. 2.

Morrison, D., & Barko, I. (2009). Dagelet and Dawes: Their meeting, their instruments and the first scientific experiments on Australian soil. *Historical Records of Australian Science, 20*, 1–40.

Nathan, D., Rayner, S., & Brown, S. (eds.). (2009). *William Dawes' notebooks on the Aboriginal language of Sydney, 1790–1791*. School of Oriental and African Studies. http://www.williamdawes.org

National Library of Australia. (n.d.) *Who was the first child born to 'free' parents in NSW?* Retrieved January 6, 2020, from https://www.nla.gov.au/faq/who-was-the-first-child-born-to-free-parents-in-nsw

*Nepean Times*. (3 February 1883). *Death of Mrs Oakes*. P. 6.

Nicolas, P. H. (1845). *Historical Records of the Royal Marine Forces* (Vol. 1). Thomas and William Boone.

Orchiston, W. (2016). *Exploring the history of New Zealand Astronomy: Trials, tribulations, telescopes and transits*. Springer.

Pembroke, M. (2013). *Arthur Phillip, Sailor, Mercenary, Governor*. Hardie Grant Books.

Phillip, A. (1790a). *The voyage of Governor Phillip to Botany Bay: with an account of the establishment of the colonies of Port Jackson and Norfolk Island ... To which are added the journals of Lieuts. Shortland, Watts, Ball, and Capt. Marshall*. John Stockdale. Retrieved September 30, 2020, from http://gutenberg.net.au/ebooks/e00101.html

Phillip, A. (1790b). Letter to Lord Sydney, dated 12 February 1790. In: *Historical Records of Australia*, Series 1, Vol. 1 (1788–1796), 140–147.

Project Gutenberg Australia. (2019). *The First Fleet—Journals, books, letters and orations by people on the first fleet*. Retrieved November 13–16, 2020, from http://gutenberg.net.au/first-fleet.html

Saunders, S. D. (1990). *Astronomy in colonial New South Wales, 1788–1858*. Ph.D. thesis. University of Sydney. Retrieved January 6, 2021, from http://hdl.handle.net/2123/1988

Southwell, D. (1788). Letter to Mrs. Southwell, dated 5 May 1788. *Journal and Letters of Daniel Southwell ("The Southwell Papers")*. Retrieved November 6, 2020, from http://gutenberg.net.au/ebooks12/1204411h.html

Southwell, D. (1790). Letter to Mrs. Southwell, dated 14 April 1790. In: *Journal and Letters of Daniel Southwell ("The Southwell Papers")*. Retrieved November 6, 2020, from http://gutenberg.net.au/ebooks12/1204411h.html

Steele, J. M. (2005). *The Aboriginal language of Sydney*. MA thesis. Macquarie University.

Stephens, P. (1786). Letter to Lieutenant-General William Sydney Smith, Marines, Portsmouth Division, dated 25 October 1786. The National Archives, Admiralty 2/1177, folio 521.

Tench, W. (1979). *Sydney's first four years, being a reprint of 'A narrative of the Expedition to Botany Bay; with an account of New South Wales, its Productions, Inhabitants, &c., to which is subjoined, A List of the Civil and Military Establishments at Port Jackson' and 'A Complete Account of the Settlement at Port Jackson in New South Wales, including an accurate description of the Situation of the Colony; of the Natives; and of its Natural Productions'*. Library of Australian History/Royal Australian Historical Society Retrieved November 12, 2020, from http://gutenberg.net.au/ebooks/e00083.html

Thomas, S. N. A. (2011). William Dawes in Antigua. *Journal of Colonialism and Colonial History, 12*(1). https://doi.org/10.1353/cch.2011.0011

Twiss, W. (1786). Letter to Brook Watson, dated 24 October 1786; forwarded to Evan Nepean, 2 November 1786. The National Archives, Home Office 42/10, folio 393.

White, J. (1790). *Journal of a Voyage to New South Wales with sixty-five plates of nondescript animals, birds, lizards, serpents, curious cones of trees and other natural productions.* Retrieved November 5, 2020, from http://gutenberg.net.au/ebooks03/0301531.txt

Wood, G. A. (1924). Lieutenant William Dawes and Captain Watkin Tench. *Journal of the Royal Australian Historical Society, 10*, 1–24.

Wright, H. (1926). The Memorial of Lieutenant William Dawes, 1826. *Journal of the Royal Australian Historical Society, 12*, 227–230.

Wright, H. (1927). Further note on Lieutenant William Dawes. *Journal of the Royal Australian Historical Society, 13*, 63–64.

# Chapter 2
# Scientific Beginnings: William Dawes' Observatories in Context

## 2.1 An Encounter with the French

Captain and Governor-designate Arthur Phillip and William Dawes arrived in Botany Bay[1] on H.M.A.T. *Supply* on 18 January 1788, followed by the fastest convict transports—the *Alexander*, *Friendship* and *Scarborough*—the next day. The remainder of the First Fleet, including H.M.S. *Sirius*, arrived on the 20$^{th}$. On 19 January, Dawes was assigned to the party tasked with exploring the shores of Botany Bay in search of a suitable site for the new settlement, as we learn from Lieutenant Philip Gidley King's private journal:

> ... at 3 the boats were hoisted out & Governor Phillip & some Officers belonging to y$^e$ *Supply* with Lieut Dawes & Myself, landed on y$^e$ North side of y$^e$ [Botany] Bay & just looked at the face of the Country, which is as Mr. [James] Cook remarks very much like the Moors in England, ...
>
> As preparations were made in the Morning to explore & examine all y$^e$ North side of y$^e$ Bay, Major Ross was added to our party which consisted of y$^e$ Governor, Lieuts [Henry Lidgbird] Ball, Dawes, [John] Long & myself in three boats, & left y$^e$ Brig, at 11 o'clock.
>
> ... at 10 o'clock the Governor, Lieut Governor, & Capt Hunter went over to the south side of y$^e$ bay a little within point Sutherland, & I was ordered with Lieut Dawes of the Marines, to explore all y$^e$ South side of y$^e$ Bay, & trace the two inlets on the South side as high as possible. (King, 1787–1790: entry 19 January 1788)

This chapter is based on de Grijs and Jacob (2021).

[1] When Cook first arrived in the body of water now known as Botany Bay—called *Yarra* by the resident Bidjigal, Gweagal and Kameygal people of the Sydney Aboriginal language group (West, ca. 1882; Attenbrow, 2009)—he initially named it 'Stingrays Harbour': "The great quantity of these sort of fish found in this place occasioned my giving it the name of Stingrays Harbour" (Cook's log, 6 May 1770). Some time after his departure, however, he changed his mind: "The great quantity of plants [the expedition's botanists] Mr. [Joseph] Banks and Dr. [Daniel] Solander found in this place occasioned my giving it the name of ~~Botanist~~ Botany Bay" (Cook's correction; Beaglehole, 1968).



R. de Grijs and A. Jacob, *William Dawes*, Springer Biographies,
https://doi.org/10.1007/978-3-031-38774-6_2

The settlers soon realised, that Botany Bay was not their hoped-for site to establish a permanent settlement. On 21 January, an exploration party led by Phillip sailed from Botany Bay and discovered that Port Jackson,[2] some 12 km further north, provided adequate shelter, deep water for the ships' anchorage, fresh water and ideal conditions for agriculture (e.g., Parker, 2009):

> 21st at 6 in the Morning Governor Phillip Capt Hunter & ye Judge Advocate with the Masters of the *Sirius* & *Supply* & some petty officers, set off in three boats to examine Port Jackson & Broken Bay & Mr Dawes with myself & a petty Officer went away in two boats to explore the upper part of ye [Botany] Bay in which we found some very large & deep coves & some extensive lagoons but no fresh water we returned onboard again at about eight in the evening, & on the 23rd in the Evening Governor Phillip & his party returned. Orders were immediately given to get the Transports & Victuallers in readiness to sail, as Port Jackson was judged a very proper place to form an establish[ment] in. (King, 1787–1790: entry 21 January 1788)

In Phillip's words, Port Jackson represented "... the finest harbour in the world, in which a thousand sail of the line may ride in the most perfect security" (Phillip, 1790: 55). Surgeon-general Arthur Bowes-Smyth said,

> The finest terras's [*sic*], lawns and grottos, with distinct plantations of the tallest and most stately trees I ever saw in any nobleman's ground in England, cannot excel in beauty those wh[ich]. Nature now presented to our view. The singing of the various birds among the trees, and the flight of the numerous parraquets, lorrequets, cockatoos, and macaws [macaw habitats do not exist in Australia], made all around appear like an enchantment; the stupendous rocks from the summit of the hills and down to the very water's edge hang'g [*sic*] over in a most awful way from above, and form'g [*sic*] the most commodious quays by the water, beggard [*sic*] all description. (Fidlon & Ryan, 1979; Hill, 2015: 107)

Following the exploration party's return to Botany Bay on 23 January 1788, the convoy prepared to depart for Port Jackson. However, "... to the infinite surprise of everybody, we saw two large ships in the offing" (Bowes-Smyth, 1790: 87). Unbeknownst to the British, a French scientific expedition led by Jean-François de Galaup, comte de Lapérouse (1741–1788; Marchant, 1967), had arrived just outside Botany Bay. The French were equally surprised, as we learn from Lapérouse's journal:

> We spent the whole of the 24th [January 1788] plying in sight of Botany Bay, but that day we had a spectacle; this was the British Squadron at anchor with their pennants and ensigns which we could plainly distinguish. (Dunmore, 1995; transl.)

The French expedition counted 10 scientists among its complement of 114, including Dagelet, whom we will meet again shortly (see also Barko, 2007). In fact, the arrival of the French frigates, *Astrolabe* and *Boussole* ('Compass'), added urgency to Dawes' and Captain Phillip's orders to establish an observatory in the new colony. Lapérouse's Royal instructions of 15 February 1785 included such orders explicitly:

---

[2] Port Jackson encompasses Sydney Harbour, Middle Harbour, North Harbour and the Lane Cove and Parramatta Rivers. It was named by Cook as the H.M.B. *Endeavour* sailed by its entrance: "... at noon we were ... about 2 or 3 miles from the land and abrest [*sic*] of a bay or harbour within there appeared to be a safe anchorage which I called Port Jackson." Cook named the harbour system after Sir George Jackson (1725–1822), Lord Commissioner of the British Admiralty and Judge Advocate of the Fleet.

> Immediately upon arriving in a harbour, he will select an appropriate site on which to erect the tents and the observatory, and will set up a guard. ... Separately from observations relating to the determination of latitudes and longitudes, for which every known and practicable method will be used, and those needed to assess the declination and inclination of the dipping needle [a freely suspended magnetic needle used to determine the local magnetic inclination], he will ensure that any celestial phenomenon which may be visible be observed; and on every occasion he will give the astronomers all the help and facilities necessary for the success of their work. (Dunmore, 1994: cxlii–cxliii; transl.)

Despite gale-force conditions, the First Fleet weighed anchor on 25 January 1788, *en route* to Port Jackson. The site the exploration party had selected for settlement, some 10 km into the natural harbour, was known as War-ran (Warrane) by the Indigenous Gadigal people. Today, the inlet adjacent to Sydney's central business district is known as Sydney Cove. The cove featured sheltered deep water close to the shore and a fresh-water tributary—now known as the Tank Stream—where "... ships can anchor so close to the shore that, at small expense, quays may be built at which the largest ships can unload" (Phillip, 1790: 55). Indeed, today the Overseas Passenger Terminal at Sydney's Circular Quay routinely hosts some of the world's largest cruise ships.

The focus of this chapter is on Dawes' Observatory in New South Wales, the developments leading up to its foundation (Sect. 2.2) and the role it played in the social fabric of the new colony (Sect. 2.3). While some of this story is reasonably well known, our aim here is to provide a comprehensive account of the history and operations of the first permanent[3] British astronomical observatory in the southern hemisphere. We have uncovered a number of new insights, and we have also established the observatory's site beyond any reasonable doubt based on a careful assessment of the historical record (Sect. 2.4). Our historical research has been based on a wide range of records. Among the online documents we consulted were the full set of papers of the Board of Longitude and eyewitness accounts of the First Fleet's voyage. In addition, we explored the online and physical holdings of the State Library and State Archives of New South Wales, the library of the Royal Australian Historical Society (RAHS), all in greater Sydney, and the National Library of Australia in Canberra.

## 2.2 An Observatory for the Colony

Dawes was keen to start construction of the settlement's observatory. However, from his letters to Maskelyne we learn that his duties as Marine officer on the *Sirius* interfered (Dawes, 1788a). His application for a shore assignment was initially refused,

[3] The adjective 'permanent' is appropriate in this context, for two reasons: (*i*) literature references to Dawes' observatory commonly identify the structure as 'permanent'; and (*ii*) although the actual building fell into disrepair only a few years after its construction, here the contrast of note is between tent observatories and facilities made of wood or bricks and mortar. The latter are longer-lived and hence 'permanent' is a more pertinent description.

although he was promised that he would be considered for the first land-based vacancy to arise (Mander-Jones, 1966; Chap. 1).

Meanwhile, the French had built a stockade, an observatory and a garden to grow fresh produce on the headland now known as La Perouse, on Botany Bay's northern shore.[4] On behalf of Governor Phillip, Dawes and King paid their French counterparts a visit on 2 February 1788, offering them any assistance they might need. They left Sydney Cove at 02:00 that morning for an eight-hour sailing and rowing voyage (Fidlon & Ryan, 1980: 37–40; Clarke, 2015: 33), battling strong southerly winds on their way to Botany Bay:

> ... & the following day at 2 in y$^{e}$ Morning Lieut Dawes of y$^{e}$ Marines & myself [King] sett off in a Cutter for Botany Bay, to visit Monsieur De La perouse on the part of Governor Phillip & to offer him whatever he might have occassion [*sic*] for, we got down to y$^{e}$ harbours mouth at day light, finding a light air from y$^{e}$ Southward, we were obliged to row all y$^{e}$ way & arrived onboard y$^{e}$ *Boussole* at 10 o'clock in y$^{e}$ Morning where we were received with the greatest politeness & attention by Monsieur de la perouse & his Officers, after delivering my Message to him, he returned his thanks to y$^{e}$ Governor for his attention to him, & made y$^{e}$ same offers which he had received, & added that as he should be in France in 15 Months & having Stores &c. enough onboard for three Years he should be happy to oblige Mr Phillip with any that he might want — ... (King, 1787–1790: entry 1 February 1788)

The party stayed at the temporary French settlement until 5 o'clock the following morning and returned to the *Sirius* by 7 p.m. on 3 February, "... having been obliged to row all the way against the wind and a great swell" (King, 1787–1790: entry 3 February 1788). We learn from King's private journal that the French observatory was already set up for astronomical observations to be taken:

> After dinner I attended y$^{e}$ Commodore & other Officers onshore where I found him [Lapérouse] quite established, having thrown round his Tents a Stoccade, guarded by two small guns in which he is setting up two Long boats which he had in frames, an observatory tent was also fixed here, in which were an Astronomical Quadrant, Clocks &c. under the Management of Monsieur Dagelet Astronomer, & one of y$^{e}$ Académie des Sciences at Paris.[5] (King, 1788)

The quadrant, one of five made by Claude Langlois (ca. 1700–1756; the most highly regarded scientific instrument maker in France between about 1730 and 1756), was the most important of Dagelet's instruments. It was on loan from Paris Observatory. In addition, the French observatory at Botany Bay was equipped with a meridian

[4] Dagelet's letter of 3 March 1788 (see below) contains the geographic coordinates of the French observatory (see also Barko, 2007; Morrison and Barko, 2009), which Dagelet obtained using the lunar distance method and de Lalande's astronomical tables (see, e.g., de Grijs, 2020). In doing so, he was assisted by De Roux d'Arbaud, his former student at the Royal Military Academy in Paris, whom he had selected in preference to another student, Napoléon Bonaparte (see, e.g., Bartel, 1954: 133–134). In addition, King recorded in his journal that the expedition's priest, Claude-François Joseph Louis Receveur (1757–1788), "... was buried near where the French had their observatory ..." (Morrison and Barko, 2009: 27–28, note 13). This gravesite is located a short walk east–northeastwards from the entrance of today's La Perouse Museum.

[5] Dagelet had been a member of the French Académie des Sciences since 1785 and a Professor of Mathematics at the Royal Military Academy in Paris since 1777.

telescope of a design by Alexis-Marie Abbé de Rochon (1741–1817), an invariable pendulum for gravity measurements similar to the pendulum Charles-Marie de La Condamine (1701–1774) had used in the Americas—which was 'invariable' since it had a fixed pendulum length—three astronomical pendulum clocks made by the celebrated clockmaker Jean-André Lepaute (1720–1789), Borda reflecting circles (see, e.g., de Grijs, 2020) and English-manufactured sextants (see, e.g., Barko, 2007).

However, the observing conditions at Botany Bay were far from ideal. On 6 February 1788, Dagelet wrote to Marie Jean Antoine Nicolas de Caritat (1743–1794), Marquis de Condorcet and permanent secretary of the Académie Royale des Sciences, "I am blinded by the bites of flies which weigh me down in my wretched observatory" (Morrison & Barko, 2009: 28, note 14).

Dawes and Dagelet, the more senior scholar, thus met in person for the first—and only—time on 2 February 1788. As men of science, their conversation naturally turned to Dawes' assignment to establish an observatory in the new settlement and to Cook's nautical and astronomical observations. The French unreservedly praised the accuracy of Cook's measurements, with Lapérouse adding magnanimously, "Mr. Cook did so much that he left me nothing to do but to admire his work" (Barton, 1889).

The French and English contingents maintained cordial relations during Lapérouse's six-week sojourn on the shores of Botany Bay. Although Dagelet never visited Dawes at his Port Jackson base, Lieutenant Charles-Marie Fantin de Boutin (born ca. 1760) visited Sydney Cove and reported back to Dagelet. De Boutin had inspected the foundations of Dawes' observatory, most likely around 21 February, on a visit with Father Jean-André Mongez (1750–1788), the French expedition's chaplain (for the relevant journal entries, see Protos, 1988). Dagelet did not undertake the arduous journey to the Sydney Cove settlement because of his comparatively poor health, as we learn from the eminent Parisian astronomer Joseph Jérôme Lefrançois de Lalande (1732–1807):

> In his tenth and last letter [now lost], written on 1 March 1788 from Botany Bay in New Holland, he wrote to me that he had received the visit of Mr Doves [Dawes], an English astronomer, who was preparing to set up an observatory and to whom d'Agelet's [Dagelet's] experience and advice were not unuseful. He wanted to visit him in his turn at Port Jackson, but as one had to follow a track where each traveller needed to cut his way through mountains, cliffs, woods and marshes, La Pérouse thought his weakened constitution would not withstand it. He himself thought that before taking to the sea again, he should temper his astronomer's zeal. We do not know whether his health was restored and whether it held up as we have not received any subsequent news from this ill-fated expedition. (de Lalande, 1803: 713)

On 3 March 1788, Dagelet sent Dawes a letter[6] that offered advice on how to set up his observatory, which was partly based on de Boutin's description of the most likely site: "M. Boutin has easily explained to me the plan you have chosen and that you are currently having carried out" (Morrison & Barko, 2009: 19; transl.).

[6] Although Dagelet's letter implies that it was delivered by an English seaman ("I profit from the presence of your seamen to send you my farewells."), there are no records of any English visit to Botany Bay at this time. Instead, in an excerpt from the logbook of the *Alexander*, we read "4 March 1788 – Several French officers came from Botany Bay" (Morrison and Barko, 2009: 27, note 7).

Dagelet's letter, written a week before the Lapérouse expedition left their base at Botany Bay, included suggestions about the observatory's layout, the placement of Dawes' instruments and ideas for future instruments—including, ambitiously, "… a sector [transit circle] with a radius of 6–8 feet …" (1.8–2.4 m). He also provided ideas as to the types of observations Dawes might pursue, encouraging him to

> … devote time to the comparative research of the right ascension of the Sun and some stars which provide the setting of coelum Australis … If you were to observe some of Venus's conjunctions, either superior or inferior, you know perhaps that I have studied its movements and intend to return to this research one day. (Morrison & Barko, 2009: 18; transl., original emphasis)

Dagelet particularly encouraged Dawes to attempt "… the measurement of a degree of the [terrestrial] meridian in this hemisphere" (Barko, 2007), a standardised measurement of length based on the Earth's circumference that had been an important objective of the Académie Royale des Sciences for more than a century (e.g., Mouton, 1670: 432). In this same context, Dagelet provided Dawes with a set of geographic coordinates for his temporary observatory on the shores of Botany Bay which he had obtained with the help of his assistant, Roux d'Arbaud (e.g., Barko, 2007; see below).

Dagelet urged Dawes to never "… forget to note down the smallest variations or slightest changes, or the improvements you make in your instruments … [which] would help you avoid all the 'ifs' and 'buts' of men who lack trust" (cited by Saunders, 1990: 107–108).[7] He offered that if Dawes was keen to compose a scientific contribution on the movements of the Sun and the stars, he would be pleased to present it to the Académie Royale des Sciences on Dawes' behalf.

Despite their brief acquaintance, both men clearly got along well (e.g., Barko, 2007). In a letter from Dagelet to his mentor, de Lalande, which he asked Dawes to send via Maskelyne, Dagelet sounded hopeful, advising that

> … at Botany Bay he had come across an English astronomer furnished with instruments who was preparing to carry out numerous observations and we may have the satisfaction of correspondence with our Antipodes. (Barko, 2007: 21)

It took until March 1788 before Dawes would be assigned to a substantial shore-based role in Governor Phillip's quest to secure the territory for the British Crown. He was appointed[8] as the settlement's engineer, artillery officer[9] and surveyor, roles

[7] Despite Dawes' careful and systematic observations, in a letter to Maskelyne on 18 June 1792, he expressed distress that his measurements had been "… so little successful …".

[8] Dawes (1788e) admitted to Maskelyne that when he was asked to take on the roles of engineer and artillery officer, he complained to Governor Phillip of the excessive workload compared with his Marine duties. Phillip was "… very highly offended and several letters past [*sic*] between us". Although the disagreement dissolved of its own accord, it set the tone of the men's tense relationship during Dawes' tenure in New South Wales (see also Chap. 4).

[9] One of Dawes' first tasks was to transfer the new settlement's ordnance from the *Sirius* to the shore, including the *Sirius*' guns. Wood (1924) explained, "… so eight of them were landed on the west point before the Observatory, and Lieutenant Dawes threw up a small breastwork in front of them, and afterwards built a platform for them" (see also Collins, 1798: 41, 166, 173 and 189).

that allowed him shore access, although he remained part of the *Sirius*' complement of Marines until 23 May 1788.[10] Nevertheless, already in February 1788 do we learn of Dawes' efforts to commence construction of an observatory:

> Among the buildings that were undertaken shortly after our arrival must be mentioned an observatory, which was marked out on the western point of the cove, to which the astronomical instruments, which had been sent out by the Board of Longitude for the purpose of observing the comet which was expected to be seen about the end of this year. The construction of this building was placed under the direction of Lieut. Dawes, of the Marines, who, having made this branch of science his peculiar study, was appointed by the Board of Longitude to make astronomical observations in this country. (Collins, 1798: 15)

Hunter's first map of Sydney Cove (see Fig. 2.1), dated 1 March 1788, already shows the observatory's site and an anticipated outline of a two-building structure. The observatory building appears large compared with other contemporary buildings. In comparison with modern maps, the location of the Governor's residence is shown close to its actual position, whereas the topography of Sydney Cove maps well onto the modern shoreline (see below). Meanwhile, land-based astronomical observations were obtained from a portable tent observatory similar to that carried by Cook and his astronomer, Charles Green (1734–1771), on the H.M.B. *Endeavour*'s round-the-world voyage of 1768–1771 (e.g., Haynes et al., 1996: 31–32): see Fig. 2.2. Assisted by four marines and a few convicts, construction of a wooden observatory had well and truly commenced by 30 April 1788:

> I have ... with the assistance of 4 Marines of my own party & 3 Convicts whic[h] the Gov$^{r}$ has allowed me, cleared a point of land of trees, and am now getting on as fast as possible, with an Observatory which I hope will be completed and the instruments in it by July sometime. This has not however been done without a good deal of my own and my servants' bodily labour which shall when necessary be cheerfully employed in the same cause. The situation struck me at first sight to be so eligible and all the necessary materials so conveniently at hand, that I did not hesitate a moment to determine on setting about a permanent one at once a sketch will say more than many words. (our emphasis: see Fig. 2.5; Dawes, 1788a)

Dawes had been given permission to build the structure on the thickly wooded, rocky promontory less than a kilometre from the settlement on the western side of Sydney Cove. It was known to the Gadigal people as Tar-ra or Tjara (Dara).[11] The site of Dawes' Observatory was separated from the main settlement by a track along the waterfront. At Dawes' request, Hunter "... was pleased to honour this Point [Tar-ra] by calling it Point Maskelyne" (Dawes, 1788e). However, the area was colloquially referred to as Dawes' Point, a name which remains in use today. (We will discuss the observatory's location in more detail in Sect. 2.4.) The Gadigal people considered

---

[10] In his private journal, Bradley (1969: 108) writes, "Friday. 23$^{rd}$. [May 1788] ... Lieu$^{t}$. Dawes was this day discharged from the *Sirius* to the Battalion.".

[11] Steele (2005: 83) and Gibson (2010: 24) point out that 'dara' in the local Indigenous language may simply be a suffix meaning 'here'. In his *Notebooks* on the Sydney Aboriginal language, Dawes writes, "They were not speaking of *Dara*, for since, I have heard them repeat *dara* in the same word when I think they could not refer to that place. It seems to me to be peculiarly used when it is spoken as of rowing to a certain place to bring another back with you. But this is mere conjecture" (Dawes' emphasis; Dawes, 1787–1788: *Notebook A*, 17).

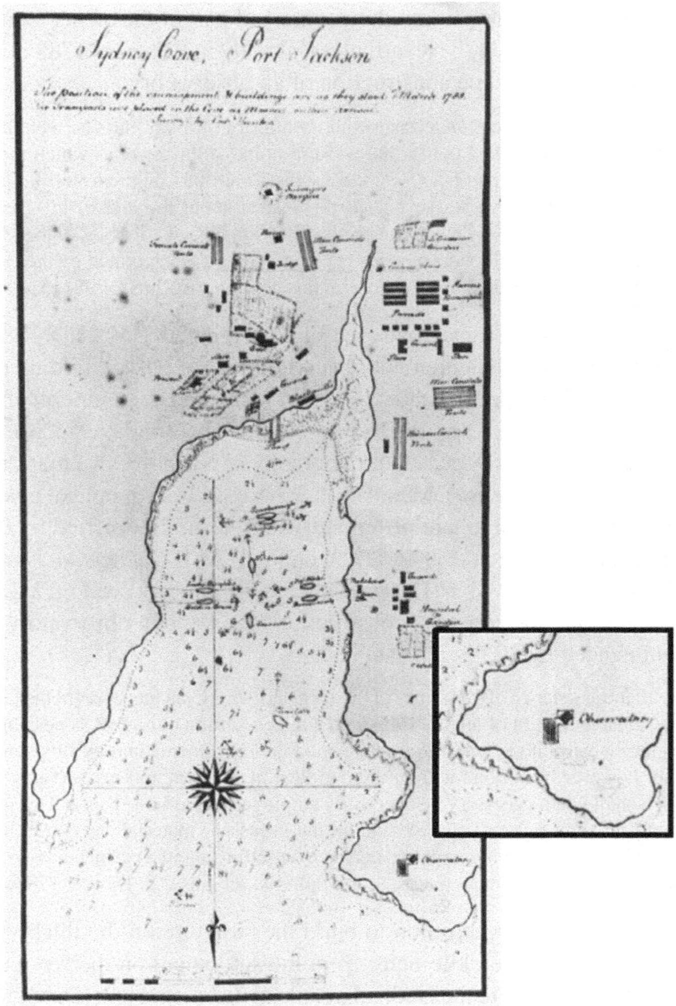


**Fig. 2.1** "Sydney Cove, Port Jackson, surveyed by Captain Hunter, 1 March 1788" (State Library of New South Wales; FL6068816; out of copyright)

the headland on the western side of Sydney Cove a safe and welcoming location to share friendship and knowledge (Moran & McAllister, 2020; see also Chap. 3).

Construction of the observatory proceeded well those first few months, despite a chronic shortage of materials, particularly linseed oil and white lead for paint, nails and sailcloth (Laurie, 1988). In fact, Phillip confided in Dawes that if he had known the difficulties they would face in sourcing building materials, he would have opted to bring an expensive, pre-fabricated observatory building along on the voyage from England (Dawes, 1788b: 283v–284r):

**Fig. 2.2** Typical tent observatory used by shipboard astronomers in the seventeenth and eighteenth centuries (Wales & Bayly, 1777: Plate II; out of copyright)

> Lamp oil will be absolutely necessary as the allowance of candle is very small & they will very soon all be expended ... On Monday last I came to reside at the Point, which is about ½ a mile from the Camp ... the Gov$^{r}$ has paid great attention to the Observatory, but very little has been in his power, on account of the inconceivable number of things to do & the very great scarcity of carpenters, & indeed of good working men of any trade, he desires me to tell you (with his best compliments) that if he could have foreseen the difficulty of getting a permanent observatory (for it is a permanent one) erected, he would much rather have been at the expense of bringing a framed one out with him … (original emphasis)

Dawes managed to motivate his construction crew by providing his Marines with shoes, rum and water "… now & then …" during the worst of the summer heat:

> ... in the excessive hot weather we had in Febr$^{y}$ March & April, it was absolutely necessary to give them some Rum & water now & then, besides which as the brush wood tore their shoes very much I gave each a pair of shoes, in return they wrought almost miraculously as is testified with astonishment by all who have seen what they have done ... (Dawes, 1788b: 283v)

On 10 July 1788, he sent an expense claim for £10/10/- (10 guineas) "... in favor of Mr. John Palmer[12] ..." to Maskelyne to cover these costs (Dawes, 1788b). By July 1788, Tench reported that the observatory was

> ... nearly completed, and when fitted up with the telescopes and other astronomical instruments sent out by the Board of Longitude, will afford a desirable retreat from the listlessness of a camp evening at Port Jackson. (Tench, 1789: Ch. XVI)

Despite the delay incurred by a lack of suitable building materials and qualified tradesmen in the colony, Dawes moved into the new observatory's living quarters in the first week of July. Governor Phillip provided an update on the observatory's construction in his letter to Lord Sydney of 9 July 1788:

> On the point of land which forms the west side of the cove an Observatory is building, under the direction of Lieutenant Dawes, who is charged by the Board of Longitude with observing the expected comet. (Phillip, 1788)

In his letter to Maskelyne of 10 July 1788, Dawes (1788b) was clearly excited about the progress he had made. He told Maskelyne that he was considering extending his time in the colony by signing on for another three years. However, having said so, just two days later he placed a number of conditions on his request for an extension. Most importantly, he declared that he would need to be given a more permanent appointment at the observatory to supplement his Marine income:

> With respect to remaining in this Country I have to observe that I shall, during the three Years for which I offer'd my Services, pay all possible attention to the observatory; & that, from a wish to make every return in my power, for the very polite and friendly manner in which you have been pleased to treat me; but, as staying here without any certainty of some addition to my income would be doing myself and Family a material injury; I propose, unless I receive some assurance of such addition, to return at the end of that term. It is not that I wish an allowance from the B$^{d}$. of Longitude in particular, but if any appointment can be thought of in any department, which may tie me to this Country, I shall with the greatest pleasure attend to the observatory to the utmost of my power without thinking of any stipulation with the board of Longitude in particular. Nevertheless, as I have had the happiness of being in some measure the means of establishing a permanent observatory in this country, I should, if a fix'd observer be thought necessary be equally happy to fill that place. If not, I will (while I can with common justice remain in this country) attend to the observatory as strictly as if I were paid a salary for that business, ... (Dawes, 1788c)

Although Dawes did not state it directly at this time, later that year he expressed his conviction that he had given the observatory as much as he could: "... no application shall be wanting in me ..." (Dawes, 1788e). He hoped that a properly qualified successor could be found by the time his six years of service were up. Clearly, he had meanwhile decided to sign on for a second three-year term. In fact, the historical record offers us a useful insight into his thinking: in a letter to his father he admitted that the "... principal reason ..." to sign on for a second term was

[12] John Palmer (1760–1833), purser on H.M.S. *Sirius*, was appointed Commissary of New South Wales, a role in which he was responsible for the colony's supplies.

> … the uncertainty of any other person being appointed to succeed him in the astronomical way, and the great loss, in his opinion, it would be to the science were the opportunity to be missed of making proper use of those valuable instruments. (Bladen & Britton, 1893: 423)

Given the observatory's almost-operational state by mid-July 1788, Dawes arranged for his instruments to be transferred to shore, unpacked and ready for use by the first week of August. He was keen to start his search for Maskelyne's comet (Dawes, 1788a). Routine scientific operations commenced on 14 September 1788.

## 2.3 Completion of the Observatory and Routine Operations

Dawes' timber and canvas observatory was, as expected for a building constructed during the colony's first year of settlement, small and far from impressive. Nevertheless, its completion represented a milestone for the young colony, given that it was the first building dedicated to astronomy and meteorology in New South Wales.

Dawes' observatory consisted of two buildings made of timber, one of which was octagonal, nine feet (2.7 m) in diameter, with a white conical canvas roof and built on a "… very large and firm …" rock (James, 2012):

> I have got an exceeding good stone cut into the form of a frustum of an octagonal pyramid whose base is 2 ft 5 in [~74 cm] and top 1 ft 5 in [~43 cm] this to be placed on the centre of the octagonal room & the quadrant on the top of it. The roof is to turn round on three or more rollers. *e*, A stair case of communication between the upper & lower room. *f*, the proposed place for the astron[omical] clock. The roof of the lower room is to be so constructed and to make good against the side of the upper room and the ridge of it is to be several inches below the top of the quadrant. (Dawes, 1788a)

The revolving canvas roof on the octagonal building included a retractable shutter. The roof's apex was located approximately a foot (~30 cm) from the vertical position, so that astronomical observations could also be made at zenith (James, 2012). The canvas roof was nailed to vertical wooden poles that rested on cannon balls in a gouged wooden track to allow for rotation (Haynes et al., 1996: 31–32). It is possible that the tent observatory used during the voyage and upon their arrival in Port Jackson was repurposed to provide the canvas roofing material (Kerr, 1986).

Dawes selected a large outcrop of either bedrock or a large 'floater' as the solid foundation for his quadrant room. The quadrant required a foundation that was stable against vibrations, "… on a good stone … an octagonal pyramid …", so as to work effectively. This thus determined the final site selection for the observatory (Saunders, 1990) at the northeastern end of the Tar-ra headland, northwest of what is now Campbells Cove.

A second, rectangular building measuring 12 by 16 feet ($3.7 \times 4.9\ m^2$) resembled a lean-to shed and was constructed against that rock (Dawes, 1788a; see also Fig. 2.4, top). It served as Dawes' study and living quarters. A staircase connected both buildings (see Fig. 2.3, right). Although the only historical drawing of Dawes' observatory is a sketch included in his 30 April 1788 letter to Maskelyne, the building

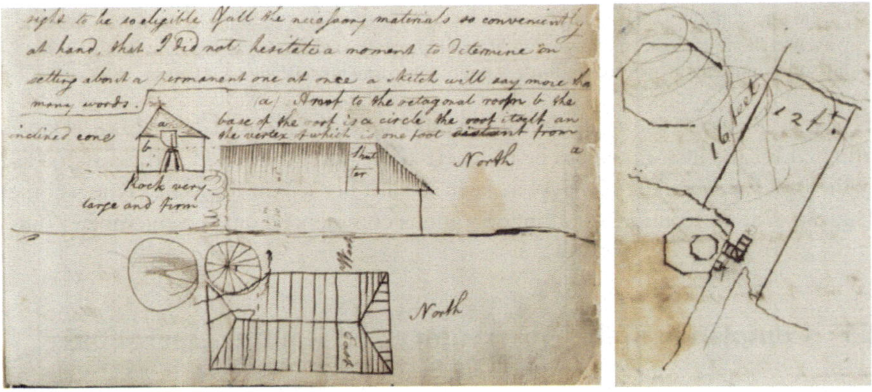


**Fig. 2.3** (*Left*) Sketches by William Dawes of his cottage and observatory on Point Maskelyne. Note that the slightly faded circle at bottom left is unlikely meant to be part of the final sketch. (*Right*) Observatory building orientation with respect to the bedrock at Point Maskelyne. Notice the staircase connecting the large rectangular building with the octagonal building on top of the bedrock. The small square to the left of the staircase may be an indication of the clock's location. (Cambridge University Library; Board of Longitude Papers, RGO 14/48, folios 281r–v; out of copyright)

was carefully reconstructed in the now-defunct historical theme park, 'Old Sydney Town', near Gosford (New South Wales): see Fig. 2.4 (bottom).

Dawes used the larger building to store his portable instruments, except for his thermometers, which were most likely stored in the well-ventilated observatory building (McAfee, 1981). The larger and more sensitive instruments, including his quadrant and the Shelton pendulum clock, were fixed in place. The historical record indicates that the quadrant was already put to practical use well before the observatory had been completed. It was clearly the most important instrument among the fledgling observatory's equipment, since it was used to accurately determine local time, longitude and latitude, and to calibrate the colony's clocks.

Dagelet referred to Dawes' quadrant in his letter of 3 March 1788, knowledge he most likely obtained following de Boutin's visit on 21 February: "I find that your q.c. [*quart de cercle*; quadrant] is perfectly well placed and leaves nothing to be desired from any point of view" (Morrison & Barko, 2009). The earliest observations with the quadrant at Sydney Cove may have been carried out by Bradley of the *Sirius*, rather than by Dawes (e.g., Johnson, 1998; Morrison & Barko, 2009): see Sect. 2.4.

In his letter of 17 November 1788, Dawes informed Maskelyne about the arrangements he had made for the Shelton clock,

> On the 6$^{th}$ Sept.$^{r}$, I got the clock scrwed [*sic*] to the frame which is let into the niche in a very large solid stone and there wedged exceedingly firm and propped in front by a short, stout piece of plank to another very large stone so that I believe it is impossible to fix a clock up much better. (Dawes, 1788e)

Dawes expressed his satisfaction with the clock's firm foundation once again in a letter of 16 April 1790. He told Maskelyne that the clock had been "... admirably

**Fig. 2.4** (*Top*) Sketch of Dawes' observatory at Sydney Cove (*Credit* Rod Bashford; McAfee, 1981; no copyright restrictions); (*bottom*) Reconstruction of Dawes' observatory at Old Sydney Town (after Jackson, 2018: Fig. 7; no known copyright restrictions)

well fixt [*sic*] … into a niche in the solid rock which has never moved since the foundation of the world & the feet of the frame are leaded into a rock below the former …" (Dawes, 1790a; original emphasis).

Dawes devoted himself to 'Observation' (for details, see Chap. 3), which encompassed taking standard astronomical, meteorological, magnetic and tidal measurements, drawing maps and recording the Sydney Aboriginal language in collaboration with his Gadigal language partner, a young girl called Patyegarang (e.g., Gibson, 2010; Nathan et al., 2009; Pybus, 2009; Thomas, 2013). On 14 September 1788, he commenced routine astronomical and weather observations at his observatory, thereby taking on the role of the colony's unofficial meteorologist as well[13] (Russell, 1877: 2; McAfee, 1981; Orchiston, 1989; Derrick, 2019; see Chap. 3). As such, he truly deserves the epitaph of Australia's first on-shore meteorologist.[14]

However, Governor Phillip continued to order Dawes to take on ever more tasks in support of the fledgling colony, which eventually led to his frustrations getting the better of him (see also Chap. 4). In a strongly worded letter written in July 1790, he told Maskelyne that "… the very few men of business in this country have always each of them more to do than he can by any efforts perform properly", adding that "I do more real business than any besides [the chaplain] in the country" (Dawes, 1790b). Most damningly, however, as well as unexpectedly indiscreetly, he complained to Maskelyne about Governor Phillip's apparent disinterest in the observatory's operations (e.g., Clarke, 2015: 42). Yet, as Governor, Phillip was ultimately in charge of the colony's day-to-day operations, and so Dawes had no choice but to act upon Phillip's practical orders—to the detriment of his scientific interests. This internal conflict may have exacerbated the men's apparently newly developed mutual antipathy. It eventually led to a complete breakdown of their personal relationship, which we will explore in more detail in Chap. 4.

In addition to his duties as officer of the Marines and his roles as artillery officer and the colony's engineer, Dawes was tasked with designing the batteries for the colony's defence. Moreover, the alleged ineffectiveness of the settlement's Surveyor-General, August(us) Alt (1731–1815), significantly added to Dawes' workload, since he was expected to come through as the settlement's *de facto* surveyor. His main and lasting achievements in this latter area—likely undertaken with the astronomical quadrant as his principal tool—include his surveying and planning of the Government farm. In addition, he surveyed[15] and planned the first streets and allotments in Sydney and Parramatta, the colony's second settlement, presently a major Sydney suburb (e.g., Haynes et al., 1996: 31–32). Figure 2.5, *A Survey of The Settlement in New South Wales, New Holland, by Lt. William Dawes, engraved by A. Dulon & L. Poates*, is the only remaining Crown plan (survey chart) of Sydney by Dawes himself (reproduced in 1886).

Dawes made his observatory into a welcoming place for everyone. It soon became the settlement's intellectual and cultural centre. Lieutenant Daniel Southwell (ca. 1764–1797), an officer from the *Sirius*, was clearly impressed: "He has a great share

[13] Dawes' meticulous meteorological work set the standard for the implementation of meteorological duties at later state observatories, which provided a straightforward public justification for the associated costs (Haynes et al., 1996).

[14] The earliest meteorological measurements in Sydney Cove were taken by Bradley on board H.M.S. *Sirius*. He recorded temperature and atmospheric pressure at noon each day during the First Fleet's voyage. He continued measuring atmospheric pressure until February 1788 and temperature until early October 1788. From 20 to 25 January 1788, shipboard pressure and temperature observations were also measured by John White (Ashcroft, 2016; Ashcroft et al., 2014).

[15] Initially, when surveyed in the weeks after the arrival of H.M.S. *Sirius*, one of the islands in Sydney Harbour was named 'Dawes Island'. In 1850, the island at the confluence of Iron Cove and the Parramatta River was renamed 'Spectacle Island' by John Skinner Prout (1805–1876), the artist and mapmaker. The "… two flat rock platforms, covered with scrub and divided by an isthmus 55 yards … partially dry at low tide …" (Bradley, 1788), resembled a pair of spectacles.

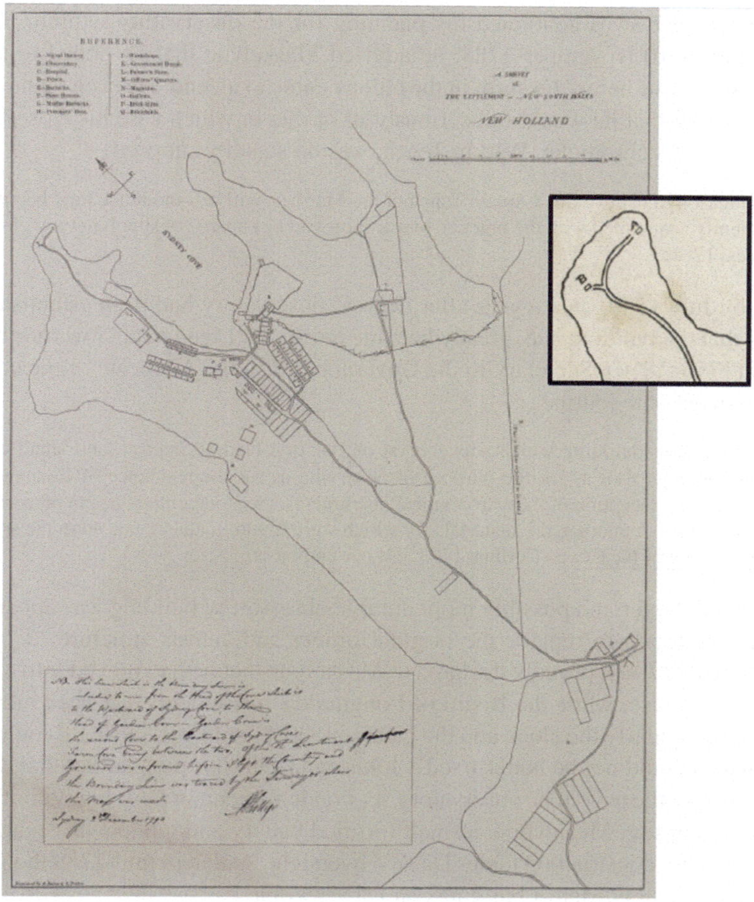


**Fig. 2.5** "A Survey of The Settlement in New South Wales, New Holland, by Lt. William Dawes, engraved by A. Dulon & L. Poates (Sydney, ca. 1886)." Dawes' observatory is building 'B'. (National Library of Australia; out of copyright)

of genuine knowledge, studious yet ever cheerful and the goodness of his disposition renders him esteemed and respected by all who know him" (cited by James, 2012).

On the other hand, from Mrs. Elizabeth Macarthur (1766–1850), the wife of John Macarthur (1767–1834), Lieutenant on the Second Fleet, we learn that Dawes was "… so much engaged with the stars that to mortal eyes he [was] not always visible …" (Macarthur-Onslow, 1973: 28)—this despite having agreed to act as her astronomy and botany instructor. In Dawes' (mental) absence, Tench often provided much-needed social interaction, leading Elizabeth Macarthur to refer to both men as "… those gentlemen and a few others [who were] chief among whom we visit …" (*Ibid.*).

Meanwhile, Dawes continued his planning for the observatory's future. In the same letter of 10 November 1788, he advised Maskelyne that he hoped to find a suitable successor before his term in the colony came to an end. He even planned for the eventuality that he would fall seriously ill or die, in which case his close friend and the colony's chronicler, Watkin Tench, was on standby support:

> I have reason to believe that Captain Tench of the Marines, will in a moderate time become sufficiently acquainted with the practice of astronomy to be capable of supplying my place. (Dawes, 1788e)

Yet, by July 1789, only a year after Dawes' observatory had been established, it appears that the building had already become too small. Lieutenant Governor David Collins (1756–1810), Secretary to the Governor and Judge Advocate of the colony, wrote in his private journal,

> The observatory building which was erected on our first landing being found small and inconvenient, as well as for the purpose of observing as for the residence of Lieutenant Dawes and the reception of the astronomical instruments, the stone-cutters began preparing stone to construct another, the materials for which were found in abundance upon the spot, the west point of the Cove. (Collins, 1798: 61; our emphasis)

A second, larger and possibly more durable observatory building was apparently under construction to replace the original timber and canvas structure. By April 1790, Dawes (1790a) proudly declared to Maskelyne that he had moved into a more comfortable house, while the Board of Longitude's instruments had been mounted securely, particularly the clock and the quadrant: "[i]f could you see it… [you would agree that] it could not be better fixed". Johnson (1998) has pointed out that during construction of the new stone observatory, its wooden precursor most likely continued to see frequent use. Meanwhile, a small fortified battery and gun powder magazine were also under construction under Dawes' oversight, so that in mid-1789 there may have been a small cluster of buildings on Dawes Point.

Most views of Dawes Point between 1788 and the early nineteenth century only show a single building on site, often associated with the guns and flagstaff of Dawes' Battery (e.g., McCormick, 1987). A view of the headland from 1793 or 1794 carries the title "… taken from the Flagstaff opposite the Observatory", suggesting that Dawes' stone observatory building was still known by that designation and probably still in use, likely as a guardhouse for the garrison or a signal station. An undated painting, possibly completed around 1813, by John William Lewin (1770–1819), the colony's first professional artist, titled "Fish catch and Dawes Point, Sydney Harbour", shows a single rectangular building, a rare representation of a stone building at a lower elevation situated North–South against the bedrock: see Fig. 2.6.

An additional clue as to the existence of a stone observatory was unearthed in October 1925, when the original battery established by Dawes was demolished to facilitate construction of the southern approaches to the Sydney Harbour Bridge. A stone marked 'R. R. 1789' (see Fig. 2.7, right) located in the internal stone wall of the old garrison building on the site may have originated from Dawes' second, stone

**Fig. 2.6** "Fish catch and Dawes' Point, Sydney Harbour", ca. 1813; John William Lewin. *Inset* enlargement of the buildings at Dawes' Point (Art Gallery of South Australia; out of copyright)

observatory building.[16] On 19 February 1926, Kathleen Butler, Secretary to John Job Crew Bradfield (1867–1943), Chief Engineer of the Sydney Harbour Bridge, advised the Principal Librarian of the Public Library of New South Wales (see Fig. 2.7, left) that this particular stone was

> … probably part of the first Observatory, being afterwards incorporated in the Military buildings, which consisted of four rooms with stone walls nearly three feet thick. It was

[16] If, as for his first wooden observatory, Dawes received help from a number of convicts and Marines in constructing his stone observatory, and assuming that 'R. R.' are the initials of the labourer who cut the stone in 1789, this possibly refers to Robert Ruth (b. 1764), a convict transported on the *Charlotte* (https://hmssirius.com.au/robert-ruth-convict-charlotte-1788/) or, perhaps less likely, to Robert Ryan (1758–1823), a Marine assigned to the *Prince of Wales* (https://peopleaustralia.anu.edu.au/biography/ryan-robert-31076). Both were used to manual labour. Ruth was a seaman by trade and, later in life, employed as a labourer for 12 months by the settler William Mitchell, former H.M.S. *Sirius* crew. Ryan was a nailer by trade and familiar with farm work. As a Marine officer, Major Robert Ross would not have been employed in construction. Other individuals on the First Fleet with the same initials included a number of ships' crew, who were either dead already by 1789 (Robert Ranson; *Alexander*) or otherwise fully occupied (Robinson Reid; *Supply*, built the Rose Hill Packet, which launched in October 1789). Richard Richardson, a marine on H.M.A.T. *Supply*, had been assigned to the shore-based marine contingent in November 1788, which presumably kept him fully occupied as well.

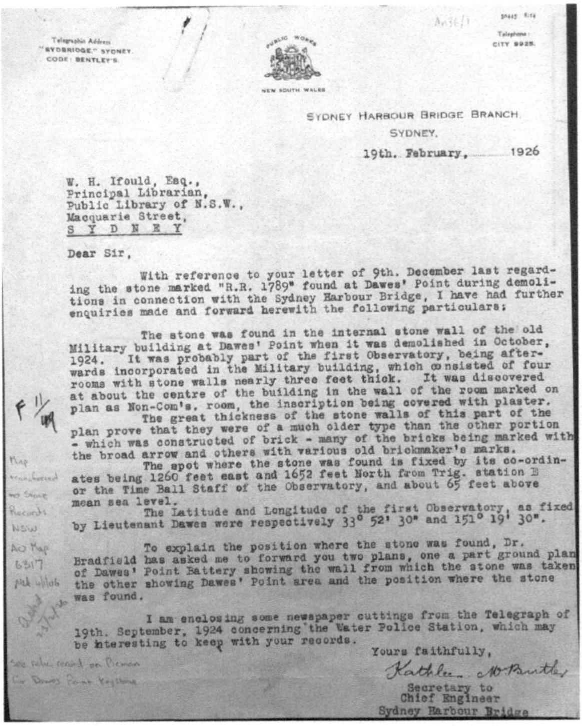

Telegraphic Address "SYDBRIDGE," SYDNEY. CODE: BENTLEY'S.

PUBLIC WORKS NEW SOUTH WALES

Telephone: CITY 9925.

SYDNEY HARBOUR BRIDGE BRANCH
SYDNEY.
19th. February, 1926

W. H. Ifould, Esq.,
Principal Librarian,
Public Library of N.S.W.,
Macquarie Street,
S Y D N E Y

Dear Sir,

With reference to your letter of 9th. December last regarding the stone marked "R.R. 1789" found at Dawes' Point during demolitions in connection with the Sydney Harbour Bridge, I have had further enquiries made and forward herewith the following particulars:

The stone was found in the internal stone wall of the old Military building at Dawes' Point when it was demolished in October, 1924. It was probably part of the first Observatory, being afterwards incorporated in the Military building, which consisted of four rooms with stone walls nearly three feet thick. It was discovered at about the centre of the building in the wall of the room marked on plan as Non-Com's. room, the inscription being covered with plaster.

The great thickness of the stone walls of this part of the plan prove that they were of a much older type than the other portion - which was constructed of brick - many of the bricks being marked with the broad arrow and others with various old brickmaker's marks.

The spot where the stone was found is fixed by its co-ordinates being 1260 feet east and 1652 feet North from Trig. station E or the Time Ball Staff of the Observatory, and about 65 feet above mean sea level.

The Latitude and Longitude of the first Observatory, as fixed by Lieutenant Dawes were respectively 33° 52' 30" and 151° 19' 30".

To explain the position where the stone was found, Dr. Bradfield has asked me to forward you two plans, one a part ground plan of Dawes' Point Battery showing the wall from which the stone was taken the other showing Dawes' Point area and the position where the stone was found.

I am enclosing some newspaper cuttings from the Telegraph of 19th. September, 1924 concerning the Water Police Station, which may be interesting to keep with your records.

Yours faithfully,
Kathleen McButler
Secretary to
Chief Engineer
Sydney Harbour Bridge

**Fig. 2.7** (*Left*) Letter regarding building stone 'R. R. 1789' found at Dawes Point from the Secretary to the Chief Engineer Sydney Harbour Bridge to the Principal Librarian, Public Library of New South Wales, dated 19 February 1926. (State Library of New South Wales; reproduced with permission) (*right*) Building stone 'R. R. 1789' (State Library of New South Wales. Call number: XR 9; record: 16AJ2aqn; no copyright)

> discovered at about the centre of the building in the wall of the room marked on the plan as the Non-Com's room, the inscription being covered with plaster.
>
> The great thickness of the stone walls of this part of the building prove that they were of a much older type than the other portion – which was constructed of brick – many of the bricks being marked with a broad arrow and others with various old brickmaker's marks. (Butler, 1926)

As we have seen already, Dawes and Governor Phillip did not get along. Dawes felt that he was unreasonably held back in his astronomical aspirations, while for Phillip juggling the young settlement's safety, security and practical day-to-day operations were at the forefront of his mind. Their mutual dislike and disagreement came to a head by the end of 1790. (We will explore their mutual antipathy in detail in Chap. 4.) Nevertheless, Dawes was keen to remain in the colony for an additional three-year term. However, even before Phillip formally notified him of his decision, Dawes expressed his doubts as to the potential success of his application: "I think it at least ten to one that I shall return with the Marine detachment" (Dawes, 1790b). And indeed, Phillip did not accept Dawes' reasons for his offending behaviour. In

December 1791, he left for England on the H.M.S. *Gorgon*, a 44-gun fifth-rate Royal Navy 'Man of War' in the Third Fleet.

Upon his departure from New South Wales, Dawes took with him the instruments, including the astronomical clocks, he had borrowed from the Board of Longitude. K1 returned to England in 1792 on the *Supply* (Hawkins, 1979). Apparently, Dawes' back-up plan to have Tench step into the role of the colony's astronomer was not activated. The building structure fell into disuse and disrepair. The observatory building collapsed. However, Collins (1798) reports that by the end of 1791 the rectangular wooden building had been appropriated as a guard room, a platform for a flagstaff, and that a cannon had been installed just behind it (Kerr, 1986). Sydney's scientific beginnings had come to a halt to give way to Britain's military might.

The last record of Dawes' observatory is found in the diary of John Crosley (1762–1817), the astronomer on board H.M.S. *Providence*, which called into Sydney on 28 August 1795 during her four-and-a-half-year round-the-world voyage known as the 'Vancouver Expedition':

> I went on shore and examined the place where Mr. Dawes' observatory was built but found nothing standing but the uprights which supported the roof and the pillar on which he placed his quadrants. (cited by James, 2012)

Nevertheless, European maps and charts of Sydney Cove continued to include Dawes' observatory until at least 1798: see Fig. 2.8 for an example.

## 2.4 A Memorial at Odds with the Historical Record

The precise location of Dawes' observatory is still a matter of debate. Here, we will shed further light on that debate. We aim to provide a comprehensive assessment of the most likely site.

Let us start by offering an assessment of the various geographic longitude and latitude determinations for the observatory available in the historical record. McGuffie and Henderson-Sellers (2012) noted that all accounts, except for Dawes' (1788e) own determination, cite a latitude of 33° 52′ 30″ South, a location that is some 2.2 km south of the actual location. Dawes' own latitude determination is, however, quite precise; his geographic position determination is included as a mere postscript, an afterthought, in his letter to Maskelyne of 10 November 1788:

> P.S. I have assumed the Longitude of the Observatory at $10^{h}$. 05 E. The Latitude I found roughly by the astronomical quadrant mounted on the stump of a tree saw'd off, in the open air to be 33.° 52' 20"; but by the zenith distances accompanying this I take it to be about 33.° 51' 10" … The longitude I have assumed is what $Capt^{n}$. Hunter & Mr. Bradley made it by a mean of upwards of 300 [lunar] Distances. (Dawes, 1788e)

Morrison and Barko (2009) suggest that Dawes' measurements were obtained as early as February or March 1788; they certainly predate the establishment of Dawes' Observatory given that the quadrant was positioned on what is likely to have been a substantial "… stump of a tree saw'd off …" that could support the instrument's

**Fig. 2.8** "A new chart of the Eastern Coast of New Holland from South Cape to Cape York, Comprehending Anthony van Diemen's Land, Furneaux's Land, and New South Wales, discovered by Tasman, Furneaux and Cook &ca. in the Years 1642, 1770 and 1773. London, Published by Laurie & Whittle, 53 Fleet Street, 17th July 1798" (page 105; National Library of Australia; out of copyright)

solid cast-iron base—and not on the bedrock where Dawes would eventually fix it firmly in place. The historical record includes a second measurement by Dawes, in the form of a carefully drawn sketch map of Sydney Cove published in July 1788, with Hunter, as part of *A Voyage of Governor Phillip to Botany Bay* (London, 1789; see Fig. 2.9). The latitude included on this map, 33° 52′ 30″ South, is slightly less accurate, although the longitude cited is somewhat more accurate (see below). It is, therefore, likely that Dawes and Hunter determined their geographic positions independently.

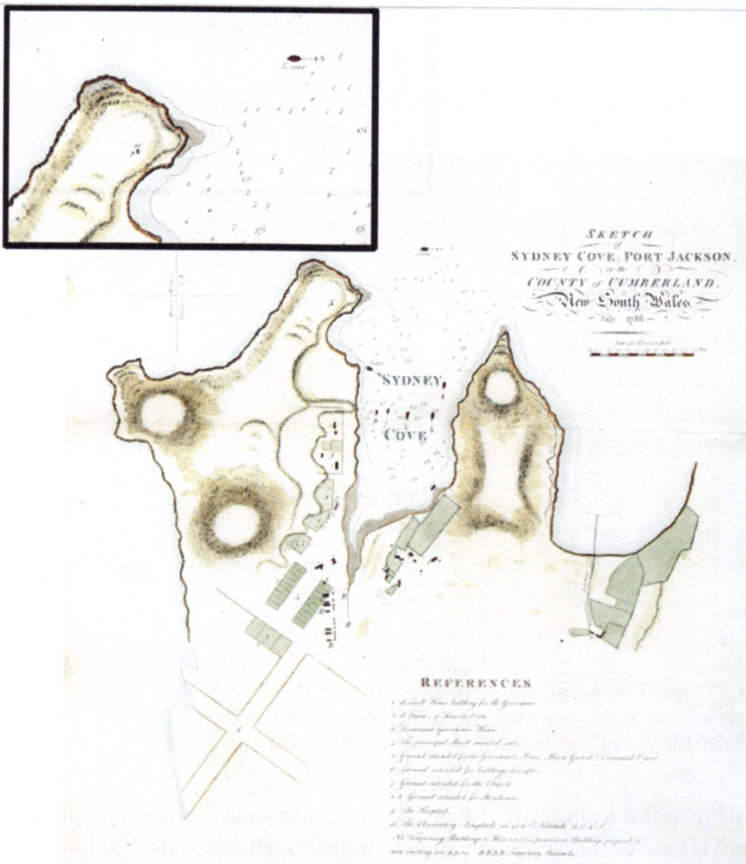


**Fig. 2.9** "Sketch of Sydney Cove, Port Jackson, in the County of Cumberland, New South Wales, July 1788. Captain John Hunter, William Dawes". (*Source* William Dawes, 1880; State Library of Victoria: ref. H24525; out of copyright)

As already indicated, Dawes had adopted the longitude determinations of Bradley and Hunter. By the end of April 1788, Bradley had independently determined the observatory's latitude at 33° 52′ 30″ South based on "… 3 meridian altitudes of the Sun with the astronomical Quadrant …"; his longitude measurement, based on 176 lunar distances, resulted in 151° 20′ East (Bradley, 1969; Dawes, 1788e). Bradley's *Chart of the Coasts and Harbours of Botany-Bay, Port Jackson and Broken Bay on the coast of New South Wales as Survey'd by Capt.$^{n}$ John Hunter of H.M.S.* Sirius

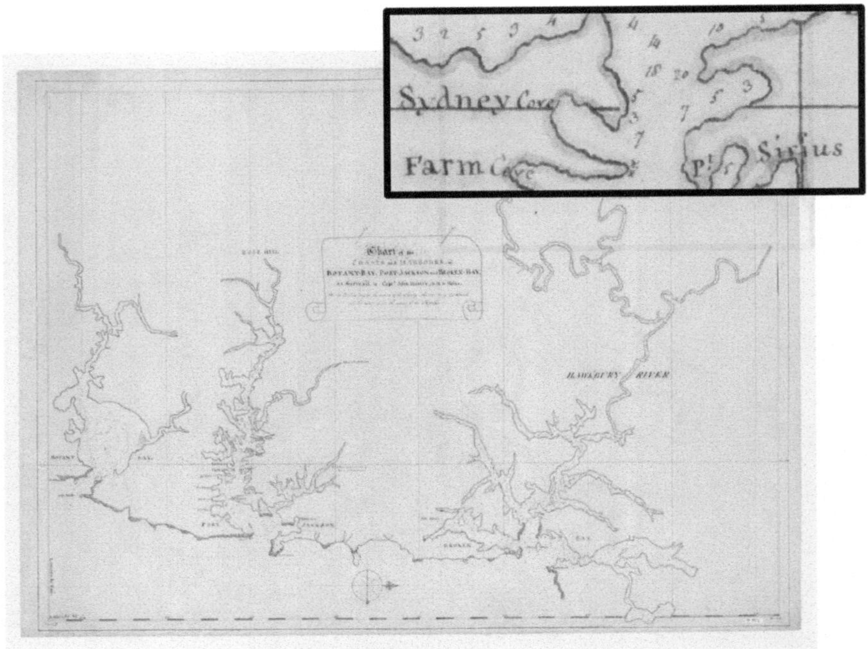


**Fig. 2.10** "Chart of the Coasts and Harbours of Botany-Bay, Port Jackson and Broken Bay on the coast of New South Wales as Survey'd by Capt$^{n}$. John Hunter of H.M.S. *Sirius* (1789?)." (Mitchell Library, State Library of New South Wales; out of copyright)

(1789?)[17] includes a single meridian, the first Australian meridian published, determined at Dawes' observatory (see Fig. 2.10, inset). Hunter took 130 lunar distances and obtained a longitude within a mile of Bradley's result (Dawes, 1788e).

It is unknown what type of observations Dawes obtained himself at this time (see also Dawes, 1788a). Dawes' own map of March 1791, *A map of all those parts of the territory of New South Wales which have been seen by any person belonging to the settlement established at Port Jackson.../faithfully constructed ... and respectfully inscribed to Capt Twiss ... By ... William Dawes*, is the only other contemporary map that includes a meridian: see Fig. 2.11. Dawes included an explicit note regarding his meridian, whose extension also appears to go through the location of Dawes' observatory: "N.B. The longitude of the observatory is determined from the observations of Capt. Hunter and Lieut. Bradley".

Meanwhile, Phillip's journal includes a position determination for the observatory that is almost certainly a transcription error: "The longitude of this observatory is

---

[17] The record accompanying this map in the collection of the Mitchell Library (State Library of New South Wales) states, "Hunter and Bradley made many charts and surveys during [their] eleven months' enforced stay on Norfolk Island [2 October 1788–8 May 1789]; this chart may have been amongst the number. ... The data gathering for [this] chart may have taken place during five months of 1788, eight in 1789, and two in 1790" (https://search.sl.nsw.gov.au/permalink/f/lg5tom/SLNSW_ALMA21147000280002626).

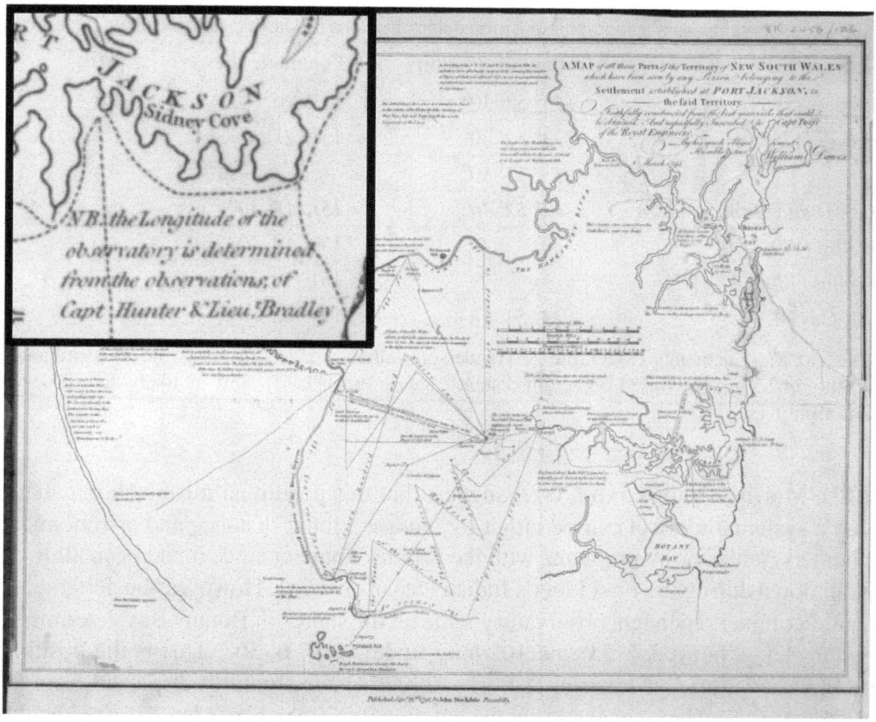


**Fig. 2.11** "A map of all those parts of the territory of New South Wales which have been seen by any person belonging to the settlement established at Port Jackson…/faithfully constructed … and respectfully inscribed to Capt. Twiss … by … William Dawes, March 1791." (National Library of Australia; out of copyright)

ascertained to be 159°19′ 30″ east from Greenwich" (Phillip, 1790). This is most likely meant to be 151° 19′ 30″ East, since Collins' (1798: 15) published longitude for the observatory, included as a journal entry for February 1788, is 151° 19′ 30″ East, as is the longitude cited on the Dawes and Hunter map of July 1788. Dawes' own measurement, 10 h and 5 min East of Greenwich, translates to a longitude of 151° 21′ East. Table 2.1 includes an overview of all contemporary geographic position determinations of Dawes' Observatory, approximately in order of publication.

The most popular contemporary geographic position determination, that of Dawes and Hunter (1788; rendered in bold italic font in Table 2.1), places the observatory some 11 km east of the actual position on Dawes Point/The Rocks and around 3.8 km off the coast of New South Wales (McGuffie & Henderson-Sellers, 2012). At the latitude of Sydney, 33° 52′ 08″ South, an East–West discrepancy of 11 km corresponds to 7.1 minutes of arc in longitude (the actual longitude difference between the Dawes and Hunter determination and the modern value is 6′ 55″), or a timing error of 29 s. Contemporary time measurements, including time differentials with respect to a reference meridian, were much more accurate than this, however (e.g., de Grijs,

**Table 2.1** Contemporary geographic position determinations of Dawes' Observatory

| Source | Latitude (South) | Longitude (East) | Notes |
|---|---|---|---|
| Bradley (1969) | 33° 52′ 30″ | 151° 20′ | (a) |
| Dawes (1788e) | 33° 52′ 20″ | 151° 21′ | (b) |
| | 33° 51′ 10″ | | (c) |
| ***Dawes and Hunter*** (1788) | ***33° 52′ 30″*** | ***151° 19′ 30″*** | |
| Phillip (1790) | | 159° 19′ 30″ | (d) |
| Collins (1798) | | 151° 19′ 30″ | |
| Google (2020) | 33° 51′ 18″ | 151° 12′ 35″ | |

*Notes* (a) Most accounts (McGuffie & Henderson-Sellers, 2012); (b) Based on measurements with the quadrant; (c) Based on zenith distance measurements; (d) '159' is likely an erroneous transcription (see text)

2017). Morrison and Barko (2009) suggest that this positional mismatch may have been a systematic offset caused either by incorrect lunar-distance and astronomical ephemeris tables or by problems with the K1 chronometer since it had been allowed to run down during the First Fleet's Indian Ocean crossing. However, the geographic location of the French tent observatory's site on the shores of Botany Bay determined by Dagelet—"latit[ude]. 33° 59′ 10″ long[itude]. 149° 6′ 30″. This is the position of our observatory." (Morrison & Barko, 2009: 4)—is similarly displaced from its modern position. This suggests that the positional mismatch was less affected by the (in)accuracy of K1 and more so by problems related to the accuracy of the contemporary almanac tables. Both Dagelet and Dawes would most likely have been aware of these issues.

At the present time, the headland to the west of Sydney Cove includes the Sydney suburbs of The Rocks on the eastern side and Dawes Point on the western side of the promontory's ridge. The area of historical interest is no longer contained by the current boundary of the Dawes Point suburb, which was formally established in 1993 (Jackson, 2018b). Since Dawes' return to London in 1792, the area known as Dawes Point gradually moved (north-)westwards from the original 'Point Maskelyne' to the present-day location of the Pier One hotel and Ives Steps at Walsh Bay. Except for Dawes' and Hunter's July 1788 map showing the entire headland, there are no historical records of the astronomer having been active on the western side of the area.

The shifting boundaries of the Dawes Point area coincided with the developing fortifications at the site (Jackson, 2018a), *viz.* Dawes' battery (closed in 1916) and Fort Phillip (present-day Sydney Observatory, established in 1858). The southern pylon of the Sydney Harbour Bridge is located at the northern extremity of the promontory. The area coincident with the location of Dawes' Observatory, today's Dawes Point (Tar-ra) Park, part of the Hickson Road/Dawes Point Reserve, is located northwest of a small inlet known as Campbells Cove, near the present-day Park Hyatt Hotel.

Nevertheless, a number of contradictory identifications of the Observatory's location appear in the literature. Johnson (1994), an archaeologist with the Sydney Cove Authority (New South Wales Government), declared that Dawes' Observatory would have been located on the promontory's highest point, a conclusion echoed by Johnson (2003) and Jackson (2018a). The former publication implies that a single building located on the promontory's ridge and shown in a drawing from the early 1790s (Johnson, 2003; his Fig. 2) may have been Dawes' observatory. However, this view is at odds with Dawes' own description of the observatory and its site, including the structure's dual-level configuration. Wood's (1924: note 9) suggestion that the *Sirius*' guns may have been located *in front of* the observatory, with respect to the waterline, may also have contributed to the confusion. Contemporary maps of Dawes Battery (for a typical example, see Fig. 2.12) show the guns arranged in a semi-circle near the ridge of the headland (close to the waterline) and in front of a steep incline, facing East towards the Sydney Harbour entrance.

Johnson (1994) refers specifically to a bronze plaque, located on Pier No. 6, the first pier 150 feet (46 m) southwest of the southern pylon of the Sydney Harbour Bridge (*Daily Telegraph*, 1926; *Sydney Morning Herald*, 1932), as his position reference. As we will see below, this south-facing tablet implies that the original Observatory was located some 90 feet (27 m) to the west of the pillar of interest: see Fig. 2.13 (left). Between February and September 1995, the Sydney Cove Authority undertook archaeological excavations near the area suggested by Johnson's (1994) report, although without locating any remains of the Observatory. Johnson (2003) suggested that this failure could be related to the site's subsequent use as stone quarry around

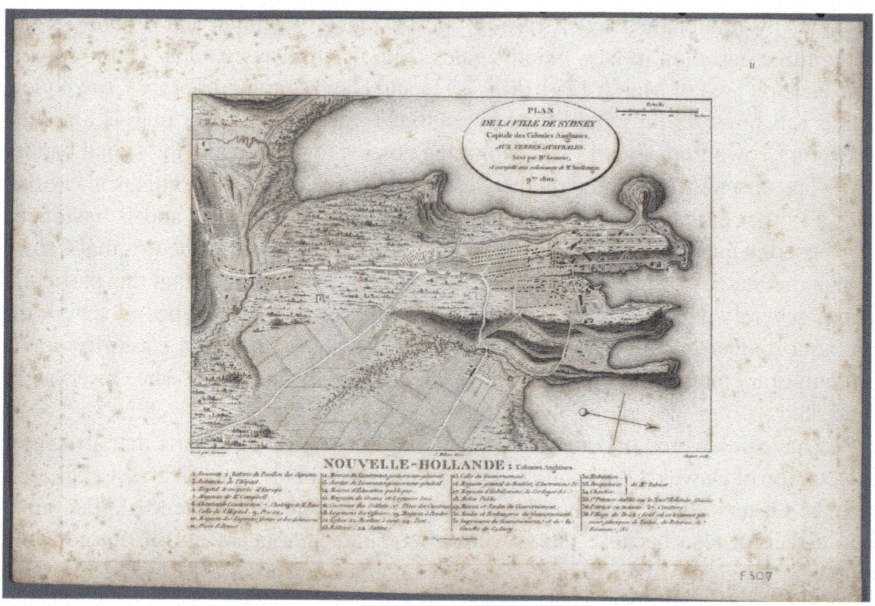


**Fig. 2.12** French map of Sydney, 1802 (National Library of Australia; out of copyright)

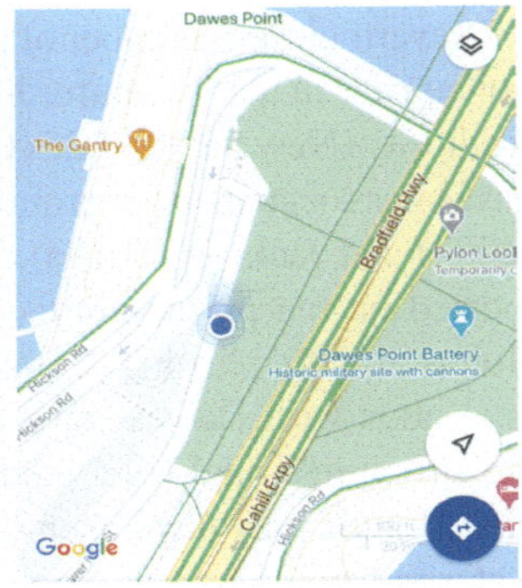


**Fig. 2.13** (*Left*) Dawes' memorial plaque attached to the Sydney Harbour Bridge (Photo Author supplied); (*right*) Blue dot: location of Dawes' observatory implied by the text on the plaque (Photo: Author supplied; map reproduced under 'fair use' principles)

1819. However, as we will show below, the memorial plaque's indicative position of Dawes' Observatory (see Fig. 2.13, right) is most likely incorrect. In addition, Collins (1798: 61) implies that Dawes and his men were quarrying stone for the second incarnation of the observatory, suggesting that the Observatory's site was not coincident with but somewhat removed from the main quarry location.

To uncover the full extent of the historical record, let us first briefly summarise what we have learnt from contemporary maps and Crown plans. The majority of contemporary maps showing Dawes' observatory were drawn by either Hunter or Dawes, including Hunter's first map of the settlement dated 1 March 1788 (Fig. 2.1), a joint map by Hunter and Dawes from July 1788 (Fig. 2.9), Hunter and Bradley's chart of the three harbours in New South Wales (Fig. 2.10), Dawes' chart of the greater Port Jackson area of March 1791 (Fig. 2.11) and Dawes' contemporary Crown plan, which was however only published by 1886 (Fig. 2.9). This latter map is a tracing based on Crown Plan CP1-172. It shows the observatory ('B') and the signal battery ('A'). The observatory is sited on the western side of Dawes Point. This is in conflict with all other early and contemporary maps, unless the labels A and B have been transposed, which we suspect may have happened. We inspected the original Crown Plan at the State Archives of New South Wales, where we noticed that this particular map is severely damaged at the location where its key would have been located (the key is no longer present). The sketch map of Fig. 2.9 is the most carefully drawn contemporary map, although it does not scale to modern shorelines. The Observatory is marked '*p*': 'proposed or now building'.

We have uncovered two additional contemporary maps and charts that also show the location of Dawes' Observatory, a survey map by Hunter from 1788 (see Fig. 2.14: note the Observatory's dual-building structure; this is likely an intended plan rather than a record, also because the Governor's residence appears out of place) and a convict's drawing of the Sydney Cove area dated 16 April 1788 (Fig. 2.15). This latter map is usually attributed to Francis Fowkes (fl. ca. 1788–ca. 1800). Fowkes is sometimes tentatively identified with the 'Port Jackson Painter' whose sketches of

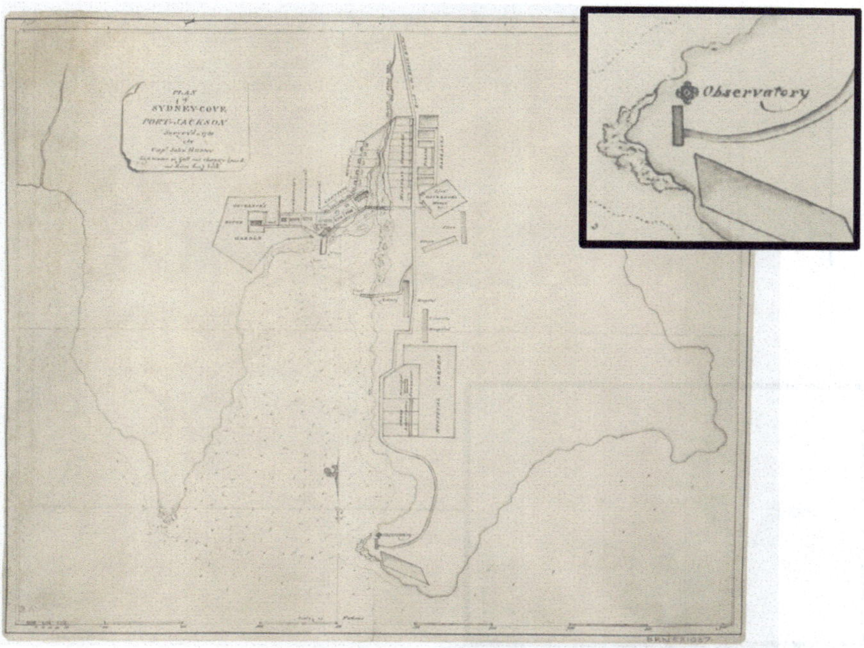


**Fig. 2.14** "Plan of Sydney-Cove, Port-Jackson survey'd in 1788 by Capn. John Hunter" (Mitchell Library, State Library of New South Wales; out of copyright)

the early colony continue to provide valuable insights into the pre-colonial era.[18] The latter drawing, although more conceptual than geographically correct, is nevertheless important, since it represents independent evidence as to the Observatory's location.

All contemporary maps and charts show the site of Dawes' Observatory on the eastern side of the headland, situated against a rocky bluff. Figure 2.16 provides an overview of the locations implied by the historical maps and charts discussed in this chapter. Panels (a) through (c) show representative examples of contemporary maps overlaid on a modern outline of Sydney Cove, including (a) Hunter's first survey map of Sydney Cove of 1 March 1788 (Fig. 2.1), (b) his planned survey of 1788 (Fig. 2.14) and (c) Hunter and Dawes' careful sketch from July 1788 (Fig. 2.9). Figure 2.16d shows the positions of Dawes' Observatory and the meridian determined with the Observatory's fixed quadrant according to the maps and charts discussed in this chapter. Note that whereas we have attempted to reconcile Dawes' survey map, Fig. 2.5, with the modern map, the match is tentative at best, since the contemporary survey's scaling of identifying features varies across the original map.

---

[18] The notion of a 'Port Jackson Painter' was first suggested by the art historian Bernard Smith in 1960 (see Smith and Wheeler, 1988: 208–210). Historians tend to identify the Port Jackson Painter with Fowkes, the colony's first Provost-Marshall Henry Brewer (1739–1796) or the convict Thomas Watling (ca. 1767–1797).

**Fig. 2.15** "Sketch and description of the settlement at Sydney Cove, Port Jackson, in the County of Cumberland taken by a transported convict on the 16th of April 1788, which was not quite 3 months after Commodore Phillips's landing there" (Mitchell Library, State Library of New South Wales; public domain)

To further support our conclusion of the observatory's location on the eastern side of the promontory, in Fig. 2.17a we provide a wide-angle view of the site's present-day appearance. The memorial plaque is visible on the pier on the left of the photograph. We have indicated the area covered by Dawes' Battery by means of a light green outline, a location supported by both archaeological records and today's restored battery. Beyond the battery, the yellow arc traces the edge of the bedrock; the yellow arrow indicates a solitary cannon at that position. As we have seen, Dawes' letters to Maskelyne imply that he built his observatory against the edge of the bedrock. Foreshortening renders a proper assessment of the distance (20–30 m) between Dawes' Battery and the edge of the bedrock difficult, however. Therefore, Fig. 2.17b shows an enlarged image of the same area but from a different vantage point, clearly showing this distance.

We are fairly confident in our determination of the most likely Observatory site, because (i) it would not be prudent to build an observatory by the shoreline; (ii) Dawes' (1788a) letter to Maskelyne implies that he constructed the rectangular building against a large block of rock and the octagonal building on top of it; (iii) we have established that the facility was built on the eastern side of the headland; and (iv) it was likely built close to the stone quarry later identified by Johnson (2003). However, we caution the reader that we do not have enough physical or documentary

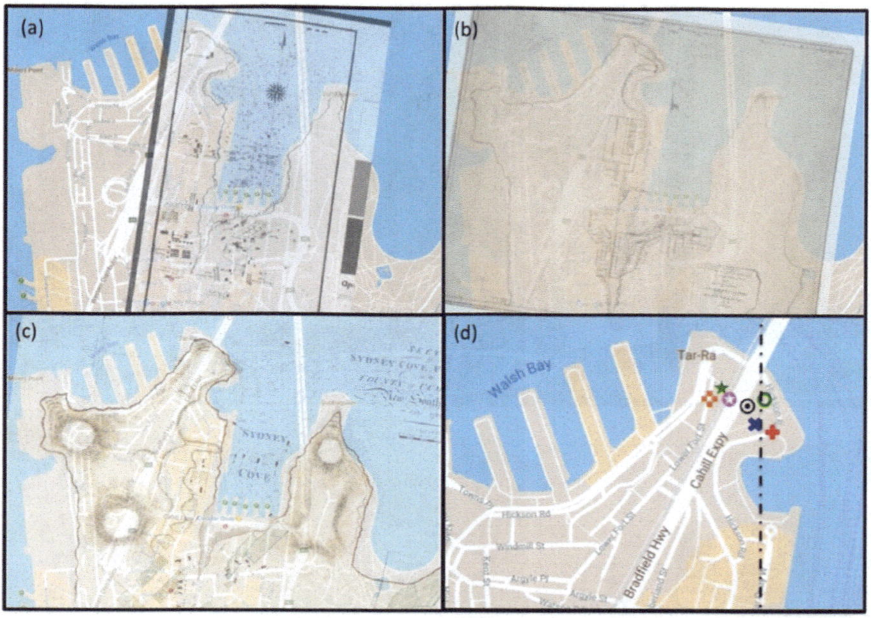


**Fig. 2.16** **a–c** Examples of matching the contemporary record (Figs. 2.1, 2.14 and 2.9, respectively) with a modern map (© Google 2020; permissible use; each panel is 1,900 × 1,400 m$^2$). **d** Locations of Dawes' Observatory implied by the records discussed in this chapter. *Black* ⊙: Hunter's map of 1 March 1788 (Fig. 2.1); *red* ✚: Hunter's, 1788 planned survey map (Fig. 2.14); *blue* ✖: Hunter and Dawes' 1788 map of Sydney Cove (Fig. 2.9); *dark green*: Locations of the Observatory ('B'; ★) and Dawes' Battery ('A'; ○) according to Dawes' Crown Plan (Fig. 2.5); *orange* ✥: Location indicated on the Sydney Harbour Bridge's memorial plaque; *pink* ✪: Approximate location of the archaeological excavations described by Johnson (2003). The black dash-dotted line corresponds to the best match to the meridian line shown in Fig. 2.10. Panel **d** covers an area of approximately 850 × 650 m$^2$. The convict's sketch shown in Fig. 2.15 is not directly transposable

evidence to state categorically where the building was located exactly. Any physical evidence has been erased by the later fortification work (which likely remodelled the hillside) and the construction of the Sydney Harbour Bridge (see below).

Yet, the evidence of a site on the eastern side of the promontory appears overwhelming, and so we were rather perplexed by the indication of a western site on the memorial plaque attached to the Sydney Harbour Bridge. Therefore, we decided to explore the origin of the citation on that tablet.

The memorial plaque was formally unveiled on 16 June 1932 (*Sydney Morning Herald*, 1932), but the idea to establish a plaque as a memorial of the local history goes back to a letter from Engineer Bradfield to the Secretary of the RAHS, Karl Reginald Cramp (1878–1956), of 23 March 1926: see Fig. 2.18. Bradfield advised the

Fig. 2.17 (*Top*) Today's appearance of Dawes' Battery under the southern approach to the Sydney Harbour Bridge. The memorial plaque is visible on the pier on the left of the photograph. The light green outline reflects the location of the original battery, while the yellow arc traces the edge of the bedrock. The yellow arrow indicates a solitary cannon at the edge of the bedrock (Photo Author supplied); (*bottom*) Overview of the area of interest, highlighting the non-negligible distance between the location of Dawes' Battery, from which this photo was taken, and the edge of the bedrock (at the location of the cannons in the middle distance), where the astronomer established his Observatory (Photo: Author supplied, May 2023)

RAHS that "[t]he Minister of Public Works[19] wishes to commemorate this historical battery [Dawes' Battery] and desires a brass plaque suitably inscribed to be erected" (Bradfield, 1926). He proceeded to ask the Society to provide a suitable inscription. From the RAHS Annual Report of 1926 (RAHS, 1927), we learn that the President

[19] In 1926, the New South Wales Secretary (Minister) for Public Works and Minister for Railways was The Hon. Mr. Martin Matthew Flannery, M. L. A. (Member of the Legislative Assembly of New South Wales).

and Secretary of the Society, Captain James Henry Watson (1841–1934) and Cramp, reached agreement on the proposed wording (see also *Daily Telegraph*, 1926; *Sydney Morning Herald*, 1927; RAHS, 1928, 1931):

> At Dr. Bradfield's request, a suitable inscription for a brass tablet to be affixed to one of the granite-faced piers of the North Shore Bridge [the Sydney Harbour Bridge], was suggested by your President and Secretary. The proposed inscription is worded as follows:–
>
> On this site
> Lieutenant William Dawes (Artillery Officer)
> erected a breastwork in 1788,
> replaced by a stone battery in 1791,
> known as Dawes Battery,

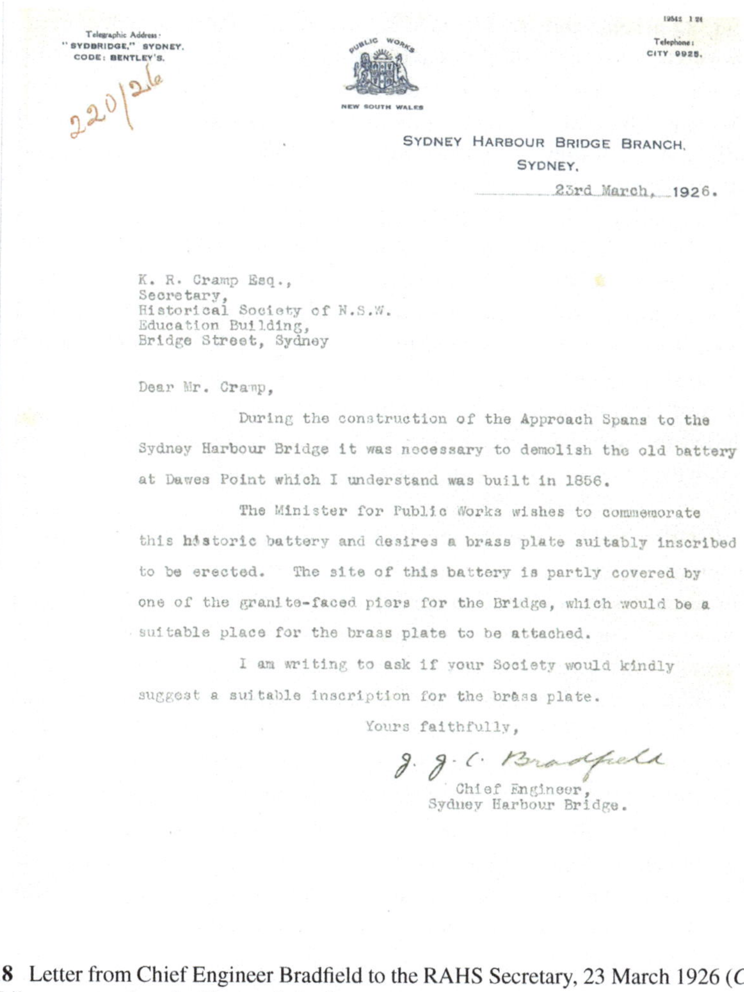

12542 1 24

Telegraphic Address: "SYDBRIDGE," SYDNEY. CODE: BENTLEY'S.

220/26

PUBLIC WORKS
NEW SOUTH WALES

Telephone: CITY 9925.

SYDNEY HARBOUR BRIDGE BRANCH,
SYDNEY,
23rd March, 1926.

K. R. Cramp Esq.,
Secretary,
Historical Society of N.S.W.
Education Building,
Bridge Street, Sydney

Dear Mr. Cramp,

During the construction of the Approach Spans to the Sydney Harbour Bridge it was necessary to demolish the old battery at Dawes Point which I understand was built in 1856.

The Minister for Public Works wishes to commemorate this historic battery and desires a brass plate suitably inscribed to be erected. The site of this battery is partly covered by one of the granite-faced piers for the Bridge, which would be a suitable place for the brass plate to be attached.

I am writing to ask if your Society would kindly suggest a suitable inscription for the brass plate.

Yours faithfully,

J. J. C. Bradfield

Chief Engineer,
Sydney Harbour Bridge.

**Fig. 2.18** Letter from Chief Engineer Bradfield to the RAHS Secretary, 23 March 1926 (*Courtesy* RAHS Library; reproduced with permission)

the whole being remodelled and enlarged by
Governor Denison, 1856.

— feet[20] west of this tablet,
the first Observatory in Australia
was erected
under the direction of Lieutenant Dawes,
1788

It is thus clear that the original wording proposed by Watson and Cramp (the emphasis is ours) was already mistaken. The location implied by the plaque affixed to the Sydney Harbour Bridge coincides almost exactly with that reported for the stone marked 'R. R. 1789' by Kathleen Butler in her letter of 19 February 1926 to the Principal Librarian of the Public Library of New South Wales (Fig. 2.7, left), that is, 1260 feet (384 m) east and 1652 feet (503.5 m) north of the time-ball staff at Sydney Observatory. Butler was Bradfield's Confidential Secretary, and so it is likely that her letter closely represents Bradfield's own words (Engineering Heritage New South Wales, 2023; W. Phippen, personal communication, May 2023). We therefore suspect that Bradfield himself may have been the source of the incorrect identification; after all, Bradfield's letter to Cramp was sent on 23 March 1926, shortly after Butler's report on the building stone. Butler's letter includes a reference to two plans that included an indication of the location where the stone had been found; a marginal note implies that those plans had been transferred to the State Archives of New South Wales. Our inquiries at the State Archives and Records of New South Wales have only managed to uncover one of those plans: see Fig. 2.19 for an enlargement of the area of interest, excerpted from the original $73.5 \times 96.5$ cm$^2$ document covering the entire Dawes Point promontory. A grainy photocopy of a sizeable section of this plan was included in the 1981 Annual Report of the Archives Authority of New South Wales, when the State Archives acquired it (Archives Authority of New South Wales, 1982: 9).

Note that Bradfield's letter to the RAHS states that "[d]uring the construction of the Approach Spans [now known as Bradfield Highway] of the Sydney Harbour Bridge it was necessary to demolish the old battery at Dawes Point". Indeed, Fig. 2.20 shows one of two enormous pits—40 feet (12.2 m) wide, 90 feet (27.4 m) long and up to 30 feet (9.1 m) deep—known as 'skewbacks', at the southern end of the Sydney Harbour Bridge span (NSW Education Standards Authority, 2014). Their excavation, in preparation of the Harbour Bridge's pylon construction, may have contributed to the removal of any remains of Dawes' Battery and his observatory. Although the skewbacks did not extend as far from the southern end of the headland as the likely location of Dawes' Observatory, other contemporary photographs in the Mitchell Library's collection clearly show that the entire area under the southern approach to the Harbour Bridge was deeply excavated and thoroughly disturbed during the Bridge's construction. Moreover, one should keep in mind that Dawes' Observatory,

[20] The distance of 90 feet indicated on the tablet already appeared in the *Daily Telegraph* of 29 May 1926.

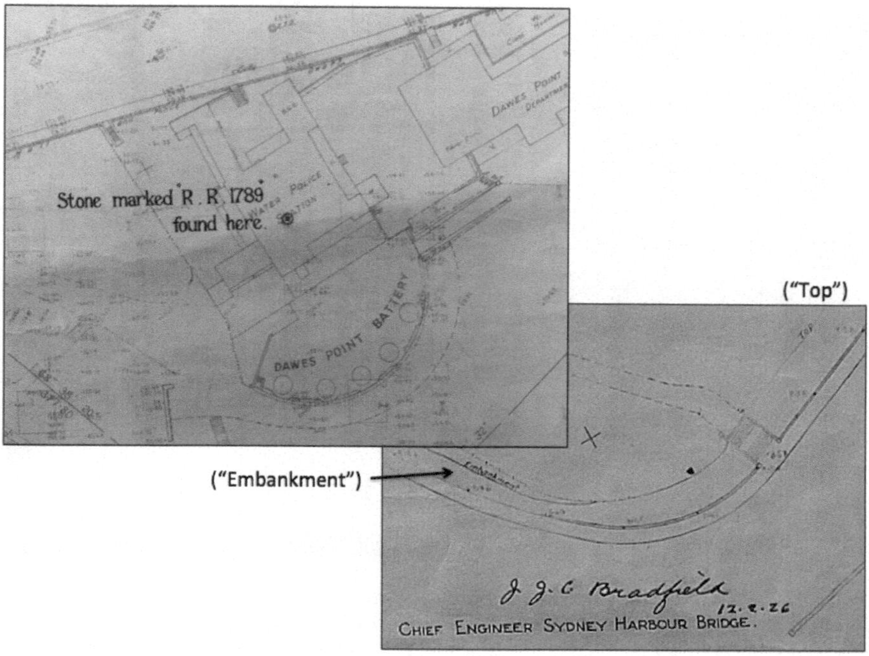


**Fig. 2.19** "Plan of Dawes Point showing the location of stone marked RR1789 found during the excavation of Sydney Harbour Bridge in 1925. Signed by J. J. C. Bradfield." (sections showing the location of the former Water Police Station and the Hickson Road area near Dawes' original observatory) (State Archives and Records of New South Wales; Map AO 6317, F11/49; public domain)

a wooden structure constructed during the first year of British settlement, was unlikely built on deep and solid foundations. Without access to proper tools and professional construction expertise, even the stone building that was under construction just a year later was likely still a simple structure without deep foundations, given the thin layer of top soil on bedrock that was prevalent around Sydney Cove (e.g., Hughes, 1987: 107). It is therefore not surprising that the 1995 archaeological excavation (Johnson, 2003), even if it had been undertaken in the correct location, was not successful in locating any remains of Dawes' Observatory.

## 2.5 Sydney's Scientific Beginnings: Conclusion

One aspect we have not yet considered is the evidence provided by the morphology of the landscape. Let us therefore explore this latter aspect as additional, although perhaps less conclusive, evidence. While landscape artists usually tried to reproduce an area's main features accurately, some measure of creative license often crept in.

**Fig. 2.20** (*Left*) Sydney Harbour Bridge, excavation of the South pylon's southeast 'skewback'. (Department of Public Works, 1927: 57); (*right*) Harbour Bridge Chief Engineer John Bradfield, construction supervisor Lawrence Ennis, structural engineer Ralph Freeman and Kathleen Butler, Bradfield's confidential secretary, in the excavation for the southwest skewback, 22 April 1926 (*Sydney Harbour Bridge Photographic Albums*, 1923–1933, Vol. 2; State Records of New South Wales; public domain)

In this context, we are most interested in exploring the shape of the original headland and particularly in the location of any obvious vertical rock faces. We have to cast our net more widely than simply focusing on the period from 1788 to 1791, however.

The Sydney harbourside topography is shaped by the nature of the so-called Sydney Basin Hawkesbury sandstone it is composed of. The rock strata is horizontal in the harbour region, and weathering causes horizontal and vertical fractures. This results in the topography of low, stepped headlands, with a short shoreline cliff backed by an approximately horizontal—sometimes gently sloping—terrace followed by one or more additional vertical terraces and steps. Higher headlands rise in a steeper set of terraces and steps up to a plateau. In reality, this terrace and step topography tends to be more chaotic and broken. From a distance, the features can be disguised and 'rounded' by vegetation. At the shoreline, a broad tidal terrace is common. Today, most shorelines have been modified by quarrying or infill behind rock retaining walls, although the natural form can still be seen at Ball's Head and in large parts of Cremorne Point, Bradley's Head and Mrs Macquarie's Point.

Figure 2.21 shows the original morphology of the harbour shore at Lavender Bay (the image is reversed with respect to the original), near the onset of the Sydney Harbour Bridge but prior to the start of construction. As the author intimates, the site of the 'fuzzy' tree on the hill at the left of the image is where the first sod was turned on 28 July 1923, marking the start of the construction of the Harbour Bridge a few days later (Cash, 1930: 7–8). The age of the picture is uncertain, other than that the author wistfully refers to the "… old-time picturesqueness …" of 60 years ago, that is, around 1870. He proceeds (our emphasis),

> One afternoon, some years ago, I met my friend, Mr. James Blue, of honourable memory. He said, "Come inside and I will show you something that will interest you". … There came into his possession some time ago, a photograph, yellow with age, but alive in historic interest. (Cash, 1930: 9)

Our first, roughly contemporary sketch of interest is titled *A View of Dawes Battery at the Entrance of Sydney Cove*: see Fig. 2.22. The sketch was published in 1820 by Walter Preston (ca. 1787–after 1821), a convict engraver (Willetts, 2010–2020). It was based on an original drawing by Captain James Wallis (1785?–1858), commander of the convict settlement in Newcastle (New South Wales). Wallis was based in Sydney between February 1814 and June 1816. The drawing clearly shows a low-lying extension to the headland, which rises steeply to a higher-level, isolated building. This same morphology of the bedrock is confirmed in a detailed sketch by the artist and former convict John Eyre (1771–1812; Fig. 2.23), thought to have been created around 1809. Eyre's sketch is particularly valuable, because he carefully sketched the shapes of the buildings individually rather than using generic shapes. In addition, the contemporary oil painting from ca. 1800 (artist unknown) shown in Fig. 2.24 also implies a similar form of the headland.

**Fig. 2.21** View of Lavender Bay in North Sydney (reversed) prior to the construction of the Sydney Harbour Bridge. The 'fuzzy' tree on the hill at the image's top left marks the site where the Bridge's first sod was turned on 28 July 1923 (Cash, 1930: 1; out of copyright)

**Fig. 2.22** A view of Dawes Battery at the entrance of Sydney Cove, New South Wales, 1820 (National Library of Australia; out of copyright)

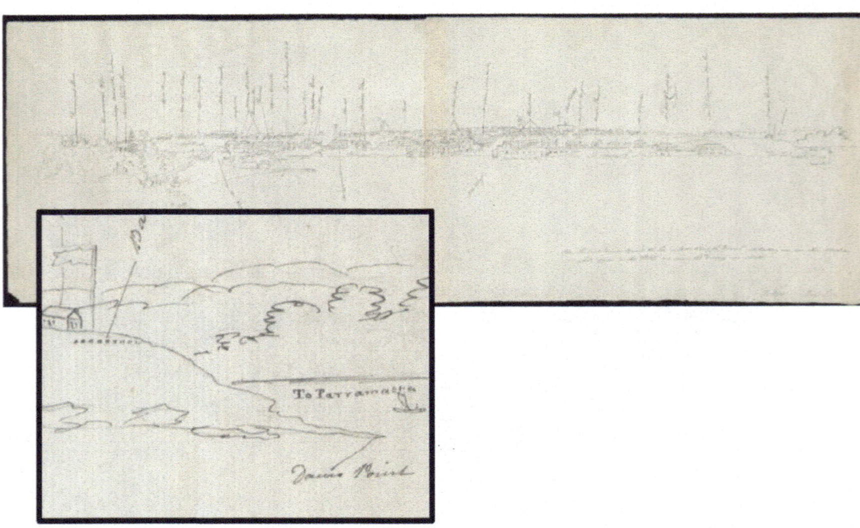


**Fig. 2.23** "Sydney Cove, east side, by John Eyre, ca. 1809" (National Library of Australia; out of copyright)

**Fig. 2.24** Sydney Cove looking to the west, ca. 1800; artist unknown (Mitchell Library, State Library of New South Wales; out of copyright)

All of these early representations viewed Dawes Point from either the East or the North East. Before concluding this section, we highlight two additional paintings, shown side-by-side in Fig. 2.25. Both view Dawes Point from due North, and both suggest that the area initially designated as Point Maskelyne was a slightly elevated area. It is tempting to suggest that Dawes' Observatory was built on its high point. However, we recommend due caution in interpreting these drawings, given the absence of such an elevated area in the side views. In addition, such a location cannot be reconciled with Dawes' (1788a) own description of the Observatory's siting against the bedrock.

A recent, independent topographical reconstruction of the Sydney Cove landscape during the first few years of British settlement (Webster, 2022: Fig. 1.2) also includes the location of Dawes' Observatory at a site closely coincident with our best estimate. The primary sources used in Webster's reconstruction included Dawes and Hunter's map of July 1788 (Fig. 2.9), as well as maps published by Mitchell et al. (1853) and Sydney Water (1994). Webster's placement of Dawes' Observatory was predominantly determined by cartographic evidence from Dawes and Hunter's map of July 1788 (Fig. 2.9) and Hunter's, 1788 plan of the Sydney Cove settlement (Fig. 2.14) (A. Webster, personal communication). For additional considerations about Dawes' Observatory and the development of Sydney Cove, see Chap. 8.

A final clue as to the location of Dawes' Observatory can be inferred from photographs taken during the demolition of buildings on Dawes Point in preparation for the construction of the Sydney Harbour Bridge. The top left and bottom

**Fig. 2.25** (*Top*) "North view of Sidney, New South Wales", ca. 1825; Joseph Lycett. (Mitchell Library, State Library of New South Wales; out of copyright); (*bottom*) "Two Forts and a Battery, 1822"; Joseph Lycett (Mitchell Library, State Library of New South Wales; out of copyright)

left panels of Fig. 2.26 show contemporary photographs of the construction site. The top left photograph is a view from 1926 showing Harbour Bridge construction Water Police building on 22 January 1925. Both photographs, but in particular the bottom left image, show what appears like a natural cliff face towards the bottom (north) of Hickson Road, behind the original retaining wall on the east. Dawes' Observatory was most likely located at the point where the rock face exhibited a sharp edge (cf. Fig. 2.3, right; see also Fig. 2.19, bottom), a location just outside of the field of view of the bottom left photograph. The two images on the right of Fig. 2.26 show the present-day landscaping of the area, from ground level (top right) and from the rooftop of the Park Hyatt Hotel (bottom right). The bottom panel of Fig. 2.19 includes an indication of the sharp 'embankment' alongside Hickson Road, which appears to have leveled off near the top of the flight of stairs allowing access from street level to Dawes' Battery (in reference to the embankment, note the label 'Top' north of the stairs).

**Fig. 2.26** (*top left*) Sydney Harbour Bridge construction activity in 1926. (State Library of New South Wales, reference 130044, call number PXD 747; public domain). (*bottom left*) Demolition of the Sydney Water Police building on 22 January 1925. (State Archives and Records of New South Wales, NRS-12685-1-[4/8725]-4/8725-52; public domain). Both photographs, but in particular the bottom left image, show what appears like a natural cliff face towards the bottom (north) of Hickson Road, behind the original retaining wall on the east. (*right*) Present-day landscaping of the area, from ground level (*top right*) and from the rooftop of the Park Hyatt Hotel (*bottom right*; with acknowledgement of access provided by the hotel's reception staff). (Photos: Author supplied, May 2023)

In summary, our best estimate of the location of Dawes' first Observatory is that it was sited approximately at the location of the (black) artefact on the Butler map (Fig. 2.19). That location, straddling the edge of the bedrock, is almost identical to the planned observatory site indicated on Hunter's 1788 survey map (Fig. 2.14).

The overall picture we have gained of Dawes' efforts in establishing an observatory in the new colony of New South Wales is one that is more intricate than most popular and many scholarly studies suggest. Dawes was not the first European astronomer to work his craft in the territory of New Holland. As we saw already, prior to his first tent-based observations, Lapérouse's astronomer, Dagelet, had already established the French expedition's portable observatory on the northern shores of Botany Bay. And, of course, Cook and Green had also carried out astronomical observations while at Botany Bay in 1770 (e.g., David, 1984). Even the concept of 'Dawes' Observatory', that is, the first British permanent astronomical building in the southern hemisphere, is fraught. As we have seen, after having established a timber and canvas structure consisting of two buildings by July 1788, there is evidence suggesting that the initial structure proved too confined and a second observatory, presumably made of stone, appears to have been under construction just a year after observations began from the original building.

## References


Archives Authority of New South Wales. (1982). Annual Report for 1981. Sydney, D. West (Government printer). https://media.opengov.nsw.gov.au/pairtree_root/3d/71/91/7b/d6/de/a2/f8/fa/97/cc/6c/67/b3/00/e7/obj/document.pdf Accessed 16 May 2023.

Ashcroft, L. (2016). *Australia's meteorological anniversary*. https://lindenashcroft.com/2016/01/26/australias-meteorological-anniversary/. Accessed 5 Oct 2020.

Ashcroft, L., Gergis, J., & Karoly, D. J. (2014). A historical climate dataset for southeastern Australia, 1788–1859. *Geoscience Data Journal, 1*, 158–178.

Attenbrow, V. (2009). Aboriginal placenames around Port Jackson and Botany Bay, New South Wales, Australia: Sources and uncertainties. In H. Koch & L. Hercus (eds.), *Aboriginal placenames naming and re-naming the Australian landscape*. Aboriginal History Monograph 19, Ch. 1. Canberra ACT, ANU E Press and Aboriginal History Incorporated. http://press-files.anu.edu.au/downloads/press/p17331/html/ch01.xhtml. Accessed 29 Sept 2020.

Barko, I. (2007). Lepaute Dagelet at Botany Bay (26 January–10 March 1788). *Explorations, 43*, 21–40.

Bartel, P. (1954). *La Jeunesse inédite de Napoléon d'après de nombreux documents*. Éditions Amiot-Dumont.

Barton, G. B. (1889). *History of New South Wales from the records. I. Governor Phillip 1783–1789*. Sydney, Charles Potter/London, Trübner and Co. http://gutenberg.net.au/ebooks12/1204171h.html. Accessed 2 Oct 2020.

Beaglehole, J. C. (ed.). (1968). *The Journals of Captain James Cook on His Voyages of Discovery. I. The Voyage of the* Endeavour *1768–1771*. Cambridge UK, Cambridge University Press. P. ccix.

Bladen, F.M., and Britton, A. (eds), 1893. *Historical Records of New South Wales*, 2. Mona Vale NSW, Lansdown Slattery & Co.

Bowes-Smyth, A. (1790). *Journal of a Voyage from Portsmouth to New South Wales and China – in the Lady Penrhyn, Merchantman – William Cropton Sever, Commander by Arthur Bowes, Smyth, Surgeon – 1787–1788–1789.*

Bradfield, J. J. C. (1926). Letter to the Secretary of the RAHS, dated 23 March 1926. RAHS 220/26.

Bradley, W. (1788). Cited by Keats (1997)

Bradley, W. (1969). *A voyage to New South Wales: The journal of Lieutenant Bradley RN of HMS Sirius*. Trustees of the Public Library of New South Wales.

Butler, K. M. (1926). Letter from the Secretary to Chief Engineer Sydney Harbour Bridge to the Principal Librarian, Public Library of New South Wales, dated 19 February 1926. State Library of New South Wales. Call number: An 36, record: 9arpR5Gn.

Cash, F. (1930). *Parables of the Sydney Harbour Bridge. Setting forth the preparation for, and progressive growth of, the Sydney Harbour Bridge, to April, 1930*. Sydney, S.D. Townsend & Co.

Clarke, R. M. (2015). *Vanguards of empire: The lives of William Dawes, Watkin Tench and George Worgan*. Ph.D. Thesis, The Australian National University.

Collins, D. (1798). *An account of the English colony in New South Wales: With remarks on the dispositions, customs, manners, &c of the native inhabitants of that country to which are added, some particulars of New Zealand, compiled, by permission, from the Mss. from Lieutenant-Governor King*, vol. 1. London, T. Cadell Jun. and W. Davies. http://adc.library.usyd.edu.au/data-2/colacc1.pdf. Accessed 30 Sep 2020.

*Daily Telegraph*, (29 May 1926). *The Harbor Bridge*, p. 20.

David, A. C. E. (1984). Cook and the cartography of Australia. *The Globe, 22*, 47–59.

Dawes, W. (1787). *The notebooks of William Dawes on the Aboriginal language of Sydney*. London, School of Oriental and Asian Studies 1787–1788. https://www.williamdawes.org/. Accessed 30 Sept 2020.

Dawes, W. (1788a). Letter to Nevil Maskelyne, dated 30 April 1788. RGO 14/48, folio 281r.

Dawes, W. (1788b). Letter to Nevil Maskelyne, dated 10 July 1788. RGO 14/48, folios 283–284.

Dawes, W. (1788c). Letter to Nevil Maskelyne, dated 12 July 1788. RGO 14/48, folio 285v.

Dawes, W. (1788d). Letter to Nevil Maskelyne, dated 1 October 1788. RGO 14/48, folio 287.

Dawes, W. (1788e). Letter to Nevil Maskelyne, dated 17 November 1788. RGO 14/48, folios 290–292.

Dawes, W. (1788–1791). *The meteorological journal of William Dawes*. Bureau of Meteorology, Australia, pp. 336–339. http://www.bom.gov.au/climate/early-data/sources/0_metadata.pdf. Accessed 5 Oct 2020.

Dawes, W. (1790a). Letter to Nevil Maskelyne, dated 16 April 1790. RGO 14/48, folio 300v.

Dawes, W. (1790b). Letter to Nevil Maskelyne, dated 26 July 1790. RGO 14/48, folio 302r.

Dawes, W. (1792). Letter to Nevil Maskelyne, dated 18 June 1792. RGO 14/48, folio 307r.

de Grijs, R. (2017). *Time and time again: Determination of longitude at sea in the 17th century*. Institute of Physics Publishing.

de Grijs, R. (2020). A (not so) brief history of lunar distances: Longitude determination at sea before the chronometer. *Journal of Astronomical History and Heritage, 23*, 495–522.

de Grijs, R., & Jacob, A. (2021). Sydney's scientific beginnings: William Dawes' observatories in context. *Journal of Astronomical History and Heritage, 24*, 41–76.

de Lalande, J. J. Le Francais. (1803). *Bibliographie astronomique, avec l'histoire de l'astronomie depuis 1781 jusqu'à 1802*. Imprimerie de la République.

Derrick, G. M. (2019). *Comparison of the First Fleet Sydney temperature records with modern data, 1788 to 2018*. https://carbon-sense.com/wp-content/uploads/2019/01/comparison-of-first-fleet-and-modern-temperatures.pdf. Accessed 5 Oct 2020.

Department of Public Works. (1927). *Report of the Department of Public Works for the year ended 30th June, 1926, together with photographs, plans and diagrams*. Alfred James Kent.

Dunmore, J. (1994). *The Journal of Jean-François de Galaup de la Pérouse 1785–1788*, 1. London, Hakluyt Society.

Dunmore, J. (1995). *The Journal of Jean-François de Galaup de la Pérouse 1785–1788*, 2. London, Hakluyt Society.
Engineering Heritage New South Wales, (2023). Kathleen Muriel Butler. https://ehnsw.mywikis.wiki/wiki/Person:Butler_Kathleen Accessed 13 May 2023.
Fidlon, P. G., & Ryan, R. J. (Eds.). (1979). *The Journal of Arthur Bowes Smyth, Surgeon, Lady Penrhyn, 1787–1789.* Australian Documents Library.
Fidlon, P. G., & Ryan, R. J. (Eds.). (1980). *The Journal of Philip Gidley King: Lieutenant, R.N. 1787–1790.* Australian Documents Library.
Gibson, R. (2010). Patyegarang and William Dawes: the space of imagination. In T. Banivanua Mar & P. Edmonds (eds.), *Making settler colonial space: Perspectives on race, place and identity* (pp. 242–254). Palgrave Macmillan.
Hawkins, J. B. (1979). Observatories in Australia 1788–1830. *The Australasian Antique Collector, 19*, 98–104.
Haynes, R., McGee, R., Haynes, R., & Malin, D. (1996). *Explorers of the southern sky: A history of Australian astronomy.* Cambridge University Press.
Hill, D. (2015). *First Fleet Surgeon: The Voyage of Arthur Bowes Smyth.* Canberra ACT, National Library of Australia.
Hughes, R. (1987). *The Fatal Shore. A history of the transportation of convicts to Australia, 1787–1868.* Collins Harvill.
Hunter, J., & Dawes, W. (1788). *Sketch of Sydney Cove, Port Jackson, in the County of Cumberland, New South Wales, July 1788.* J. Stockdale.
Jackson, D. (2018a). *Dawes point: Geographical review.* http://walshbayhistory.net/wp-content/uploads/2018/09/dawes-point-geography-review-19-feb-2018.pdf. Accessed 5 Oct 2020.
Jackson, D. (2018b). Walsh Bay, Before 1800: Arrivals. *Our historic shore of Sydney Harbour.* http://walshbayhistory.net/timeline/before-1800-arrivals. Accessed 5 Oct 2020.
James, A. (2012). The William Dawes observatory. *Southern astronomers and Australian astronomy.* http://www.southastrodel.com/Page031b.htm. Accessed 28 Sept 2020.
Johnson, A. W. (1994). *Archaeological assessment, Dawes Pt. Battery.* Sydney Cove Authority.
Johnson, A. W. (1998). *Dawes Point battery archaeological excavations 1995. Dawes Point, the Rocks, Sydney. I. Introduction and History.* Sydney Cove Authority.
Johnson, A. W. (2003). Showdown in the Pacific: A remote response to European power struggles in the Pacific, Dawes Point Battery, Sydney, 1791–1925. *Historical Archaeology, 37*, 114–127.
Keats, M. (1997). A cautionary tale: A study of the Macro Bivalve and Gastropod Molluscan Fauna of Spectacle Island, Sydney Harbour. *Wetlands (Australia), 16*, 72–82.
Kerr, J. (1986). The Architecture of Scientific Sydney. *Journal and Proceedings of the Royal Society of New South Wales, 118*, 181–193.
King, P. G. (1787–1790). *The Journal of Philip Gidley King, Lieutenant, R.N.* Australian Documents Library (1980). http://purl.library.usyd.edu.au/setis/id/kinjour. Accessed 12 Nov 2020.
King, P. G. (1788). *Journal.* State Library of New South Wales Heritage Collection. Sydney, Mitchell Library, State Library of New South Wales. Microfilm CY 298, frames 1–236.
Laurie, P. S. (1988). William Dawes, and Australia's first observatory. *Quarterly Journal of the Royal Astronomical Society, 29*, 469–482.
Macarthur-Onslow, S. (1973). *Some early records of the Macarthurs of Camden.* Adelaide, Rigby.
Mander-Jones, P. (1966). Dawes, William (1762–1836). *Australian Dictionary of Biography*, 1. http://adb.anu.edu.au/biography/dawes-william-1968. Accessed 28 Sept 2020.
Marchant, L. R. (1967). La Pérouse, Jean-François de Galaup (1741–1788). *Australian Dictionary of Biography*, 1. http://adb.anu.edu.au/biography/la-perouse-jean-francois-de-galaup-2329/text3029. Accessed 5 Jan 2021.
Maskelyne, N. (1786). Advertisement of the Expected Return of the Comet of 1532 and 1661 in the Year 1788. *Philosophical Transactions of the Royal Society of London, 76*, 426–431.
Maskelyne, N. (1787). *The Nautical Almanac and Astronomical Ephemeris for the Year 1767*. W. Richardson and S. Clark.

McAfee, R. J. (1981). Dawes's meteorological journal. Bureau of Meteorology, *Historical Note* No. 2.

McCormick, T. (ed.). (1987). *First views of Australia 1788–1825: A history of early Sydney.* Chippendale NSW, David Ell Press, in association with Longueville Publications.

McGuffie, K., & Henderson-Sellers, A. (2012). Interdisciplinary climate: The case of the first 50 years of British observations in Australia. *Weather, Climate, and Society, 4*, 118–131.

Mitchell, T., Lowry, J. W., Nicholl, H., Boone, W. (Firm). (1853). *Trigonometrical survey of Port Jackson.* Engraved by Lowry, J. W. London, T. & W. Boone. http://nla.gov.au/nla.obj-231443454. Accessed 20 Sept 2022.

Moran, A., & McAllister, J. (2020). Patyegarang was Australia's first teacher of Aboriginal language, colonisation-era notebooks show. *ABC/Walking Together*. https://www.abc.net.au/news/2020-03-11/patyegarang-and-how-she-preserved-the-gadigal-language/12022646. Accessed 29 Oct 2020.

Morrison, D., & Barko, I. (2009). Dagelet and Dawes: Their meeting, their instruments and the first scientific experiments on Australian soil. *Historical Records of Australian Science, 20*, 1–40.

Mouton, G. (1670). *Observationes diametrorum solis et lunae, huic adjecta est brevis dissertatio de Nova Mensurarum Geometricarum Idea.* Paris.

Nathan, D., Rayner, S., & Brown, S. (eds.). (2009). *William Dawes' notebooks on the Aboriginal language of Sydney, 1790–1791.* London.

NSW Education Standards Authority. (2014). *Building the Bridge—Engineering Studies—Concrete.* https://sydney-harbour-bridge.nesa.nsw.edu.au/engineering-studies/concrete.php. Accessed 16 Oct 2020.

Orchiston, W. (1989). William Dawes: Australia's first meteorologist. *Quarterly Journal of the Royal Astronomical Society, 30*, 359–360.

Parker, D. (2009). *Arthur Phillip: Australia's first governor* (pp. 115–116). Woodslane Press.

Phillip, A. (1788). Letter to Lord Sydney, dated 9 July 1788. http://firstfleet.uow.edu.au/s_jul9.html. Accessed 2 Oct 2020.

Phillip, A. (1790). *The voyage of Governor Phillip to Botany Bay: With an account of the establishment of the colonies of Port Jackson and Norfolk Island ... To which are added the journals of Lieuts. Shortland, Watts, Ball, and Capt. Marshall.* London, John Stockdale. http://gutenberg.net.au/ebooks/e00101.html. Accessed 30 Sept 2020.

Protos, A. (1988). *The road to Botany Bay: The story of Frenchmans Road Randwick through the Journals of Lapérouse and the First Fleet Writers.* Randwick and District Historical Society.

Pybus, C. (2009). 'Not fit for your protection or an honest man's company': A transnational perspective on the saintly William Dawes. *History Australia, 6*(1), 12.1–12.7.

Royal Australian Historical Society. (1927). Tablets. *Twenty-Sixth Annual Report and Statement of Accounts for 1926*, 387.

Royal Australian Historical Society. (1928). Historical tablets erected by the Royal Australian Historical Society. *Twenty-Seventh Annual Report and Statement of Accounts for 1927*, 8.

Royal Australian Historical Society. (1931). Answers to questions, 1930. *Thirtieth Annual Report and Statement of Accounts for 1930*, 486.

Russell, H. C. (1877). *Climate of New South Wales.* Charles Potter.

Saunders, S. (1990). *Astronomy in Colonial New South Wales.* Ph.D. Thesis, The University of Sydney.

Smith, B., & Wheeler, A. (1988). *The art of the First Fleet & other early Australian drawings.* Oxford University Press.

Steele, J. M. (2005). *The Aboriginal language of Sydney.* MA Thesis, Macquarie University, Sydney.

*Sydney Morning Herald.* (16 February 1927). *Historical Society. Erection of Tables* (p. 16).

*Sydney Morning Herald.* (17 June 1932). *Dawes Point. Bronze Tablet Erected* (p. 8).

Sydney Water. (1994). *The Tank Stream.* SW94 09/10. https://www.sydneywater.com.au/content/dam/sydneywater/documents/tank-streamheritage-fact-sheet.pdf. Accessed 20 Sept 2022.

Tench, W. (1789). *A narrative of the expedition to Botany bay, with an account of New South Wales, its productions, inhabitants, &c. To which is subjoined, A List of the civil and military establishments at Port Jackson*. J. Debrett. http://setis.library.usyd.edu.au/ozlit/pdf/p00039.pdf. Accessed 1 Oct 2020.

Thomas, S. (2013). A transnational perspective on William Dawes' treatment of women. *History Australia, 10*(1), 187–204.

Wales, W., & Bayly, W. (1777). *The original astronomical observations, made in the course of a voyage towards the South Pole, and Around the World* ... Strahan.

Webster, A. (2022). *The foundation of Australia's capital cities. Geology, landscape, and urban character*. Rowman & Littlefield.

West, O., & (n.d. ca. 1882). Our Harbour and Ocean Bays. *Old and New Sydney, 6*, 22–32.

Willetts, J. (2010–2020). *Free settler or felon. A search through the records for Walter Preston*. https://www.jenwilletts.com/walter_preston.html. Accessed 6 Oct 2020.

Wood, G. A. (1924). Lieutenant William Dawes and Captain Watkin Tench. *Journal of the Royal Australian Historical Society, 10*, 1–24.

# Chapter 3
# William Dawes: The Scientist, The Explorer, The Linguist

## 3.1 William Dawes, The Scientist

### *3.1.1 Astronomy*

Upon the termination of Dawes' appointment in the colony of New South Wales and following a resolution by the Board of Longitude, on 3 June 1795 Maskelyne sent five of Dawes' "… folio manuscripts [journals] of Astronomical Observations made at Port Jackson and in the Voyage thither …" to William Wales (1734?–1798), Secretary of the Board of Longitude, for reanalysis and completion. Wales died on 29 December 1798, before he had had the chance to complete and publish his work. Dawes' astronomical journals have not been seen since (e.g., Bosloper, 2010).

And so, when Henry Chamberlain Russell (1836–1907)—Government astronomer at Sydney Observatory—first discussed Dawes' astronomical contributions in 1882, he was generous in his assessment, given the lack of information:

> … we are not told what was done at Dawes' Point Observatory … no doubt a great deal of useful work in determining the position and other things [was] committed to Lieutenant Dawes' care … (Russell, 1882: 1–2)

In his Presidential address of 1951 to the Royal Society of New South Wales, Harley Wood (1911–1984)—Director of Sydney Observatory—was similarly kind to Dawes, citing his high workload as a reason for the absence of any tangible observations:

> Dawes ... could scarcely have been expected to undertake other astronomical work since, as well as being charged with the care of a small garrison, he was a very active member in the survey and exploration of the young colony. Besides other maps prepared later, he made a sketch of an area of the southern shore of Sydney Harbour as early as July, 1788, on which his observatory is marked [Fig. 2.9], its position being noted as longitude 151°19′30″ and latitude –33°52′30″. (Wood, 1951: 3)



R. de Grijs and A. Jacob, *William Dawes*, Springer Biographies,
https://doi.org/10.1007/978-3-031-38774-6_3

By 1958, Wood's opinion on the matter had remained unchanged, as shown by his reference to Dawes' astronomical achievements in his centennial history of Sydney Observatory:

> No astronomical work is known to have been done by Dawes, which is scarcely surprising since, as well as being charged with the care of a small garrison, he was an active member in the exploration and preliminary survey of the young colony. (Wood, 1958: 3, note 12)

Yet, these early assessments of Dawes as resident astronomer in New South Wales do not do proper justice to the man's achievements. With our straightforward modern access to the Board of Longitude papers, we can now obtain a much better-informed picture of his astronomical contributions. As we saw in Chap. 2, Dawes started his search for Maskelyne's comet as early as 7 August 1788 (Dawes, 1788b), following the inauguration of his Observatory:

> The Instruments were got on shore from the *Sirius* on the 31$^{st}$ July; the 7$^{th}$ August I looked out all night for the Comet (having previously computed its place on different suppositions of its perihelion distance [the comet's closest distance to the Sun] as hereafter explained) and from that time to the present, have not omitted looking carefully for it every favourable opportunity, but have not yet seen anything of it; the weather indeed has been so constantly cloudy for these three weeks past, particularly at night, that I should not wonder on the next clear night if I were to discover it pretty near us, and we have had sometimes so long a series of cloudy weather that I have doubted whether it were not possible for it to have passed entirely without being seen; on the other hand when we have clear weather it generally lasts as long, with some intervals of cloudy [weather], but without being excessively so, or for a long time together. (Dawes, 1788b)

He sent Maskelyne regular updates about his observational progress, although the variable weather presented a serious challenge: "... there cannot be a more unfavourable country for observation. I have been 10 & 15 days without being able to get altitudes of sun or stars ... I have seen nothing of the expected comet or any other" (Dawes, 1788b). Meanwhile, on 1 October he advised Maskelyne that he

> ... had been so much employed in matters quite foreign to Astronomy that [he had] just had time to settle the going of the clock,[1] and look out for a few Phenomena [and he had] constantly, when the weather would permit search'd for the Comet ... (Dawes, 1788a)

Despite his failure to observe the comet, he discovered a number of new nebulae, observed a solar eclipse and took observations of the Moon's parallax and Jupiter's satellites (James, 2012). By mid-November, Dawes had designed his own tracking approach using Maskelyne's orbital predictions, combined with his own calculations:

> I find Barker's Treatise [Barker, 1757] so very easy and intelligible that I calculate the place on different suppositions of the time from the Perihelion, with great ease and I believe, accuracy, but however I do not confine myself to the track calculated, but sweep a little on each side of it. I do not make any use of the orbit projected as I don't think it accurate enough, but I compare notes with it now and then to prevent mistakes. I have seen nothing of the Comet yet but have discover'd several Nebula [*sic*] exactly, or very nearly in the track, the places of these I have settled, sufficiently to find them and do it more accurately when time

---

[1] Dawes reported that the clock's rate was decreasing by 36 s of sidereal time per day (Dawes, 1788a; see below).

will permit. I compute the Comet's Geo[centric]. Lon[gitude]. and Lat[itude]. to every tenth day in the Year, and on different suppositions of time from Perihelion and put them in a Table … and so as always to keep ten Days ahead in the computations. These lon[gitudes]. and lat[itudes]. I mark in pencil on the Celestial Charts (allowing for the precession[2]) and then I see where to look for the Comet in the heavens. … This [letter] goes by the *Sirius* to the Cape [of Good Hope],[3] from when [*sic*] Col. [Robert] Gordon will send it to England. (Dawes, 1788a)

However, despite his 'sweeping' approach, he could easily have missed seeing the comet if his calculations were not sufficiently accurate (e.g., Saunders, 1990: 101). In his letter of 17 November 1788, Dawes elaborated further on his method:

I endeavour to keep ten days at least forward of the present time always computed, and I find Mr. Barker so clear and easy that I think it much less trouble to calculate the whole by figures than to make use of the orbit at all, unless just to see when to add and when to subtract to get [the angle] at the Earth etc. I subtract 1°23′ from the Lon[gitude]. for the precession and then mark the place on the chart in pencil where the comet would be supposing it 200 Days from the perihelium, then make a mark for 180 Days etc. all along the line for that day and night [I] look along that track in the heavens and on intermediate nights I look in the intermediate spaces between the tracks, but without confining myself particularly to the very spot calculated for, but take in a degree or more round about.

Yet, despite valiant efforts, Dawes did not manage to observe Maskelyne's comet. Neither did any astronomers anywhere else, for that matter.

In addition to these purely scientific pursuits, he also worked at establishing the Observatory's precise longitude and latitude (for details, see Chap. 2), for positional reference purposes, he tracked changes in the rates of both K1 and the other timing devices present in the colony, and he observed eclipses—solar, lunar and those of Jupiter's satellites. He reported that it was "… impossible to fix a clock up much better …" (Dawes, 1788b). He supported this confident statement by a range of comparisons between his timekeepers, obtained between 8 and 30 September 1788.

Dawes also attempted to take solar altitudes and measure lunar distances to independently calibrate the running of his timepieces. Combined with careful measurements of 'equal altitudes' of the Sun on either side of local noon, a standard timing approach, he concluded that K1 was losing, on average, 4.77 s each day (Dawes, 1788b). However, as his workload increased, he eventually only managed to take solar altitude measurements irregularly, in intervals of sometimes five, seven, ten or 20 days, or even longer, which was clearly unsatisfactory (Dawes, 1790b). He managed to take lunar distance observations even less frequently; and indeed, those measurements he had obtained "… were very defective, owing principally or entirely to my want of practice; this cause as it is in my power to destroy it, so I trust it will not continue long …" (Dawes, 1788b). Similarly, he felt that he was not sufficiently

[2] Precession refers to the gradual, long-term movement of the Earth's axis with respect to the positions of the fixed stars.

[3] With the colony in desperate need of provisions, H.M.S. *Sirius* was sent to the Cape of Good Hope. She subsequently completed a full circumnavigation of the globe before returning to Sydney Cove.

trained to obtain solar zenith distances to obtain accurate angular measurements, given that the solar limb appeared hazy where it touched the micrometer wire of his measurement device (Dawes, 1788b).

In November 1788, Dawes reported to Maskelyne that he was pleased with his observations of the eclipses of Jupiter's Galilean moons, and of stellar zenith distances, particularly of Fomalhaut, the brightest star in the constellation *Piscis Austrinus* (Southern Fish). However, by July 1790 he estimated that he had observed only one quarter of all visible eclipses of the Jovian moons, as well as a small number of lunar occultations of fixed stars.[4] At that time, he had also managed to observe two solar eclipses—the latest as recently as 14 May 1790—and one lunar eclipse, although "… imperfectly, for want of skill, having never practised these observations before" (Dawes, 1790b).

At Maskelyne's request, Dawes also attempted to compare the declination (position) of Mars with respect to the positions of several fixed stars, but the resulting measurements were woefully inaccurate because of his instruments' unstable foundation and problems with movement of the Observatory's floor stones (Dawes, 1790b). Determined to improve his skills and extend his astronomical observations, in April 1790 he requested to be provided with a transit instrument (for a contemporary example, see Fig. 3.1). Following joint efforts by Maskelyne and Benjamin Dawes, in a letter of 27 January 1791 Sir Philip Stephens (1723–1809), Secretary of the Admiralty, wrote to the commander of H.M.S. *Gorgon*, Captain John Parker (ca. 1749–1794),

> Dr. Maskelyne, the Astronomer Royal, having acquainted my Lords Commissioners of the Admiralty his letter of yesterday's date [26 January 1791], that he is desirous of sending to Lieutenant Dawes at Port Jackson, for the use of the Observatory there, the *Nautical Almanacs* for 1793 to 1795, tables required to be used therewith, and a transit instrument. I am commanded by their Lordships to signify their direction to you to receive the same on board the ship you command, and deliver them upon your arrival at Port Jackson agreeable to their address. (Bladen & Britton, 1893: 431)

Although the instrument arrived by the end of 1791, it was never unpacked. Instead, it returned forthwith—on the *Gorgon*—to England, this time accompanied by Dawes himself (Dawes, 1792).

### *3.1.2 Gravity Measurements*[5]

In addition to Dawes' 'standard' astronomical duties and his daily meteorological observations, we learn from his correspondence with Nevil Maskelyne that he also intended to take gravity measurements (e.g., Forbes, 1975: 172; Morrison & Barko, 2009; Bosloper, 2010): "I shall also pay particular Attention to the Experiment of Gravity, … The Rates of the Clock … shall be properly attended to …"

---

[4] Lunar occultations refer to instances when the Moon passes in front of certain stars.

[5] This section is based on de Grijs (2022).

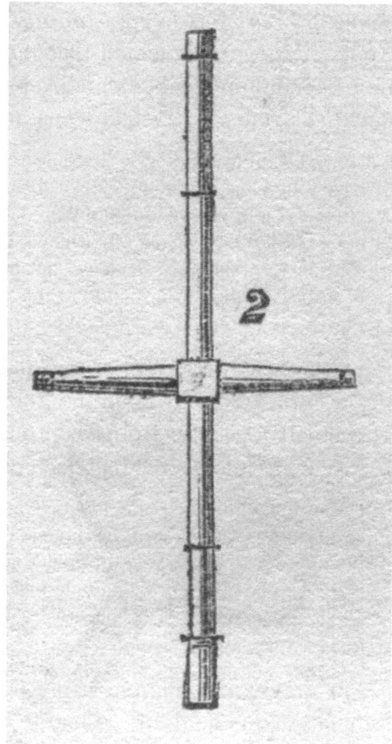


**Fig. 3.1** Contemporary example of a transit instrument: William Bradley's 8-foot transit device at Greenwich Observatory (1750). (*Gentleman's Magazine*, March 1866; out of copyright)

(Dawes, 1787a). Dawes' Shelton regulator clock was clearly meant to provide gravity measurements, as we learn from his letter to Maskelyne of 1 October 1788, sent from their final destination in New South Wales:

> I fix'd very firmly the Clock in a niche of a solid Rock, on Sat$^{y}$. the 6$^{th}$ of Sept$^{r}$. and by Altitudes taken at two or three days interval found it to be losing at the rate of about 36″00 on sidereal time in one sidereal day but this Rate seems to be increasing ... for in these last 11 Days the Clock has lost after the Rate of 37″25 on sidereal time in one sidereal day ... (Dawes, 1788a)

State-of-the-art gravity experiments in the late eighteenth century consisted of an extended series of daily clock-rate measurements of a carefully calibrated pendulum clock, compared with local time measurements based on astronomical observations. Dawes' measurements formed part of a comprehensive British research programme undertaken in the 1760s to 1780s of gravity measurements around the world. These included measurements obtained on H.M. Bark *Endeavour* by Charles Green (1734–1771) at Botany Bay during Cook's first voyage to the Pacific and Australia (1768–1771), and by Bayly and Wales on H.M.S. *Resolution* in New Zealand, in 1773, during Cook's second voyage to the Pacific (1772–1775) (Bosloper, 2017).

At the time of the First Fleet's voyage, it had already been known for more than a century that pendulum swing periods vary with geographic latitude. As a case in point, in his *Philosophiæ Naturalis Principia Mathematica* (1687), Isaac Newton (1642–1726/7) commented that the French astronomer Jean Richer's (ca. 1630–1696) pendulum clock—which kept perfect time in Paris (latitude 48.86°N)—went slow in Cayenne (French Guyana; latitude 04.94°N):

> Now several astronomers, sent into remote countries to make astronomical observations, have found that pendulum clocks do accordingly move slower near the Equator than in our climates [at moderate northern latitudes]. And, first of all, in the year 1672, Mr. Richer took notice of it on the island of Cayenne; for when, in the month of August, he was observing the transits of the fixed stars over the meridian, he found his clock to go slower than it ought in respect of the mean motion of the Sun at the rate of $2^{m}$ $28^{s}$ a day. (Newton, 1687; see Fig. 3.2)

Dawes was rather concerned that, upon his arrival at Botany Bay, the gradually accumulating errors in the Shelton pendulum clock's rate—possibly resulting from small deviations from isochronism[6]—would preclude him from performing the gravity experiments that had been proposed by Maskelyne:

> I wish to be informed how long a Time after our Arrival at Botany Bay, you think will be sufficient for making the Experiment of Gravity by the going of the Clock? it may reasonably be supposed to keep a very different Rate from that which it had at Greenwich, and I presume were it to be allowed to continue that Rate a long Time; the allowing for the Error of the Clock when that Error becomes very great will be rather inconvenient. (Dawes, 1787b: ff. 255r–v)

For gravity calculations using pendulum clocks, the length of the temperature-compensated pendulum rod had to be precisely set and calibrated, and it had to be reproducible anywhere, to within a tolerance of 10 μm. One also needed to measure the pendulum's half-amplitude—known as the 'arc from the vertical' or the 'arc of vibration'—so as to apply an amplitude-dependent correction to the clock-rate measurements. Maskelyne was well aware of these requirements, and he also knew how to calculate the amplitude-dependent correction; he made sure that all of his observers determined the 'arc from the vertical' of their pendulum clocks during their voyages, for a predetermined pendulum length that was set and calibrated at Greenwich.

As we saw already, Dawes' instruments included a Shelton astronomical regulator clock (see Fig. 3.3), a highly precise, tall 'grandfather clock' equipped with a temperature-compensated grid-iron pendulum (see Fig. 3.4) of precisely known length. Such regulator clocks were equipped with a so-called 'seconds' pendulum, which "... escaped dead seconds in the manner of the late Mr. [George] Graham" (Wales & Bayly, 1777) and completed a full swing period in 2 s. They were calibrated to swing exactly 86,400 times during a mean solar day (24 h, equivalent to 86,400 s), which could be achieved by adjusting a regulator nut, that is, a fine-motion screw, at the bottom of the pendulum bob (see, e.g., the inset in Fig. 3.3). These regulator nuts generally worked such that a full turn altered the clock rate by close to half a minute per day (Bosloper, 2010).

[6] 'Isochronism' refers to a clock's property when a pendulum's period is, at least for small swing angles, approximately independent of its (half) 'amplitude', that is, the maximum extent or angle of the swing.

**Fig. 3.2** Drawing from Richer's *Observations astronomiques et physiques faites en l'isle de Caïenne* (Paris, 1679). Most of the astronomical instruments used by Richer are shown, including one of Isaac Thuret's (1630–1706) pendulum clocks. (Artist: Sebastian Le Clerc; Creative Commons Attribution-ShareAlike License)

**Fig. 3.3** Shelton No. 35 regulator pendulum clock. (© The Royal Society. Screen captures from *Objectivity* video 57: https://www.youtube.com/watch?v=9fcWWBksfng. Reproduced for academic purposes under *YouTube*'s 'Fair Use' policy)

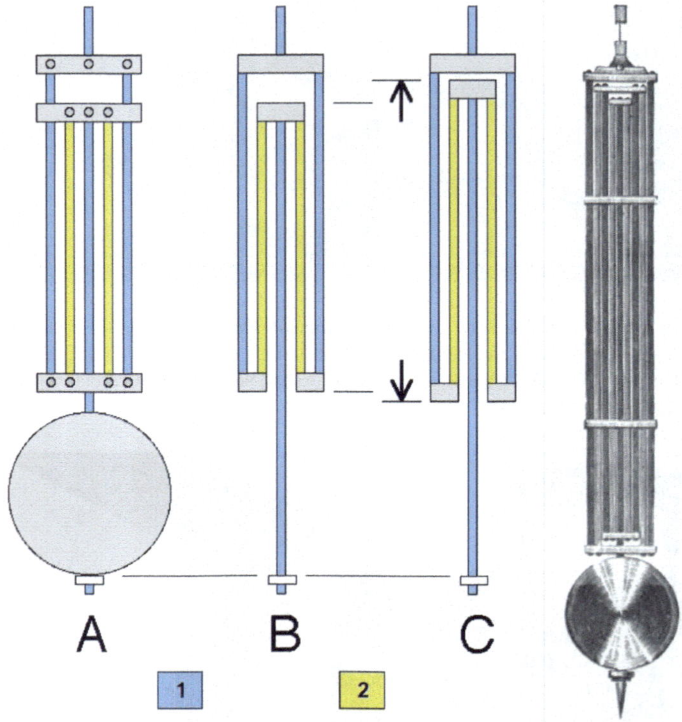


**Fig. 3.4** Operation of a temperature-compensated grid-iron pendulum. The pendulum uses rods of a high thermal-expansion metal, zinc (yellow) to compensate for the expansion of rods of a low thermal-expansion metal, iron (blue). A: General appearance; B: Schematic at normal temperature; C: Schematic at higher temperature. (Creator: Leonard G., via Wikimedia Commons; public domain) The example pendulum on the right dates from the 1880s. Note the regulator nut at the bottom of the bob. (Brockhaus, 1890: Plate 124, Fig. 7, via Wikimedia Commons; public domain)

Such astronomical regulator clocks were isochronal at their calibration location, usually Greenwich (at least initially). In 1733, James Bradley (1692–1762), Savilian Professor of Astronomy at Oxford University, had advised the Royal Society of London (Bradley, 1733) that the standard isochronal pendulum length was assumed to be 39.126 'English inches' (99.380 cm) (Bosloper, 2017: 248), a value attributed to the English clockmaker George Graham (1673/5–1751) in 1722. In 1817–1818, the British physicist and army captain Henry Kater (1777–1835) determined that the mean length of his 'reversible' solar seconds pendulum,[7] in London, at sea level, in ambient conditions at a temperature of 62 °F (17 °C), swinging in a vacuum, was 39.13858 inches (99.412 cm) from knife edge to knife edge on his device (Kater, 1818). At the time of Dawes' observations during his voyage on the First Fleet

[7] A reversible or 'Kater's pendulum' is a physical pendulum with two adjustable knife edges, purpose-built for accurate determination of the local gravitational acceleration, *g*.

and, subsequently, at Sydney Cove, Graham's value was still in use, however; it was adjusted to 39.128 inches (99.385 cm) in 1790 (*Ibid.*). We learn from a letter from Maskelyne to Henry Cavendish (1731–1810) in 1761 that the Astronomer Royal was fully aware of these developments (Maskelyne, 1761–1762). Isochronal clocks marked precise 'dead' seconds at a given location, with the timing calibration provided by either stellar transit observations (for sidereal time) or local noon determinations (for mean solar time).

The length of the pendulum could, theoretically, be set precisely to a 'reduced length' of $g/\pi^2$ metres—where $g$ is the gravitational acceleration at one's reference location, e.g., Greenwich—simply by adjusting the regulator nut *when at that reference location*. The value of the gravitational acceleration ('absolute gravity'[8]) could then be determined, in principle, by multiplying the isochronal pendulum length by $\pi^2$, *modulo* a non-linearity correction.

In 1761, before he set off on his voyage to St. Helena to observe a Venus transit, Maskelyne had compiled a list of specifications for a pendulum clock to act as a frequency counter for gravity measurements (Higgit, 2014). Most importantly, it had to be equipped with a clearly visible and audible seconds hand. The requirement for the clock to tick once a second constrained the pendulum's length. In addition, the clock had to maintain power, that is, it could not suffer a loss of time when it was rewound, which was done approximately once a month.

Finally, the pendulum had to be temperature-compensated. This was achieved by ensuring that the entire pendulum was composed of a bi-metal (see the grid-iron pendulum referenced above), an idea that had initially been proposed by Graham. It was famously applied by John Harrison in the fourth incarnation of his 'longitude clock', H4. The ratio of the metals' expansion coefficients in relation to their lengths had to be carefully balanced so as to provide full temperature compensation. In other words, expansion of one component of the bi-metal was meant to balance expansion of the other, thus keeping the pendulum length unchanged.

While Maskelyne and Dawes were planning the scientific aspects of the First Fleet's voyage, they envisioned taking pendulum clock-rate measurements at all ports of call. Upon departure, Captain Phillip had planned to call into the Santa Cruz roads at Tenerife, Canary Islands, as well as at St. Jago (Cabo Verde), St Sebastian (Rio de Janeiro) and Table Bay. As we saw in Chap. 1, the First Fleet did indeed call into Santa Cruz, St Sebastian and Table Bay, but a visit to St. Jago was abandoned at the last moment because of a sudden change in the prevailing winds that would have made entering the local harbour too dangerous.

Unfortunately, since Captain Phillip would not allow Dawes to disembark the Shelton regulator at either Tenerife or Table Bay, the Marine managed to obtain the pendulum clock's decay rate only in the harbour of Rio de Janeiro. In his letter of 3 September 1787, he went to great lengths to explain to Maskelyne his operational

[8] Strictly speaking, this gravity measurement was with respect to that at Greenwich, which became the *de facto* standard after Cavendish's publication of *G*, the 'gravitational constant', in 1798 (Cavendish, 1798).

practice in St Sebastian while also pointing out a number of mishaps that had affected his measurements:

> The Reason why the mean Rate of the Clock is deduced [only] from the 16 to the 27th [August 1787] is, that about 12 o'Clock at Night on the 17th the Shoar [shore] mark'd in the Plan with a Stroke across it, was displaced, which I did not discover 'till the next Morning when I immediately replaced it, as nearly as I could judge in the same Position that it had before; notwithstanding which I thought the mentioning of it indispensable, as also the not allowing the last Day's Rate which seems to have been affected by the Accident. I conclude that the small apparent Irregularity in the Rate of the Clock to be merely apparent, or very nearly so, and to arise principally if not entirely from the unavoidable Errors of Observation: I have therefore taken the mean Rate of the Clock –48″,067 in 24 sidereal, or 48″,20 in 24h Mean Time for the Rate of the Clock each Day in the Comparisons of the Time Keeper with it. (Dawes, 1787c; original emphasis)

Dawes also provided information about his clock's regulator nut: "… the screw is at 15 on the nut …" (Dawes, 1788a). However, Dawes only realised that the nut was set at that specific position following some confusion in Rio de Janeiro, potentially leading to erroneous results:

> Your Letter to me in which you mention the going of the Clock &c. says that the Index stood at 17 on the Nut;– when I opened the Clock Case, I found it exactly at 15 on which I immediately examined the line drawn on the Bar & found it but just appearing above the Bob, as if it (the Bob) had not been in the least lower'd; the measurement between the Bob & the upper End of the Brass Bar was accurately (as to Sense) the same as mentioned in your Letter and as I could in no wise account for the Nut having moved exactly 2 Divisions I thought there might possibly have been a mistake in taking down the Number on the Nut; I therefore set the Clock going with the Index exactly at 15 and on the next Page I have given the Results of each Day's Equal Altitudes. … On further consideration I do not think any Mistake could well have happen'd at Greenwich in noting what the Index stood at on the Nut & am rather fearfull [*sic*] that I ought to have screwed it to 17 before I set the Clock going; but in order to remedy as much as possible any ill Consequence that may have arisen from the not doing it, I intend the first Opportunity to try the Clock a Number of Days with the Index at 15 and afterwards to give it as fair a Trial as possible with the Index at 17. (Dawes, 1787c: ff. 269v, 272v)

Despite Dawes' valiant efforts, where he tried to determine whether the setting of the regulator nut was meant to be 15 or 17, in practice this must have been nigh impossible. After all, the two nut positions would move the scratch mark by no more than 4–5 hundredths of a millimetre.[9] One would need to have access to properly calibrated daily decay rates at Greenwich, for both nut settings, to reach firm conclusions as to the original nut setting.

Eventually, Dawes' most accurate clock-rate observations were obtained once they had safely gained solid ground again at their final destination in New South Wales. As we saw already, at the time Dawes wrote to Maskelyne that he had observed that the Shelton regulator suffered a sustained loss of 37.25 s per day over a period of 11 days—as it had done in Rio de Janeiro, he added (Dawes, 1788a).

---

[9] Although we can only speculate at this time, it is possible that the ship's vibrations during the voyage, as well as shocks, however minor, during the embarkation and disembarkation of the Shelton regulator at Portsmouth and in Rio de Janeiro may have caused such a minor shift in the nut's position.

This information was all that Maskelyne needed to determine the gravity at Sydney Cove. Maskelyne knew which Shelton regulator Dawes had brought along with him on the voyage, as well as the precise length of its pendulum. It appears that he had set this length to 39.111 inches (99.342 cm), appropriate for a geographic latitude of 45°. In a letter dated 16 April 1790, from Sydney Cove, Dawes recorded that "... its arc of vibration is 1°30' + and continues constantly the same ..." (Dawes, 1790a). This is the only recorded measurement of his pendulum clock's (half) amplitude obtained during the First Fleet's voyage. For isochronous operation, it applies to the entire voyage, however, as the half-amplitude is predominantly determined by a clock's escapement mechanism.

Modern analysis of Dawes' gravity measurements at Sydney Cove suggests that he indeed used a pendulum length of 39.111 inches, as opposed to the characteristic Greenwich length of 39.126 inches (Bosloper, 2010), corresponding to a difference of a full turn of the regulator nut (and a difference of 30 s a day). By analogy with the gravity measurements undertaken by William Wales on Cook's second voyage using the same Shelton regulator, it has been suggested that Dawes made a similar 'mistake' as Wales (*Ibid.*). Wales recorded that the pendulum length

> ... was always altered, in order to its being packed up, yet on setting up again, it was constantly brought back to its proper length, by means of a scratch on the rod, and the numbers on the nut. (Wales & Bayly, 1777: 131)

However, in the introductory remarks to his observations from Cook's second voyage, he concedes,

> On reconsidering the circumstances of the clock's different rates of going at the Cape of Good Hope in November 1772 and April 1775, I am rather inclined to alter my opinion (see page 131) [previous citation] and to conclude that I made a mistake in setting the pendulum to its proper length, either when here in November 1772 or at Dusky Bay [now Dusky Sound] in New Zealand, after which time it was never altered; basically as the difference corresponds nearly to that which would arise from a whole revolution of the nut which supports the ball of the pendulum, namely 28″ or 29″, increased by the same quantity that the clock had gone faster on being set up a second time both at Point Venus [in Tahiti] and Queen Charlotte's Sound [New Zealand]. (Wales & Bayly, 1777: Introduction)

Dawes provided regular updates on the clock's rate of change during part of September 1788, after their arrival in New South Wales: on 1 October 1788 he advised Maskelyne that he "... had been so much employed in matters quite foreign to Astronomy that [he had] just had time to settle the going of the clock, ..." (Dawes, 1788a). In the same letter, he reported that the clock's rate was decreasing by 36 s of sidereal time per day. He continued his measurements for most of his term in the colony, but his records were unfortunately misplaced after their transfer to William Wales back in England.

The surviving measurements from September 1788, combined with a pendulum length ($L$) of 39.111 inches and local noon determined by 'equal altitudes' of the Sun, a standard approach to time measurement, result in a gravity at Sydney Cove of $g = 9.79705$ m sec$^{-2}$, with an associated uncertainty of 25 parts per million (ppm) (Bosloper, 2010). The 1788 value resulting from Dawes' measurements is $g$

= 9.79621 m sec$^{-2}$ (Morrison & Barko, 2009: 34, Note 129). Combining this with the pendulum's arc of vibration of 1°30' leads to $g$ = 9.79705 m sec$^{-2}$, which is within 13 ppm of the modern value (*Ibid.*: Note 130), $g$ = 9.796720 m sec$^{-2}$ (Bell et al., 1973). Alternatively, based on the same pendulum length but adopting the change in the clock rate of –37.25 s (slow with respect to the daily sidereal time) leads to a pendulum swing period of $T$ = 2.00086264 s instead of exactly 2 s (see below). Combined with the simple pendulum equation, $g = 4\pi^2 L/T^2$, we obtain $g$ = 9.79621 m sec$^{-2}$ at Sydney Cove (Morrison & Barko, 2009: 34, note 129).

We can now apply the same approach to obtain an estimate of the gravitational acceleration at Rio de Janeiro as implied by Dawes' measurements of 1787. For a given daily decay rate, the corresponding pendulum swing period $T$ would become slower than the standard 2 s by a factor of 86,400/(86,400 – decay rate), where the decay rate is expressed as a positive value. Following Bosloper (2010), the gravitational acceleration can be expressed as

$$g = 4\pi^2 \frac{L(1+\Delta)^2}{T^2},$$

where $L$ = 39.111 inches = 0.9934194 m, and

$$(1+\Delta)^2 = \left(1 + \frac{1}{2^2}\sin^2\frac{\theta_M}{2} + \frac{1}{2^2}\frac{3^2}{4^2}\sin^4\frac{\theta_M}{2} + \ldots\right).$$

For the arc of $\theta_M$ = 1.5° vibration pertaining to Dawes' Shelton regulator, it follows that $(1+\Delta)^2$ = 1.000086 (Bosloper, 2010).[10] Combined with his measured decay rate of –48.067 s with respect to the daily sidereal time, and assuming that the latter rate is entirely driven by the Earth's local gravity, this yields $g$ = 9.7946 m sec$^{-2}$ for Rio de Janeiro, presumably with a similar uncertainty of 25 ppm as for Sydney Cove.

The modern value for the gravitational acceleration in Rio de Janeiro is $g$ = 9.7878 m sec$^{-2}$ (Kochsiek & Gläser, 1999: 522). The gravitational acceleration implied by Dawes' clock-rate measurements is 0.0068 m sec$^{-2}$ larger than the modern value. This difference is well in excess of the probable uncertainty estimated by Bosloper (2010), although the accuracy of our measurements does not allow us to assess the difference to the parts-per-million level. At face value, this suggests either that gravity may not have been the only force acting on the Shelton regulator in Dawes' care or that one of our assumptions could be incorrect. Incomplete temperature compensation or any number of minor mechanical effects could have contributed to a slightly increased deceleration of the pendulum bob,[11] including non-negligible

[10] Note that this equation strongly implies that small changes in the pendulum's amplitude, e.g., from $\theta_M$ = 1°30′ to $\theta_M$ = 1°40′ or $\theta_M$ = 1°45′ (e.g., Maskelyne, 1761–1762: 441; Cope, 1950: 261), will have a negligible effect on the resulting gravitational acceleration.

[11] The pendulum and bob of the No. 35 Shelton regulator operate inside a closed wooden case, so that any effects associated with air flow or air resistance can be ruled out.

friction at the pivot point, an imbalance of the escapement mechanism or mechanical energy loss in the gear train.

On the other hand, let us consider which decay rate would be expected for the modern value of $g = 9.7878$ m sec$^{-2}$. For the known values of $g$, $L$ and $(1 + \Delta)^2$, the corresponding pendulum swing period would be $T = 2.00180753$ s, corresponding to a declining clock rate of 78.1 s per day. Now recall that Dawes had measured a decay rate of –48.067 s per day in Rio de Janeiro. The difference between Dawes' measurement and the expected decay rate for the swing period implied by the modern value for the gravitational acceleration is almost exactly 30 s per day.

Although we can no longer verify the actual setting of the regulator nut used by Dawes, a difference of 30 s a day corresponds to a full turn of the regulator nut (Bosloper, 2010). Could it be that the clock's pendulum length had not been altered with respect to the characteristic Greenwich length after all?

A full turn of the nut changes the pendulum length by 0.66 mm, and one index number higher or lower on the nut only changes the pendulum length by 0.023 mm. If we now consider these implications with our current 20/20 vision, eighteenth-century observers seem to have found it hard to judge whether they were a full turn off from the proper setting of the nut, as witnessed by the errors made by Wales, Bayly and Dawes. That is most likely the reason why they also made a scratch mark on the pendulum rod just above the bob. One such scratch mark might refer to the calibration setting at Greenwich, another to an isochronal setting in a particular location where astronomical observations are timed. However, it was even harder to see if a change of two index steps on the nut—corresponding to a pendulum length change of 0.046 mm—made it disagree with the scratch mark, as Dawes noticed in Rio de Janeiro.

### 3.1.3 Meteorology

Although Dawes' opportunities for night-time observations were severely curtailed during his first few months in the new colony, he managed to maintain a detailed journal of meteorological observations (Dawes, 1791). Like his astronomical and gravity observations, his meteorological journal went missing shortly after his return to England: "[t]here seems little doubt that Dawes kept weather records but there are no records of them ..." (Russell, 1877). Covering 115 pages, it was rediscovered in the library of the Royal Society of London in 1977 by a PhD student at Macquarie University (McAfee, 1981a).

Dawes' weather observations are comprehensive, detailed and meticulous, indeed a rare historical find in the Indo–Australian region. The only comparable set of weather records from the same geographic region was obtained by the British East India Company in Madras (Chennai, India), starting in 1796 (Allan et al., 2002; Gergis et al., 2010). Dawes' weather journal is clearly organised (although not necessarily easy to read; see Fig. 3.5), containing seven days of records per page in six columns, including the day and time of the observation; the wind direction and speed; daily weather comments; barometric (atmospheric) pressure; temperature; and general comments, e.g. "Distant thunder at long intervals" (McAfee, 1981a: 2–5 November 1788).

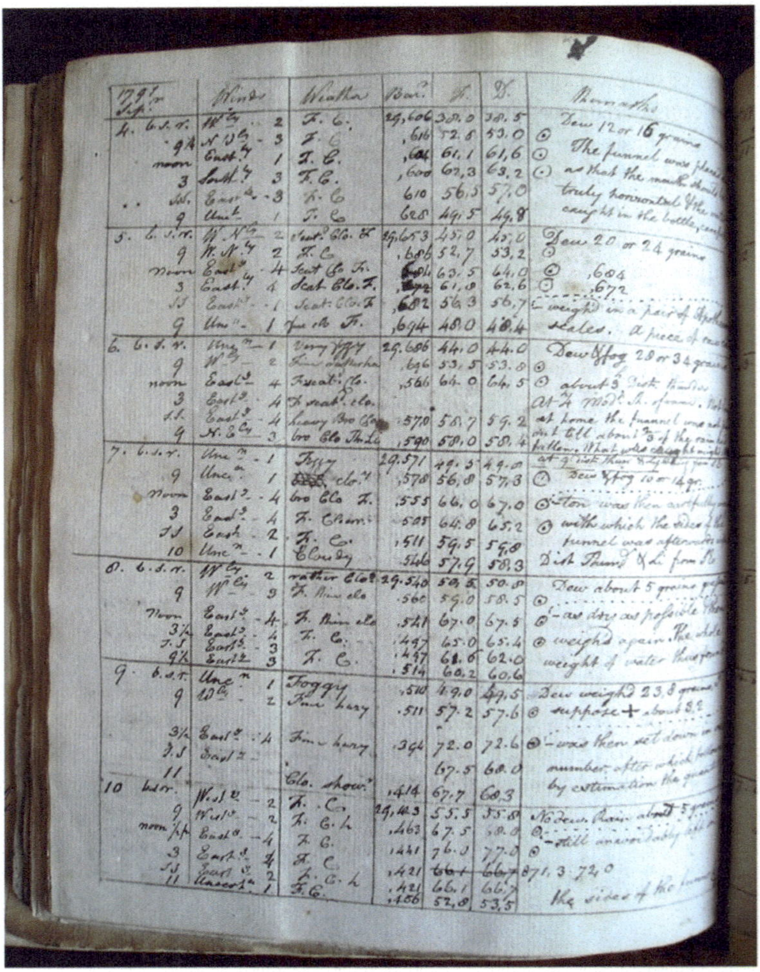

**Fig. 3.5** Page from Dawes' meteorological journal. (Photo: Dr. Joëlle Gergis; reproduced with permission)

Dawes' general comments include remarks about special qualities in the shifting wind, descriptions of the sky including cloud cover, the pressured weight of the air, and sensing and saying what type of heat or cold was prevalent (Gibson, 2012: 56). He used more than 50 expressions for different types of rain and in excess of 100 ways to describe the sky conditions (McAfee, 1981a). He also kept careful track of the properties of and changes to his instruments, realising the need to provide standard instrument exposure measurements. Some of the recorded wind directions are difficult to interpret and it is unclear which scale he used for wind force. His wind-force measurements range from 1 to 8, and so his adopted scale may have been similar to that used by most British ships at the time, which was formalised by Admiral Sir Francis Beaufort (1744–1856) in the 1830s (McAfee, 1981a: 19, 20).

Dawes' temperature measurements (recorded in degrees Fahrenheit), obtained from his observatory at Point Maskelyne, span the period from 14 September 1788 until 6 December 1791. The only gaps in his observations occurred during periods when he was otherwise engaged or away from his observatory, often on exploratory expeditions (see Sect. 3.2). On one occasion, when he was unable to take his own measurements, he recorded in his journal, "… rather doubtful being kept by a friend in my absence …" (McAfee, 1981a: 9 September 1788). He does not specify which 'friend' this comment refers to, although Watkin Tench is an obvious candidate.

Readings were taken four to six times a day, although not always consistently at the same time. In general, Dawes took his atmospheric pressure and temperature measurements roughly every four hours, usually from 8 o'clock in the morning until 8 o'clock at night. His observation times vary from 'before sunrise', 6 a.m., 'sunrise', 'after sunrise', 9–10 a.m., Noon, 2–4 p.m., 'before sunset', 6 p.m., 'sunset', 'after sunset' and occasionally at 10 p.m.–midnight.

Dawes points out that his thermometer was exposed outside, away from direct sunlight and about three feet above a stone surface (Dawes, 1791; Kington, 1997). In his private journal, John Hunter provides temperature and weather observations of which he says, "I was furnished with the following months [October 1788–January 1789] by Lieutenant William Dawes, of the marines" (Hunter, 1793: 75). He then explicitly continues, "The thermometer, as marked for these last four months, was in the open air occasionally exposed to the sun and wind". Meanwhile, Tench elaborated, "The thermometer, whence my observations were constantly made, was hung in the open air, in a southern aspect, never reached by the rays of the sun, at the distance of several feet above the ground".

On 3 July 1791, Dawes installed a second thermometer, "… a very good one by Ramsden's …", borrowed from Lieutenant James Furzer (d. 1799), quartermaster on H.M.S. *Sirius*. On 11 July, he comments that "I observe when the thermometers have been long at nearly the same height that they agree" (Dawes, 1791). He conducted such comparisons a number of times throughout 1791, concluding that both thermometers agreed with one another to within 1 °F (McAfee, 1981b: I, 121; note 3). In July 1791, Dawes recorded a detailed description of his thermometers:

> My own thermom$^{r}$. by Nairne & Blunt, small & divided only to every two degrees [°F], the whole scale from 0 to 212 being in length 6.23 inches. Both were kept constantly out of their cases so that the bulb might be fully exposed to the air, having nothing attached to it but

> the metal scale in Ramsden's (Mr Furzer's thermometers) & the wooden scale in Nairne & Blunt's. (Dawes, 1791; cited by McAfee, 1981a: 15)

To get a sense of the weather and climatological conditions the first settlers had to contend with, it is instructive to reproduce some of the early records. Hunter's four months of early temperature and weather records read as follows (Hunter, 1793: 75):

> [October 1788] The first and middle parts of this month the weather was moderate and cloudy, and the wind very variable, frequent thunder and lightning with showers of rain; the latter part was clear, fine weather in general, with distant thunder and lightning, and a few violent squalls of wind, which happened generally in the night. The Thermometer from 49° [F] to 81° [F].
>
> [November 1788] In the beginning of this month the weather was generally cloudy and hazy, the wind from the eastward; the middle part also cloudy with frequent light showers of rain, thunder, and lightning, sometimes distant and sometimes very heavy; latter part, cloudy and hazy, with violent thunder, lightning, and rain; wind from north-east to south-east; and the Thermometer from 53° [F] to 93° [F].
>
> [December 1788] The first part was cloudy and hazy, with some thunder, attended with light rain; middle, same kind of weather, with frequent and light showers of rain; latter part, moderate weather with a good deal of rain; the wind chiefly from the northward and eastward. The Thermometer from 53° [F] to 102° [F].
>
> [January 1789] During the whole of this month, the weather was cloudy and hazy, with light showers of rain, and sometimes distant thunder; the wind chiefly, though from the north-east and south-east, and during the night, westerly or land winds. The Thermometer from 63° [F] to 112° [F].

Dawes' records reveal that the summer of 1788–1789 was warm, with average to wet conditions. He recorded a seasonal maximum of 12 rain days during January 1789, including 9–12 January 1789. The following summer, 1789–1790, witnessed a transition from dry conditions to the usual warm and wet weather characteristic of Sydney summers. November 1789 was very warm. Significant rain caused flooding in Sydney in February 1790 (Ashcroft, 2016), with wet conditions prevailing during much of the first half of that year. The summer of 1790–1791 appears to have been particularly brutal, as we learn from Tench's private journal:

> My observations on this extreme heat [in December 1790] … were, that of all animals, man seemed to bear it best. Our dogs, pigs, and fowls lay panting in the shade, or were rushing into the water. … The gardens suffered considerably; all the plants, which had not taken deep root, were withered by the power of the sun. No lasting ill effects, however, arose to the human constitution; ... During the time it lasted, we invariably found, that the house was cooler than the open air, and that in proportion as the wind was excluded, was comfort augmented.
>
> But even this heat was judged to be far exceeded in the latter end of the following February [1791], when the north-west wind again set in, and blew with great violence for three days. At Sydney, it fell short by one degree of what I have just recorded [109 °F on 27 December 1790]: but at Rose Hill, it was allowed, by every person, to surpass all that they had before felt, either there, or in any other part of the world. Unluckily they had no thermometer to ascertain its precise height. It must, however, have been intense, from the effects it produced. An immense flight of bats, driven before the wind, covered all the trees around the settlement, whence they every moment dropped dead,[12] or in a dying state, unable longer to endure the

[12] Bats die from heat stress in large numbers at temperatures in excess of 42 °C (Welbergen, 2015).

burning state of the atmosphere. Nor did the perroquettes, though tropical birds, bear it better; the ground was strewed with them in the same condition as the bats.

Historical climate data suggest that the period from 1788 to 1790 was subject to a significant shift from La Niña to El Niño climate conditions (Gergis & Fowler, 2009), usually associated with extended cool and warm periods. However, modern analyses of Dawes' temperature data do not provide firm evidence of such El Niño–Southern Oscillation conditions in eastern Australia (Ashcroft, 2016; Derrick, 2019; Gergis et al., 2010). Although comparisons of Dawes' records with modern observations suggest the presence of some observational biases,[13] given the prevailing measurement uncertainties and the non-standard set-up adopted by Dawes and his contemporaries,[14] any apparent systematic seasonal temperature differences do not appear statistically significant.

In addition to his temperature readings, from 14 September 1788 until 1 July 1791 Dawes obtained atmospheric pressure measurements using the Burton barometer provided by the Board of Longitude (Dawes, 1791; McAfee, 1981a). His original measurements were recorded in inches of mercury (In Hg), which depends on the altitude of one's measurement location. Dawes' Observatory was located approximately 14 m above sea level (Dawes, 1791), and so a correction must be applied to obtain the relevant atmospheric pressure at mean sea level (Wallace & Hobbs, 2006). Dawes' barometric pressure readings are known to be slightly biased towards observations of the nearest 100$^{th}$ In Hg (Bureau of Meteorology, n.d.). However, Dawes became increasingly concerned that his instrument's readings were erroneously low (Gergis et al., 2010; McAfee, 1981b):

It was my intention to have it [the barometer] examined on its return to England, as, owing to the maker having depended upon glue to keep the cistern tight where it had been crack'd, it had lost a small quantity of quicksilver [mercury], which I had discovered by observing a number of very small globules immediately under it on the floor of the observatory. In packing it up, however, I found a large crack in the cistern, thro' which perhaps all the quicksilver would have escaped in the course of the passage [back to England] and have endangered breaking the tube; I therefore emptied it entirely and then packed it up. (McGuffie & Henderson-Sellers, 2012: 124[15]; Gergis et al., 2010: 84–85)

The imperfect performance of the Burton barometer may be behind the lack of atmospheric pressure measurements between 3 September 1789 and November 1789 (Gergis et al., 2010). Dawes eventually packed away the faulty Board of Longitude

---

[13] Dawes' data are characterised by a slight positive bias towards values ending in 8, and a small negative bias towards values ending in 0 and 5. However, given that his thermometer scale was graduated in units of 2 °F, these biases are insignificant (Bureau of Meteorology, n.d.).

[14] In addition to the non-standard instrument exposures owing to the absence of 'Stevenson screens'—so that thermometer measurements did not only record the ambient air temperature—Dawes did not use the type of 'maximum and minimum thermometers' in use today, so that the calculated maximum and minimum average temperatures used in modern analyses are only approximate. Moreover, the frequency with which he obtained his measurements is much lower than today's common practice, while inconsistencies in observing times may also contribute to the observed differences, particularly in the cool season (Derrick, 2019; Gergis et al., 2010).

[15] Text corrected with respect to earlier renditions.

barometer on 2 July 1791. It was replaced by a Ramsden barometer "… supplied by Mr. Furzer in the colony …" (Dawes, 1791):

> With respect to that of Mr. Furzer's [barometer], which he was so obliging as to spare me, it appears to be a very good one of Ramsden construction and is furnished with a gage point, tho' from the one I had before used, not having any such thing, and my then ignorance of the instrument, it was never attended to, but the bottom of the bag was let down as low as it would come and the observations made without attending to the gage point at all.

Dawes thus seems to imply that he lacked the appropriate experience to operate the instrument successfully. This appears to be borne out by the observation that barometric pressure readings taken between 12 July 1791 and 6 December 1791 seem far too low to be considered reliable (Gergis et al., 2010); the monthly averages in the period from September to December 1791 were 17.7 hPa lower than observations obtained with the Burton barometer (Bureau of Meteorology, n.d.).

Whether or not Dawes' temperature and pressure measurements were indeed reliable also depends on the instruments' placement. We do not know their exact locations, beyond the oblique references by Hunter and Tench cited earlier. However, it would be fair to assume that the thermometer and barometer were placed in the same location and attached to each other, as was standard practice at the time (McAfee, 1981a). Today, so-called 'Stevenson screens' shield meteorological instruments from the influence of direct heat radiation and provide appropriate ventilation, thus moderating the registration of extreme measurements. In the eighteenth century, meteorological set-ups were much simpler, which often explains measurement differences with equivalent modern observations (Nicholls et al., 1996; Torok & Nicholls, 1996). In any case, the absence of Stevenson screens in the collection of Dawes' meteorological data likely introduced unknown, irreducible uncertainties (McGuffie & Henderson-Sellers, 2012). As such, caution should be exercised in interpreting historical climate data sets.

From September to December 1791, Dawes also recorded rainfall volume, as well as dewfall, which he expressed in units of 'grains' of weight using a purpose-built hygrometer:

> The funnel in which the rain was caught and through which it descended into a common quart bottle was by very careful measurement known to be exactly 6 inches diam$^{r}$. at the mouth; appeared to be very accurately circular & thro'-out well made. It was constructed in London on purpose.
>
> The funnel was placed so that the mouth should be truly horizontal & the water caught in the bottle, carefully weigh'd in a pair of Apothecary scales. A piece of raw cotton was then carefully weigh'd with which the sides of the funnel was afterwards wiped as dry as possible & then weigh'd again. The whole weight of water thus found was set down in one number after which follows by estimation the quantity. (McAfee, 1981a: 15)

Although rainfall records are among the most common meteorological observations today, in Australia standard rainfall observations did not begin until the early 1820s. It is possible that the early colonists did not pay much attention to the absence of regular rainfall until water became scarce. By the time Dawes' records cease, David Collins, the colony's chronicler, raises the alarm regarding the prevailing drought conditions:

> By the dry weather which prevailed our water had been so much affected, besides being lessened by the watering of some transports, that a prohibition was laid by the Governor on the watering of the remainder of Sydney ... to remedy this evil, the Governor had employed the stone-mason's gang to cut tanks out of the rock, which would be reservoirs for the water large enough to supply the settlement for some time. (Collins, 1798: Chap. XV)

The deteriorating drought conditions led to Sydney's earliest documented water restrictions. The tanks that were cut out of the rock eventually lent their name to the trickle of fresh water flowing into Sydney Cove, which has since been known as the Tank Stream.

## 3.2 William Dawes, The Explorer

Given the significant demands on his time, particularly during the first two years of the British presence at Sydney Cove, Dawes' activities were initially largely confined to the settlement. However, by late 1789 his workload seems to have eased somewhat, and he became increasingly involved in expeditions towards the western mountains (for exploratory routes, see Fig. 3.6). During the remainder of his service in the colony, Dawes participated in no fewer than seven exploratory parties, although not all of them were successful. Russell (1882: 2) referred to Dawes' heroic attempts "… to find a road over the Blue Mountains, the great barrier to the progress of the Colony". He was clearly impressed by Dawes' persistence in exploring "… these almost inaccessible mountains [before] he gave up the enterprise …". Among the settlers, Governor Phillip himself appears to have been particularly keen to conquer the difficult terrain, likely driven by the urgent need to find sufficient arable land for cultivation and provide food for the fledgling colony:

> At the distance of sixty miles inland, a prodigious chain of lofty mountains runs nearly in a north and south direction, further than the eye can trace them. Should nothing intervene to prevent it, the Governor [Phillip] intends shortly to explore their summits; and I think there can be little doubt that his curiosity will not go unrewarded. (Tench, 1789a: 118)

On 5 July 1789, an expedition led by Phillip had reached a high hill, which he decided to call 'Richmond Hill'. In his private journal, Hunter wrote on that date,

> In the morning, we walked to the top of the hill, and we found we were not more than five or six miles from a long range of mountains, between which, and that where we stood, there is a deep valley, or low country, through which, probably, a branch of this river may run. This range of mountains we supposed to be those which are seen from Port Jackson, and called the Blue Mountains … In that range of high land there is a remarkable gully, or chasm, which is seen distinctly at a distance, and from which we appeared to be distant about five miles. The hills on each side of this gap were named by Governor Phillip; on one side the Carmarthen [including Mt Hay, Mt Irvine and others], on the other, the Lansdown hills [the Lower Blue Mountains]; and that on which we stood was called Richmond hill. (Hunter, 1793: 151)

Next, Phillip assembled a small party, including Dawes, to explore the Ca(e)rmarthen Mountains. In early December 1789 they left the fertile area around

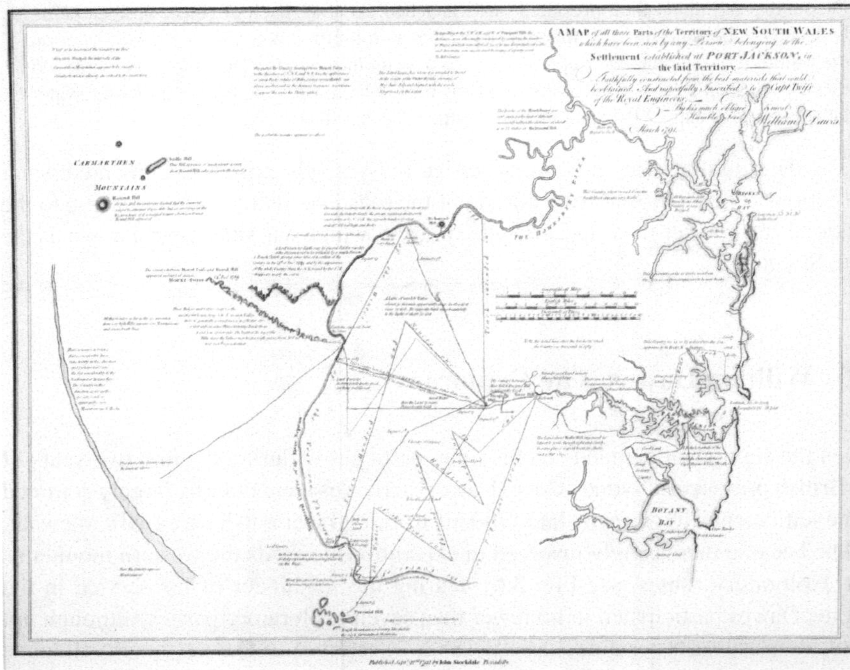


**Fig. 3.6** "A map of all those parts of the territory of New South Wales which have been seen by any person belonging to the settlement established at Port Jackson… / faithfully constructed … and respectfully inscribed to Capt Twiss … by … William Dawes, March 1791." (National Library of Australia; Map NK 2456/126; out of copyright)

the newly formed outpost at Rose Hill (Collins and Fletcher, 1975 [1798]: 72–73), which housed a small garrison commanded by Tench. However, after nine days of fruitless attempts, the men were forced to return to the settlement, empty-handed:

> *December, 1789.* At the request of his excellency, lieutenant Dawes of the marines, accompanied by lieutenant [George] Johnston and Mr. Lowes [surgeon's mate], about this time undertook the attempt to cross the Nepean river, and to penetrate to Carmarthen mountains. Having discovered a ford in the river, they passed it, and proceeded in a westerly direction. But they found the country so rugged, and the difficulty of walking so excessive, that in three days they were able to penetrate only fifteen miles: and were therefore obliged to relinquish their object. This party, at the time they turned back, were farther inland than any other persons ever were before or since, being fifty-four miles in a direct line from the sea coast when on the summit of mount Twiss, a hill so named by them, and which bounded their peregrination. (Tench, 1793: 151)

They had traversed the inaccessible terrain further than any previous expedition, to the vicinity of today's town of Linden, and their distance record would not be broken for a number of years. Dawes named their point of return 'Mount Twiss' (see Fig. 3.7), after Captain William Twiss, Royal Engineer in the British Army, who had supported his application to join the expedition to Botany Bay. The exact location of Mount Twiss remains unknown. Current maps show Mount Twiss about five kilometres north of Linden on Dawes Ridge, with Dawes Creek running northwards to the east of Dawes Ridge. These names were gazetted following a comprehensive investigation, including field work, conducted in the 1940s (Craft and Mitchell, 1941): see Fig. 3.8. However, more recent work (Paish, 1989/2017) places Mount Twiss about 3.5 km closer to Mt Hay, just east of the deep-cut and steep-sided valley containing Wentworth Creek.

The harsh terrain had defeated the expedition members this time. In fact, careful consideration of a chart drawn by Dawes in March 1791 uncovers a number of revealing statements about the difficulty of the terrain. Near Round Hill (Mt Hay), a mountain located beyond Mount Twiss, Dawes has scribbled, "… of this hill the Governor desired that the summit might be attained, if possible; but on arriving at the western brow of it [Mt Twiss], a rugged country between it and Round Hill

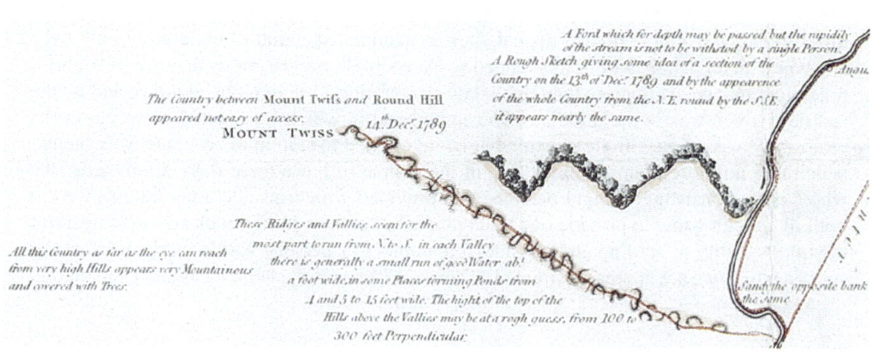


**Fig. 3.7** Enlargement of a section of Fig. 3.6, showing the 'Mount Twiss' area. (National Library of Australia; Map NK 2456/126; out of copyright)

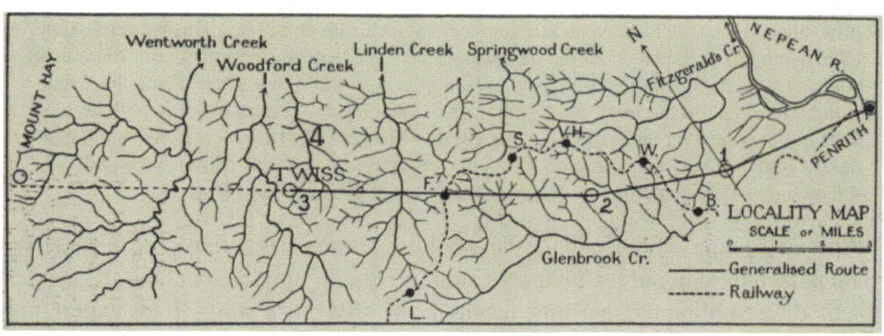


**Fig. 3.8** Locality map of the Lower Blue Mountains from Craft and Mitchell (1941: 257): "**1**: First major turn on the crest of Lapstone Hill; **2**: Second major turn on the ridge crest south of Valley Heights; **3**: Position of Mt Twiss on the heights east of Woodford Creek; **4**: Creek for which the name of Dawes' Creek is suggested." Initials indicate railways stations, **B**: Blaxland; **W**: Warrimoo; **V.H.**: Valley Heights; **S**: Springwood; **F**: Faulconbridge; **L**: Linden. (© Royal Australian Historical Society; reproduced with permission)

appeared". Elsewhere we read, "... all this country, as far as the eye can reach from very high hills, appears very mountainous and covered with trees", yet Dawes advises the reader that "... the dotted lines show where it is intended to travel in the course of the winter months ensuing, of May, June, July, and August" 1790 (Barton, 1889: Chap. 1.15).

Despite Phillip's aspirations to conquer the Blue Mountains, on 29 June 1792 the *Kentish Gazette*, a local British newspaper, reported on the party's attempts to traverse the region, lamenting that "... farther it appears impossible for any man to travel ...", as the gullies and rocky hills were "... sterile in the extreme ...", to the extent that "... a rat would starve ..."—most likely Dawes' own words (cited by Clarke, 2015: 35, note 80), given his scathing assessment in a frank letter to Maskelyne of 16 April 1790:

> I should like now to say something, which may give you an adequate idea of this country; but as it would take much time to be particular; I can only inform you that I have been about 53 miles from the sea into the country, and in the course of my journey passed thro' only about 25 miles of middling land, ...; tho' from the spot to which I attained I could see about 30 miles round, and a more dreary, dismal, barren, inanimated country I believe does not exist any where in the whole world; it is indeed so exceedingly rugged, stony and full of frightful precipices that we were more than three days penetrating 15 geographical miles, and on the last day, which was in the middle of December, and in which we exerted ourselves to the utmost, we were exceedingly mortified at being obliged to stop in the evening after having penetrated no more than 2½ miles!!! – In the course of these three days & upwards (the whole party consisting of eight persons well provided with arms and ammunition) we did not kill enough game to provide one scanty repast ... at least I think it clearly demonstrates the impossibility of settling any considerable number of persons within many miles of the spot on which we are at present situated. (Dawes, 1790a; original emphasis and exclamation marks)

Dawes estimated that they had managed to get as close as 11 miles from the mountain range (Barton, 1889: 138), yet that distance remained as yet unassailable. David Collins' account of the expedition is indeed telling:

> Early in this month, Lieutenant Dawes, with a small party, taking with them just as much provisions as they could conveniently carry, set off on an attempt to reach the western mountains by and from the banks of the freshwater river [the Nepean river], first seen some time since by Captain Tench, and supposed to be a branch of the Hawkesbury. From this excursion he returned on the ninth day without accomplishing his design, meeting with nothing after quitting the river but ravines that were nearly inaccessible. He had, notwithstanding the danger and difficulty of getting on through such a country, reached within eleven miles of the mountains, by computation. (Collins and Fletcher, 1975 [1798]: 89)

Among the educated men in the early colony, Dawes was perhaps the most appropriate person to map out the unknown country. Single-minded to a fault, he applied a method resembling the maritime technique of 'dead reckoning' to his exploration of the western expanses of the Sydney Basin, systematically counting, remembering and mapping every single step. Tench was duly impressed:

> Our method, on these expeditions, was to steer by compass, noting the different courses as we proceeded; and counting the number of paces, of which two thousand two hundred, on good ground, were allowed to be a mile. At night when we halted, all these courses were

> separately cast up, and worked by a traverse table, in the manner a ship's reckoning is kept, so that by observing this precaution, we always knew exactly where we were, and how far from home; an unspeakable advantage in a new country, where one hill, and one tree, is so like another that fatal wanderings would ensue without it.[16] This arduous task was always allotted to Mr. Dawes who, from habit and superior skill, performed it almost without a stop, or an interruption of conversation: to any other man, on such terms, it would have been impracticable. (Tench, 1793: 133, footnote)

Following their failed attempt to reach and explore the Blue Mountains, Dawes became well and truly involved in exploratory expeditions in 1790. He joined the first of three expeditions, with Tench and George Bouchier Worgan (1757–1838), former surgeon on H.M.S. *Sirius*, in August 1790. The next two expeditions followed in quick succession, later that month and in September 1790, when they were also joined by the Rev. Richard Johnson on a three-day expedition to Broken Bay. The first two expeditions, to the southwest and northwest of Rose Hill through difficult terrain, spanned a total of 12 days. Tench recorded concise summaries of these otherwise mostly unremarkable treks:

> *August, 1790.* In the beginning of this month, in company with Mr. Dawes and Mr. Worgan, late surgeon of the *Sirius*, I undertook an expedition to the southward and westward of Rose-hill [Rose Hill], where the country had never been explored. We remained out seven days, and penetrated to a considerable distance in a S.S.W. direction, bounding our course at a remarkable hill, to which, from its conical shape, we gave the name of Pyramid-hill [Mt. Hunter, near present-day Camden]. Except the discovery of a river (which is unquestionably the Nepean near its source) to which we gave the name of the Worgan, in honour of one of our party, nothing very interesting was remarked.
>
> Towards the end of the month, we made a second excursion to the northwest of Rose-hill, when we again fell in with the Nepean, and traced it to the spot where it had been first discovered by the party of which I was a member, fourteen months before; examining the country as we went along. Little doubt now subsisted that the Hawkesbury and Nepean were one river.
>
> We undertook a third expedition soon after to Broken Bay; which place we found had not been exaggerated in description, whether its capacious harbour, or its desolate incultivable shores, be considered. On all these excursions we brought away, in small bags, as many specimens of the soil of the country we had passed through, as could be conveniently carried; in order that by analysis its qualities might be ascertained. (Tench, 1793: Chap. VII)

Despite Tench's confident statement that "[l]ittle doubt now subsisted that the Hawkesbury and Nepean were one river", the matter as to whether both rivers were one and the same had actually not yet been settled. In addition, Phillip was determined more than ever that the settlers should reach the Blue Mountains. And so, on 11 April 1791, Phillip himself took command of an expeditionary force that included Dawes, Tench, Collins and his servant, three convicts ('gamekeepers'), eight soldiers including two Serjeants, and Surgeon John White. The party was completed by the addition of two Aboriginal guides, Colebe(e) (Colbee; Smith, 2008a; d. 1806?) and Ballederry (Balloderree, Boladeree; Goodman, 2020), as "... much information was expected from them".

---

[16] It was so common that wanderers got lost in the unexplored landscape, even near Sydney Cove and in view of the harbour, that annual casualty records included a column 'Lost in the Bush'.

> Baneelon wished to go, but his wife would not permit it. Colbee on the other hand, would listen to no objections. He only stipulated (with great care and consideration) that during his absence, his wife and child should remain at Sydney under our protection, and be supplied with provisions. …
>
> Every man (the Governor excepted) carried his own knapsack, which contained provisions for ten days; if to this be added a gun, a blanket, and a canteen, the weight will fall nothing short of forty pounds. Slung to the knapsack are the cooking kettle and the hatchet, with which the wood to kindle the nightly fire and build the nightly hut is to be cut down. …
>
> At twenty minutes before seven o'clock, we started from the governor's house at Rose Hill, and steered for a short time nearly in a north-east direction, after which we turned to north 34° west, and steadily pursued that course until a quarter before four o'clock, when we halted for the night. (Tench, 1793: Chap. VII)

This was the first occasion on which both Aboriginals and convicts, the latter "… considered good shots …" (Barton, 1889: Chap. 1.15), were included among expedition members. Phillip led his men from Rose Hill to the Hawkesbury opposite Richmond Hill, across the Nepean and northwestwards towards the mountains. The next day, now some 18½ miles from Rose Hill, they decided to follow the river's winding course, with one member told to count his steps so that "… they might always know their situation in the woods, and the direction it would be necessary to take when they returned across the country" (Barton, 1889: Chap. 1.15). Tench recorded in his personal journal that the Aboriginal trackers were out of their depth shortly after the party had left Rose Hill, "… being absolute strangers inland":

> Their total ignorance of the country, and of the direction in which they had walked, appeared, when they were asked which way Rose Hill lay; for they pointed almost oppositely to it. Of our compass they had taken early notice, and had talked much to each other about it: they comprehended its use; and called it "*Naa-Mòro*," literally, "To see the way"; — a more significant or expressive term cannot be found. …
>
> To convey to their understandings the intention of our journey, was impossible. For, perhaps, no words could unfold to an Indian, the motives of curiosity, which induce men to encounter labour, fatigue, and pain, when they might remain in repose at home, with a sufficiency of food. We asked Colbee the name of the people who live inland, and he called them *Boò-roo-ber-ongal*; and said, they were bad; whence we conjectured, that they sometimes war with those on the sea coast, ... (Barton, 1889: Chap. 1.15)

On the third day, they climbed a hill and thought that they could see Richmond Hill, the southern extremity of the mountain range, "… distant about eight miles, in a contrary direction from what we had been proceeding upon" (Barton, 1889: Chap. 1.15). As Tench was the first to spot their destination, Phillip decided to call their vantage point after him:

> But, before I descend, I must not forget to relate, that to this pile of desolation, on which, like the fallen angel on the top of Niphates, we stood contemplating our nether Eden, his excellency [Phillip] was pleased to give the name of Tench's Prospect Mount. (Barton, 1889: Chap. 1.15; see Fig. 3.9)

With the terrain considered impassable, however, the party was forced to turn back, thereby losing any advantage they had gained over the previous two days. Despite

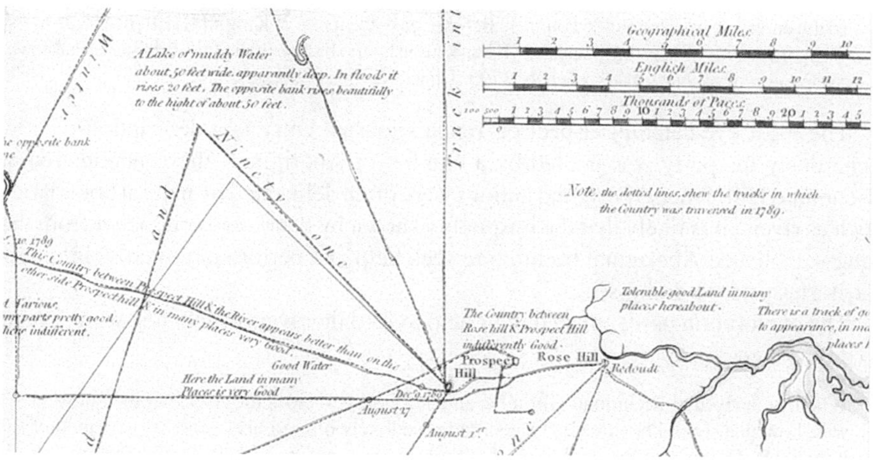


**Fig. 3.9** Enlargement of a section of Fig. 3.6, showing the Prospect Hill area. (National Library of Australia; Map NK 2456/126; out of copyright)

receiving help from "… Bèr-ee-wan, of the tribe of Boorooberongal …, a man about thirty years old …" and directions from another local Aboriginal, Gom-beè-ree,[17]

> [i]t was clear to all of them that their chance of exploring the Blue Mountains was a very poor one, seeing that they could not do a day's journey without being stopped by a creek or a gully, compelling them to go round it, and thus bringing them back to the place from which they started. (Barton, 1889: Chap. 1.15)

And so, after six days of failed attempts to penetrate the impassable wilderness and the baffling multitude of gullies and streams, Phillip took the decision to return to Rose Hill, only sixteen miles from their camp, instead of pressing on towards the Blue Mountains: "Saturday April 16$^{th}$, 1791. … We started early; and reached Rose Hill about three o'clock, just as a boat was about to be sent down to Sydney." (Barton, 1889: Chap. 1.15).

Undeterred, on 24 May 1791 Tench and Dawes, "… accompanied by a serjeant of marines and a private soldier …" (Tench, 1793: Chap. XIV), set out on yet another expedition to Richmond Hill, eventually confirming Tench's earlier hunch as to the nature of the Hawkesbury and Nepean confluence:

> The Nepean or Cowpasture River is a fine stream, rising a few miles north of Berrima and flowing in a northerly direction through a fine agricultural district into the Hawkesbury River, between Penrith and Richmond, or at the confluence of the Grose River. The Nepean is, in fact, only another name for the upper end of the Hawkesbury. (Whitworth, 1866)

---

[17] A significant outcome of this expedition was the realisation among Europeans that there were different Aboriginal languages and cultures across the country (Wilkins and Nash, 2008; see also Troy, 1994). Some indications that there was more than one Aboriginal language had come to the fore when the colonists realised that the Aboriginals in the Sydney Basin did not understand the majority of the words collected by Cook's expedition at the Endeavour River estuary in North Queensland. That list had been included among the First Fleet's general papers (Troy, 1992: 146).

> To the elevation which bounded our research, we gave the name of Knight[s] Hill [near Kurrajong], in honour of the trusty serjeant [Isaac Knight], who had been the faithful indefatigable companion of all our travels. (Tench, 1793: Chap. XIV)

The most eye-catching aspect of Tench's journal entry is a description of how generously the party was treated by a number of Aboriginals they encountered at Richmond Hill. Since Aboriginal nations were often delineated by natural boundaries such as rivers, it is likely that the hospitality shown by these men originated from the long-established Aboriginal tradition to seek help and permission to cross into new territories prior to doing so.

The exploration party enlisted their help to ford the river, which they were forced to do in stages:

> When we arrived at Richmond Hill it became necessary to cross the river; but the question was, how this should be effected? Deedora immediately offered his canoe: we accepted of it, and Mr. Dawes and the soldier putting their clothes into it, pushed it before them, and by alternately wading and swimming, soon passed. … My clothes, half our knapsacks, and three of our guns, yet remained to be transported across. These I recommended to the care of our grim ferrymen, who instantaneously loaded their boat with them, and delivered them on the opposite bank, without damage or diminution.
>
> During this long trial of their patience and courtesy, in the latter part of which I was entirely in their power from their having possession of our arms, they had manifested no ungenerous sign of taking advantage of the helplessness and dependence of our situation, no rude curiosity to pry into the packages with which they were entrusted, and no sordid desire to possess the contents of them; although among them were articles exposed to view of which it afterwards appeared they knew the use and longed for the benefit. …, produce me a brighter example of disinterested urbanity than was shown by these denizens of a barbarous clime to a set of destitute wanderers on the side of the Hawkesbury. (Tench, 1793: Chap. XIV)

Shortly after their return to the settlement, in July 1791 Dawes and Tench embarked on their final joint expedition, this time "… in search of a large river, which was said to exist a few miles to the southward of Rose Hill" (Tench, 1793: Chap. XIV):

> We went to the place described, and found this second Nile, or Ganges, to be nothing but a salt water creek, communicating with Botany Bay, on whose banks we passed a miserable night, from want of a drop of water to quench our thirst; for as we believed that we were going to a river, we thought it needless to march with full canteens.

Although Sydney's climate is mild even in winter, this expedition was beset by extraordinarily cold weather, as we learn from a note in Tench's journal:

> *Monday, July 18.* The sun arose in unclouded splendour, and presented to our sight a novel and picturesque view; the contiguous country as white as if covered with snow, contrasted with the foliage of trees flourishing in the verdure of tropical luxuriance. Even the exhalation which steamed from the lake beneath contributed to heighten the beauty of the scene. … Thermometer at sunrise 25°[F]. The following night was still colder. At sunset, the thermometer stood at 45°[F]; at a quarter before four in the morning, it was at 26°[F]; at a quarter before six, at 24°[F]; at a quarter before seven, at 23°[F]; at seven o'clock, 22°.7 [F]; at sunrise, 23°[F]; after which it continued gradually to mount, and between one and two o'clock, stood at 59°.6 [F], in the shade. … Nothing but demonstration could have convinced me that so severe a degree of cold ever existed in this low latitude. Drops of water on a tin-pot,

not altogether out of the influence of the fire, were frozen into solid ice in less than twelve minutes. Part of a leg of kangaroo, which we had roasted for supper, was frozen quite hard, all the juices of it being converted into ice. On those ponds which were near the surface of the earth, the covering of ice was very thick; but on those which were lower down, it was found to be less so in proportion to their depression; and wherever the water was twelve feet below the surface (which happened to be the case close to us), it was uncongealed. (Tench, 1793: Chap. XIV)

## 3.3 William Dawes, The Linguist

Today, William Dawes is perhaps best known more generally for Dawes' Point, the headland in central Sydney colloquially named after him, and for a set of manuscripts describing local Aboriginal language characteristics "... in the neighbourhood of Sydney".[18] These manuscripts were compiled during the first few years of the New South Wales colony and are increasingly used today to reclaim the Sydney language spoken before colonisation (Kohen, 1993; Nash, 2011; Nathan et al., 2009; Steele, 2005; Troy, 1992, 1994). They consist of three 'notebooks', two comprehensive language notebooks (*Notebooks A* and *B*; see Fig. 3.10) and a semi-official lexicon of the local Indigenous language, *Notebook C*.[19] It appears that *Notebooks A* and *B* were composed entirely by Dawes, whereas *Notebook C* resulted from the work of a number of individuals:

> Manuscript 'c' seems to have been the work of several authors as it is written in at least three different hands including both 'rough' and 'fair' scripts. [At the time] it was common for literate people to have a 'rough' hand for rapid note-taking and composing and a 'fair' or careful hand for final copy. One of the hands in the manuscript is exactly the same as Governor Arthur Phillip's rough hand [as found in] many surviving manuscripts ... [Other] evidence ... suggests that two other officers, David Collins and John Hunter, also contributed to the manuscript ... [So] it is very likely that [it] was composed by Phillip, Collins and Hunter. (Troy, 1994: 5; see also Troy, 1992)

Dawes' notebooks were lost for almost 180 years. Upon Dawes' return to England, his notebooks ended up with a London collector, the Orientalist, linguist and colonial trophy collector William Marsden (1754–1836) (Nathan et al., 2009):

> So long ago, however, as the year 1792, I received from a Mr Dawes, whose familiar habits of intercourse with the natives enabled him to acquire a competent knowledge of their tongue, a collection of memoranda that furnish a tolerable idea of their grammatical system (if such it may be termed), particularly as respects the inflections of the verbs, with their accompanying pronouns. (Marsden, 1834; cited by Smith, 2008b: 8)

---

[18] Dawes referred to his local Aboriginal visitors as 'Eora', a term which represents 'people'. In today's New South Wales, reference is routinely—and erroneously—made to the 'Eora nation' (see Foley and Read, 2020: Chap. 1).

[19] The notebooks are available in full at https://www.williamdawes.org. The online version is jointly managed by Aboriginal Affairs NSW and the Hans Rausing Endangered Languages Project at the School of Oriental and African Studies (SOAS), University of London. SOAS holds the original notebooks.

**Fig. 3.10** Dawes' *Notebooks A* and *B*. (State Library of New South Wales; no copyright)

Marsden published some excerpts from Dawes' notebooks in 1834. In 1835, he offered his entire library to King's College London. In 1916, Dawes' notebooks were eventually relegated to an archive box at the library of the newly opened School of Oriental and African Studies of the University of London, where they were re-discovered in 1972 by the Australian librarian Phyllis Mander-Jones (1972: 362). Careful analysis of the notebooks, where very few entries or pages are dated, suggests that Dawes started recording Aboriginal language aspects in November 1790; the final dated entry is from 13 November 1791.

Between November 1790 and December 1791, the relationship between the local Indigenous population and the British settlers had significantly improved, to the extent that increasing numbers of Aboriginals started to visit the settlement at Sydney Cove: "With the natives we are hand and glove. They throng the camp every day, …" (Tench, 1793: Chap. X). It seems that the local Aboriginal people had decided that it was in their best interests to understand the settlers and learn about their ways and customs.

Shortly after the arrival of the First Fleet, Governor Phillip had ordered the capture of a number of local Aboriginal people in an attempt to learn their language and to facilitate intercultural communication. Dawes would have had the opportunity to learn some of the local language at that time from men like Arabanoo (ca. 1759–1789; captured at Manly on 31 December 1788) and Woollarawarre Bennelong (ca. 1764–1813; captured on 23 November 1789). Many of the educated First Fleet officers tried their hand at learning the Sydney Aboriginal language at one time or another. Their notes are still of prime importance in the study of the languages spoken in the Sydney Basin at the time of colonisation, since other nineteenth-century records are virtually non-existent.

Many Aboriginals were initially too intimidated to enter the Sydney Cove settlement. However, Dawes' more isolated Observatory offered a safe and more welcoming venue to interact with and learn about the different cultures represented on the harbour's south shore, nurturing intercultural friendships and the sharing of knowledge. In his effort to compile his notebooks, Dawes likely benefited from suggestions by Colebe, Bennelong and Bennelong's second wife, Barangaroo (d. 1791). However, his network extended much more widely: a total of sixteen Aboriginals contributed to 125 identified records in his notebooks, whereas the young, 14 or 15-year-old[20] Aboriginal woman known by her 'skin name' Patyegarang—often endearingly referred to as 'Patye' by Dawes—supplied at least 59 (Steele, 2005: 313). Patyegarang's contributions do not commence until *Notebook B*, possibly by mid-1791 (Steele, 2005: 87). Dawes was also in active conversation with Aboriginal children who had been residing with settlers since April 1789, including Boorong (Abaroo), who had been a house guest of the Rev. Richard Johnson, and Yirînîbî, her brother, as well as Nanbarri, Anganángan, Kooroóda and, on 15 November 1790, Balûdérri. Dawes quickly became an expert speaker of the Sydney Aboriginal language, an impression that is further supported by Patyegarang's praising of his linguistic abilities: *Mr. Dawes budyiri karaga* – "Mr. Dawes pronounces well" (literally, "Mr. Dawes good mouth") (*Notebook B*: 11, 24; Smith, 2008b: 8).

On the whole, Dawes' colleagues in the colony were not aware of his private intellectual pursuits, so it should not have been a surprise that Collins appeared exasperated at the settlers' ignorance of the Sydney language as late as January 1791:

> It was also unfortunately found, that our knowledge of their language consisted at this time of only a few terms for such things as, being visible, could not well be mistaken; but no one had yet attained words enough to convey an idea in connected terms. (Collins, 1798: 148)

Collins had been collecting examples of the local language, and he was undoubtedly aware of similar efforts by Tench, Phillip, King and Southwell. King and Southwell were the most prolific collectors among the educated settlers (Steele, 2005: 60). Initially, Dawes was also determined to take a systematic approach to language acquisition, collecting neat columns of nouns and verbs. However, it appears that he lost interest in such a strictly prescribed method quite quickly. And so, mostly in his *Notebook B*, he began recording dialogues, conversational snippets, narrative miniatures and phrases that placed the Aboriginal language into the appropriate societal and cultural context, covering personalities and feelings, actions and their drivers, straying from concrete word lists to abstract concepts and connected ideas—including 'to dream', 'to think', 'to be ashamed' and 'to tell a lie' (*Notebook A*: 16, 35; *Notebook B*: 4, 26) (Gibson, 2012: 34). He recorded tense, grammar, sentence structure and the societal and cultural norms that determined how language was used. Dawes was clearly a pioneer—once again. Nevertheless, one should keep in mind that his interpretation and linguistic notes about the Sydney language were influenced by his own first language (English), by the level of fluency he was able to master and by ideas developed as his language studies proceeded (Troy, 1992):

> Although people solely look to Dawes's manuscript to assist in reconstructing language, culture and customs of coastal Sydney, we are aware that he was not a trained linguist, received information from numerous sources including travelling Europeans and [had] limited understanding of our culture and customs. (Ingrey, 2019)

[20] Oblique references in Dawes' notebooks suggest that she already practised rituals associated with menstruation (Gibson, 2012: 80). In addition, it has been recorded that she wore a 'barrin' (*Notebook B*: 6, 32), a garment worn by girls "... until they are grown into women and attached to men" (Collins, 1798: 562).

He collected phrases and dialogues about the daily activities of his visitors, including their living arrangements, but also about a range of minor cultural, often somewhat humorous misunderstandings. Tench recorded one example of a good-humoured misunderstanding, which occurred because of the settlers' determination to frame all intercultural interactions in positive terms. They knew that "[t]he epithet *Wee-ree* [represented something] bad …". To learn the local word for 'good' from their informant, Arabanoo, they therefore applied *wee-ree* to things that they knew she appreciated. And so, "… in order to provoke [her] to tell us that it was good, Arabanoo, taxed in this way, replied '*beeal*', which we translated, and adopted for 'good'" (Tench, 1793: 231). Much later, they learnt that Arabanoo "… meant no more than simply to deny our inference, and say, 'no – it is not bad'" (Tench, 1793: 231). The settlers had fallen into the obvious trap where they did not allow for ambiguity in expressions, applying linguistic (lexical and grammatical) rules strictly and without room for interpretation.

The complexity of the local Aboriginal language and its many subtleties baffled Dawes for some time. Throughout Aboriginal history, multigenerational layers of observation, learning and teaching have become embedded into Indigenous knowledge of place, which in turn becomes an experiential part of language. Given these complexities, it is thought that Dawes may have encountered the type of 'cognitive dissonance' recognised by linguists in connection with English speakers trying to understand the complexities of the Sydney language:

> English speaking people had difficulty in their efforts to find words for direction and time in the Sydney language because notions of time and space in Aboriginal languages are very different to those expressed in English. (Troy, 1993: 20)

> They use the ellipsis in speaking very freely; always omitting as many words as they possibly can, consistent with being understood. They inflect both their nouns and verbs regularly; and denote the cases of the former, and the tenses of the latter, not like the English by auxiliary words; but like the Latins by change of termination. Their nouns, whether substantive or adjective, seem to admit of no plural. I have heard Mr. Dawes hint his belief of their using a dual number, similar to the Greeks: but I confess that I never could remark aught to confirm it. The method by which they answer a question, that they cannot resolve, is similar to what we sometimes use. Let for example the following question be put — *Waw Colbee yagoono?* — Where is Colbee to-day? — *Waw, baw!* Where, indeed! would be the reply. They use a direct and positive negative: but express the affirmative by a nod of the head, or an inclination of the body. (Tench, 1793: 203)

Dawes began to learn how to use different pronouns, depending on the circumstances—modified by added suffixes, defined by vocal tricks of the larynx, the environment in which something was discussed or even by how many people were listening. He soon realised that his European nominalist principles, where the relationship between words, objects and concepts is predetermined, were not directly applicable to the Indigenous language spoken in the Sydney region. For instance when speaking in the first person plural, expressions for 'we two' and 'we all' were significantly different. At one time that difference was sufficiently important for Patyegarang to adjust her language *post facto* when she realised that an additional person had been in their presence without her having noticed him (Tench, 1793: 203; *Notebook B*: 35):

> The difference of speaking of we two and we three as above expressed was obtained 27 Nov[ember]. [1791] by Patyegaráng first speaking to me mark'd #1 and afterwards as mark'd #2, when on asking her why she did not speak in the same way the 2$^{d}$. time as the 1$^{st}$., she said it was because she had forgot that Pündül was with them, and explained herself very clearly. (*Notebook B*: 34)

A second example of the multifaceted nature of the Aboriginal mindset in relation to their environment is found in the numerous expressions Dawes recorded for the concept 'wind', including expressions related to something in the air, coming from the Sun; something with a strong smell; something cold; something associated with a particular island; or something associated with a particular time of the year when certain botanical or zoological factors dominate (Troy, 1994: 50; Gibson, 2010: 251–252).

From a cultural angle, Dawes was particularly interested in Aboriginal spiritual and religious attitudes and beliefs, perhaps because of his own deep religiosity. He tried to teach Boorong about the tenets of Christianity, but to no avail; she was too consumed by "... her levity, and love of play ..." (Tench, 1789b: 280). Elizabeth Macarthur (1766–1850), wife of the British grazier John Macarthur, observed that Dawes was of the opinion that the Aboriginals "... have the tradition of the Flood amongst them" (Tench, 1789b: 280; Macarthur Onslow, 1973: 28):

> Mr. Dawes, who has studied their [the Aboriginals'] language or jargon a good deal, has endeavoured to learn what their notions are of the Deity. It is not discovered that they worship the sun or any of the heavenly bodies, and yet they say all who die go up to the clouds. Mr Dawes thinks they have a tradition of a Flood amongst them. (Macarthur, 1791: 499)

The first page of *Notebook A* contains a few almost indistinct outlines drawn in pencil (see Fig. 3.11). However, at closer inspection and by comparison with modern maps and charts, the outline resembles what is today The Rocks, the headland jutting out from Sydney Cove that forms the southern support of the Sydney Harbour Bridge and where Dawes spent most of his free time in pursuit of astronomical breakthroughs. The word 'Dara' is barely decipherable (see Chap. 2, note 12, for context). It is located inside the outline of the headland, pointing down. With that realisation, the roughly oval outline to the right of 'Dara' becomes Goat Island, 'Me-mel':

> Bennillong [Bennelong] ... often assured me, that the island Me-mel (called by us Goat Island) close by Sydney Cove was his own property; that it was his father's ... To this little spot he appeared much attached; and we have often seen him and his wife Ba-rang-aroo feasting and enjoying themselves on it. He told us of other people who possessed this kind of hereditary property, which they retained undisturbed. (Collins, 1798: 599)

Other landmarks may also be indicated, including Balmain ('Kowang').

One of the first instances of the local Aboriginal language Dawes discusses in *Notebook A* is closely related to his own reasons for his presence in the colony: *Naa* – to see (see Fig. 3.12). He illustrates the use of the term—"I have not seen him" (*Notebook A*: 4)—through a story involving Yirîníbî in search of Boorong, anxious to see her. But at the same time, Dawes' recording of their conversation gives us insights into how Dawes grappled with the complexities of the Sydney language:

**Fig. 3.11** First page of Dawes' *Notebook A*, showing 'Dara' Point (centre left) and 'Me-mel' island (centre right). (Special Collections, SOAS Library, MS 41,645; reproduced with permission)

> ~~This~~These ~~words was~~ words were spoken to me by Yirînîbî, Booro[o]ng's Brother and he was evidently anxious in enquiring after Booro[o]ng. I have alter'd the [E]nglish signification from the top of the page in consequence of discovering the 1st person plural of the future in the verbs Wĭda & Páta. (Gibson, 2012: 43; *Notebook A*: 4)

Yirînîbî was clearly concerned about his sister, a deep concern rooted in the horrors he had seen enacted by the raging smallpox epidemic that was devastating the Aboriginal community at the time (Gibson, 2012: 33, 35).

As the settlement's resident astronomer, one would expect Dawes to have collected as much information about Aboriginal star lore as he could gather. Indeed, his notebooks contain many words for stars and planets, but nothing about their knowledge of celestial phenomena other than ephemeral concepts such as *mulumulu* (a cluster of shooting stars) and references to *yanada* (moon) *gili* (light) sparkling across the *badu* (water) (Poll, 2019). For a man immersed in the local culture, showing respect for the beliefs of his Aboriginal counterparts, this lack of depth is curious. It is hard

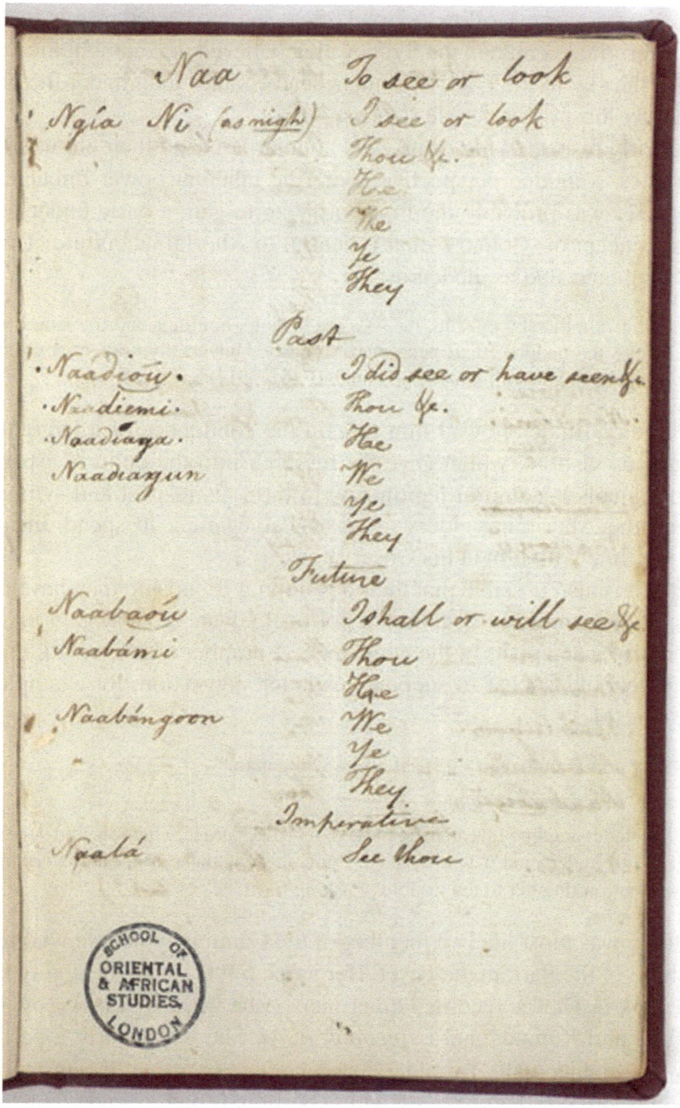
Naa — To see or look
Ngía Ni (as nigh) — I see or look
Thou &c.
He
We
Ye
They
Past
Naadiou. — I did see or have seen &c.
Nadiemi. — Thou &c.
Naadiaga. — He
Naadiagun — We
Ye
They
Future
Naabaou — I shall or will see &c
Naabámi — Thou
He
Naabóngoon — We
Ye
They
Imperative
Naalá — See thou



**Fig. 3.12** Examples of the use of *Naa*—'to see or look', in Dawes' *Notebook A*. (Special Collections, SOAS Library, MS 41,645; reproduced with permission)

to believe that Dawes would not have quizzed his Aboriginal visitors about their understanding of the heavens… We glean some more information from Collins:

> Their acquaintance with astronomy is limited to the names of the Sun and Moon, some few stars, the Magellanic [C]louds, and the Milky Way. Of the circular form of the Earth, they have not the smallest idea, but imagine that the Sun returns over their heads during the night to the quarter whence he begins his course in the morning. (Collins, 1798: 471)

This short extract from Collins' journal offers an unexpected insight, namely that the Darug Aboriginal people in the Sydney area believed that at night the Sun would return across the sky to the East, over their heads, rather than under the ground, as was believed by most other people (Norris, 2016).

Dawes clearly respected his Aboriginal counterparts and their cultural traditions and sensitivities, a unique perspective given the inherent power imbalance in the early colony. He was probably the first European to gain a basic understanding of the spiritual concept of 'Country' that is central to Aboriginal culture. Through his Aboriginal language studies, he learnt

> ... the land kinship, the sky kinship, the creative kinship, huntings and trackings; the way that land shapes the people, [that] people are Country. Universally, indigenous people all have that same metaphor with kinship and land. (Page, 2014)

Dawes' open attitude allowed him to gain the confidence of a large number of local Indigenous visitors, which gave his research into the cultural aspects of the Sydney Aboriginals a stamp of legitimacy. In turn, his respect and sympathy may have swayed the Aboriginal elders to allow Patyegarang to spend an increasing amount of her time with him at his Observatory.

Of course, it is also possible that their developing friendship may have been seen as a strategic opportunity to negotiate a political advantage and gain insights into the minds, motives and plans of the white men. A number of conversations recorded in Dawes' notebooks appear to support this latter suggestion, for example (Carter, 2007: 433, note 11),

> Dawes: "Why don't you [scorn to] speak like a white man?"
>
> Patyegarang: "*Mangabuninga bial.*"
>
> Dawes: "Not understanding this answer, I asked her to explain it, which she did very clearly, by giving me to understand it was because I gave her victuals, drink and everything she wanted, without putting her to the trouble of asking for it."

Patyegarang was most likely a member of the Cammeray people who resided on the north shore of the Parramatta River. Her name refers to the local grey kangaroo. In his *Notebook B*, Dawes recorded other names she was also known by, including Tagaran, Tuba and Kanmangnal (*Notebook B*: 4). She was clearly a partner to be reckoned with, intellectually formidable and politically astute. For instance, late in *Notebook B* Dawes asks Patyegarang why Aboriginals had wounded a white settler (*Notebook B*: 34):

> Dawes: I then told her that a whiteman had been wounded some days ago in coming from Kadi to Wáráng & asked her why the black men did it.
>
> Patyegarang: (Because they are) angry.
>
> Dawes: Why are the b[lack]. m[en]. angry?
>
> Patyegarang: Because the white men are settled here. The kamarigals [Cammeraygal] are afraid.
>
> Dawes: Why are the k– afraid?
>
> Patyegarang: *Gun-in*—because of the Guns.

The Cammeray Darug people had become angry, and the white men were to blame in their eyes, because it had become clear that they had now settled more permanently at Sydney Cove. This clearly shows that she was not afraid to pass on complex political messages, despite their developing friendship (Gibson, 2010: 246).

Indeed, over time a close relationship of trust and even intimacy appears to have developed between Dawes and Patyegarang (Collins, 1798: 147), who had become Dawes' main language tutor and cultural adviser by the second half of 1791. Much has been written about their fledgling relationship, with opinions among scholars diverging as to the nature—platonic, sensual or sexual—of their liaison (Troy, 1992; Karskens, 2009: 411–412; Nathan et al., 2009; Pybus, 2009; Gibson, 2010; Thomas, 2013). Without adding unduly to this debate, here we will focus on the contents of Dawes' notebooks as they pertain to their interactions and attempt to place these snippets of information into the context of the era. We will leave the physical details of their relationship to the reader's imagination, with the recommendation to consider Dawes' behaviour in its entirety (for a balanced discussion, see Clarke, 2015: 47–48; see also Troy, 1992).

Nevertheless, any reader of the notebooks will gain an appreciation of the warm and trusting relationship that had developed, with Dawes and Patyegarang sharing emotions, humour, intellectual pursuits and mutual respect. As a case in point, Patyegarang shared expressions that reflect intimacy, including.

> *Putuwá*—to warm one's hand by the fire & then to squeeze gently the fingers of another person (*Notebook B*: 21).

Other examples of intimate conversations include,

> At this time Patyegarang was standing by the fire … naked, and I desired her to put on her cloaths, on which she said *Goredyú tágarin* the full meaning of which is "I will or do remain longer naked in order to get warm sooner, as the fire is felt better without cloaths than if it had to penetrate thro' them." (*Notebook B*: 29)
>
> Patyegarang: *Taríadyaou*— "I made a mistake in speaking." Dawes: "This Patye said after she had desired me to take away the blanket when she meant the candle." (*Notebook B*: 30)
>
> Patyegarang: "*Nyímuŋ* candle Mr. D." — "Put out the candle Mr. D." (*Notebook B*: 34)
>
> Dawes: "*Mínyin bial naŋadyími?*" – "Why don't you sleep?"
>
> Patyegarang: "*Kandúlin.*" – "Because of the candle." (*Notebook B*: 36)

Dawes also recorded the phrase *tyarrsbabaouínıa*, an expression for playfully splashing water over a companion, just above the sound *ma·kūŋ* (*Notebook B*: 20), for 'sweetheart' or 'lover'. In addition, he learnt what it meant to wink and how to interpret feints and cues given with eyelashes and eyebrows (*Notebook B*: 18, 21, 33). He also recorded the verbs 'to embrace, to hug' and 'to make love to' (*Notebook B*: 19), whereas snippets of conversation affirm their closeness and suggest that they spent evenings and nights in each other's company (Nathan et al., 2009: vii). Notable examples include conversations that discuss skin complexion, eyes, hands, fingernails, snot, testicles, hair, sores, wounds, body lice, washing a variety of body parts and how fire and the night air affect a naked body (Gibson, 2010; 248; *Notebooks A* and *B*).

Dawes recorded Patyegarang exclaiming "I shall not become white", followed by "This was said by Patyegaráng after I had told her, if she would wash herself often, she would become white, at the same time throwing down the towel as in despair" (*Notebook B*: 19). However, we should also note that Dawes' interest in recording such personal and bodily aspects did not start with Patyegarang's presence at his Observatory. In fact, as early as 17 November 1790, the notebooks include a tale of Nanbarry and Boorong chatting about bathing (*Notebook A*: 8), sometime later followed by Dawes referring to Wauriwéeal telling him about her petticoat (*Notebook A*: 42). Moreover, Dawes recorded Patyegarang's platonic insistence, *Matarabaun nagaba*, that is, "We shall sleep separately" (*Notebook B*: 17).

Dawes' untimely departure from New South Wales in late 1791 left his studies of the Aboriginal language and culture unfinished. Whereas first drafts of a grammar and elements of a vocabulary are preserved, much is left to be uncovered, despite significant linguistic efforts in recent decades (e.g., Troy, 1992, 1994). Drawing on local Indigenous knowledge, it has become clear that some words Dawes translated in his notebooks continue to survive in today's Sydney Aboriginal language, including *gnamoro* (compass), *mari nawi* (big ship) and *nangamai* (to dream, to sleep) (Jackson, 2019).

Dawes' departure almost erased all memories of his linguistic advances. In fact, already by 1792, the sustained English–Aboriginal exchanges had created a contact-induced *lingua franca* which eventually developed into a pidgin language:

> Several of their young people continued to reside among us, and the different houses in the town were frequently visited by their relations. … and it was observed, that they conversed with us in a mutilated and incorrect language formed entirely on our imperfect knowledge and improper application of their words. (Collins and Fletcher, 1975 [1798]: 174)

And even though Hunter noted that "Mr. Collins, the judge-advocate, is very assiduous in learning the language, in which he has made a great progress …" (Hunter, 1793: 269),

> By slow degrees we began mutually to be pleased with, and to understand each other. Language, indeed, is out of the question; for at the time of writing this [September 1796] nothing but a barbarous mixture of English with the Port Jackson dialect is spoken by either party; and it must be added, that even in this the natives have the advantage, comprehending with much greater aptness than we can pretend to, every thing they hear us say. (Collins and Fletcher, 1975 [1798]: 451)

By 1814 William James Shelley (1774–1815), a former missionary to the South Pacific, lamented, "It is very remarkable that tho' this Colony has been settled nearly thirty years no one has attempted the study of the language of the Natives" (Shelley, 1814). Shelley's advice to the Church Missionary Society eventually led to the foundation of the Native Institution at Parramatta, Governor Lachlan Macquarie's (1762–1824) experimental school, in 1815. It was established to "… civilise, Christianise and educate …" Aboriginal children, train them in Western customs and equip them to undertake domestic duties (Christie, 1994; Cruickshank, 2008).

## References


Allan, R. J., Reason, C., Carroll, P., & Jones, P. (2002). A reconstruction of Madras (Chennai) mean sea level pressure using instrumental records from the late 18th and early 19th centuries. *International Journal of Climatology, 22*, 1119–1142.

Ashcroft, L. (2016). *Australia's meteorological anniversary*. https://lindenashcroft.com/2016/01/26/australias-meteorological-anniversary/ [accessed 4 September 2022].

Barker, T. (1757). *An account of the discoveries concerning comets with the way to find their orbits and some improvements in constructing and calculating their places. For which reason are here added new tables, fitted to those purposes; particularly with regard to that comet which is soon expected to return*. London, J. Whiston and B. White. Quarto.

Barton, G. B. (1889). *History of New South Wales from the Records. I. Governor Phillip 1783–1789*. Sydney, Charles Potter.

Bell, G. A., Gibbings, D. L. H., & Patterson, J. B. (1973). An absolute determination of the gravitational acceleration at Sydney, Australia. *Metrologia, 9*, 47–61.

Bladen, F. M., and Britton, A. (eds). (1893). *Historical Records of New South Wales*, 2. Mona Vale NSW, Lansdown Slattery & Co.

Bosloper, C. (2010). William Dawes' Gravity Measurement in Sydney Cove, 1788. In: *XXIV FIG International Congress 2010: Facing the Challenges – Building the Capacity. TS 6M History of Surveying*. 14pp. https://www.icsm.gov.au/sites/default/files/CongressPaper4534.pdf [accessed 3 September 2022].

Bosloper, C. (2017). The First Absolute Gravity Observations in Australia and New Zealand. In: Janssen, V. (ed.). *Proceedings of the 22nd Association of Public Authority Surveyors Conference*. Pp. 247–257. https://www.apas.org.au/files/conferences/2017/APAS2017-proceedings.pdf [accessed 3 September 2022].

Bradley, J. (1733). An account of some observations made in London, by Mr. George Graham, F.R.S. and at Black-River in Jamaica, by Colim Campbell, Esq., F.R.S. concerning the going of a clock; In order to determine the difference between the lengths of isochronal pendulums in those places, *Philosophical Transactions of the Royal Society, 38*, 302–314.

Brockhaus, F. A. (ed.). (1890). *Iconographic Encyclopedia of the Arts and Sciences – Applied Science*. Philadelphia, Iconographic Publishing Co. Vol. 6.

Bureau of Meteorology. (n.d.) *Table 1. Details of the 1788–1859 Southeastern Australia Historical Climate Data Sources: Source 2. The meteorological journal of William Dawes*. http://www.bom.gov.au/climate/early-data/sources/0_metadata.pdf [accessed 4 September 2022].

Carter, P. (2007). Public space: Its mythopoetic foundations and the limits of the law. *Griffith Law Review, 16*, 430–443.

Cavendish, H. (1798). Experiments to determine the density of Earth. *Philosophical Transactions of the Royal Society, 88*, 469–526.

Christie, M. (1994). Educating Bungelene: A case of educational colonialism. *History of Educational Review, 23*, 2–46.

Clarke, R. M. (2015). *Vanguards of empire: the lives of William Dawes, Watkin Tench and George Worgan*. Ph.D. Thesis, The Australian National University.

Collins, D. (1798). *An Account of the English Colony in New South Wales: with remarks on the dispositions, customs, manners, &c of the native inhabitants of that country to which are added, some particulars of New Zealand, compiled, by permission, from the Mss. from Lieutenant-Governor King*. Vol. 1. London, T. Cadell Jun. & W. Davies. http://adc.library.usyd.edu.au/data-2/colacc1.pdf [accessed 30 September 2020].

Collins, D., & Fletcher, B. (ed.). 1975 (1798). *An Account of the English Colony in New South Wales*. Vol. 1. Sydney, A. H. & A. W. Reed. http://adc.library.usyd.edu.au/data-2/colacc1.pdf [accessed 30 September 2020].

Cope, T. D. (1950). A Clock Sent Thither by the Royal Society. *Proceedings of the American Philosophical Society, 94*, 260–268.

Craft, F. A., & Mitchell, R. E. (1941). In Search of Dawes' Mount Twiss. *Royal Australian Historical Society Journal and Proceedings, XXVII*, 245–275.
Cruickshank, J. (2008). To exercise a beneficial influence over a man: Marriage, gender and the native institutions in early colonial Australia. In: *Evangelists of Empire? Missionaries in Colonial History*. eScholarship Research Centre, in collaboration with the School of Historical Studies, Melbourne. Pp. 115–124.
Dawes, W. (1787a). Letter to Nevil Maskelyne, dated 16 January 1787. RGO 14/48, folios 241r–v; 243r–v.
Dawes, W. (1787b). Letter to Nevil Maskelyne, dated 20 March 1787. RGO 14/48, folios 255r–256v.
Dawes, W. (1787c). Letter to Nevil Maskelyne, dated 3 September 1787. RGO 14/48, folio 270r.
Dawes, W. (1788a). Letter to Nevil Maskelyne, dated 1 October 1788. RGO 14/48, folio 287.
Dawes, W. (1788b). Letter to Nevil Maskelyne, dated 17 November 1788. RGO 14/48, folios 290–292.
Dawes, W. (1790a). Letter to Nevil Maskelyne, dated 16 April 1790. RGO 14/48, folio 300v.
Dawes, W. (1790b). Letter to Nevil Maskelyne, dated 26 July 1790. RGO 14/48, folio 302r.
Dawes, W. (1791). *Journal or meteorological records for Port Jackson, 1788–1791*. Unpublished manuscript. London, The Royal Society.
Dawes, W. (1792). Letter to Nevil Maskelyne, dated 18 June 1792. RGO 14/48, folio 307r.
de Grijs, R. (2022). Gravitational conundrum: Confusing clock-rate measurements on the 'First Fleet' from England to Australia. *Journal of Astronomical History and Heritage, 25*, 737–744.
Derrick, G. M. (2019). *Comparison of the First Fleet Sydney temperature records with modern data, 1788 to 2018*. The Saltbush Club. https://saltbushclub.com/wp-content/uploads/2019/01/comparison-of-first-fleet-and-modern-temperatures.pdf [accessed 4 September 2022].
Dow, S. (2014). First Contact. The secret history of Bangarra Dance Theatre's '*Patyegarang*'. *The Monthly, Arts and Letters*. https://www.themonthly.com.au/issue/2014/july/1404136800/steve-dow/first-contact [accessed 22 September 2022].
Foley, D., & Read, P. (2020). *What the colonists never knew*. Canberra, National Museum of Australia Press.
Forbes, E. G. (1975). *Greenwich Observatory: The story of Britain's oldest scientific institution, the Royal Greenwich Observatory at Greenwich and Herstmonceux, 1675–1975*. Taylor and Francis.
Gergis, J., Brohan, P., & Allan, R. (2010). The weather of the First Fleet voyage to Botany Bay, 1787–1788. *Weather, 65*, 315–319.
Gergis, J., & Fowler, A. (2009). A history of El Niño–Southern Oscillation (ENSO) events since A.D. 1525: Implications for future climate change. *Climatic Change, 92*, 343–387.
Gibson, R. (2010). Patyegarang and William Dawes: The Space of Imagination. In T. Banivanua Mar & P. Edmonds (Eds.), *Making Settler Colonial Space: Perspectives on Race, Place and Identity* (pp. 242–254). Palgrave Macmillan.
Gibson, R. (2012). *26 Views of the Starburst World: William Dawes at Sydney Cove 1788–91*. University of Western Australia Press.
Goodin, V. W. E. (1848). Public Education in N.S.W. before 1848. *Journal and Proceedings of the Royal Australian Historical Society, XXXVI*.
Goodman, M., (2020). *Baludarri – A Significant Figure in Parramatta History*. City of Parramatta Research and Collections. https://historyandheritage.cityofparramatta.nsw.gov.au/research-topics/aboriginal/baludarri-significant-figure-in-parramatta-history [accessed 13 July 2022].
Higgit, R. (ed.) (2014). *Maskelyne: Astronomer Royal*. London, Robert Hale. Cited by Bosloper (2017).

Hunter, J. (1793). *An historical journal of the transactions at Port Jackson and Norfolk Island, with the discoveries which have been made in New South Wales and in the Southern Ocean, since the publication of Phillip's voyage, compiled from the official papers; including the journals of Governors Phillip and King, and of Lieut. Ball; and the voyages from the first sailing of the Sirius in 1787, to the return of that ship's company to England in 1792.* London, John Stockdale.

Ingrey, R. (2019). Cited in *Sydney Morning Herald*, 14 July 2019.

Jackson, M. (2019). Cited in *Sydney Morning Herald*, 14 July 2019.

James, A. (2012). The William Dawes observatory. *Southern astronomers and Australian astronomy.* http://www.southastrodel.com/Page031b.htm [accessed 28 September 2020].

Karskens, G. (2009). *The Colony. A History of Early Sydney.* Sydney, Allen & Unwin.

Kater, H. (1818). An account of experiments for determining the length of the pendulum vibrating seconds in the latitude of London. *Philosophical Transactions of the Royal Society of London, 108*, 33–102.

Kington, J. A. (1997). The voyage of the British First Fleet from Portsmouth to Port Jackson in 1787–1788 and its impact on the history of Meteorology in Australia. In J. M. Kenworthy & J. M. Walker (eds.), *Colonial Observatories and Observations: Meteorology and Geophysics* (pp. 37–48). Department of Geography, University of Durham.

Kochsiek, M., & Gläser, M. (eds), 1999. *Comprehensive Mass Metrology.* Wiley.

Kohen, J. L. (1993). *The Darug and their neighbours: The traditional Aboriginal owners of the Sydney region.* Darug Link/Blacktown and District Historical Society.

Macarthur, E. (1791). Letter to Ms Kingdon, dated 7 March 1791. *Historical Records of New South Wales.* Vol. 2. Sydney, Charles Potter.

Macarthur Onslow, S. (ed.). (1973). *Some Early Records of the Macarthurs of Camden.* Adelaide, Rigby.

Mander-Jones, P. (1972). *Manuscripts in the British Isles Relating to Australia, New Zealand, and the Pacific.* University Press of Hawaii.

Marsden, W. (1834). *Miscellaneous Works of William Marsden.* London, Parbury, Allen, and Co.

Maskelyne, N. (1761–1762). Observations on a Clock of Mr. John Shelton, Made at St. Helena: In a Letter to the Right Honourable Lord Charles Cavendish, Vice-President of the Royal Society, from the Reverend Nevil Maskelyne, M. A. F. R. S. *Philosophical Transactions (1683–1775), 52*, 434–443.

McAfee, R. J. (1981a). Dawes's meteorological journal. Bureau of Meteorology, *Historical Note.* No. 2. Canberra, Australian Government Publishing Service.

McAfee, R. J. (1981b). *Towards a climatic history for southeastern Australia 1788–1860.* Ph.D. Thesis (unsubmitted), School of Earth Sciences, Macquarie University.

McGuffie, K., & Henderson-Sellers, A. (2012). Interdisciplinary Climate: The Case of the First 50 Years of British Observations in Australia. *Weather, Climate, and Society, 4*, 118–131.

Morrison, D., & Barko, I. (2009). Dagelet and Dawes: Their Meeting, Their Instruments and the First Scientific Experiments on Australian Soil. *Historical Records of Australian Science, 20*, 1–40.

Nash, D. (2011). *Dawes' Law generalised: Cluster simplification in the coastal dialect of the Sydney language.* In: Baker, B., Mushin, I., Harvey, M., & Gardner, R. (eds). Indigenous language and social identity: Papers in honour of Michael Walsh. *Pacific Linguistics, 626*, 159–178.

Nathan, D., Rayner, S., & Brown, S. (eds.). (2009). *William Dawes' Notebooks on the Aboriginal Language of Sydney, 1790–1791.* London, Hans Rausing Endangered Language Project and SOAS Library Special Collections.

Newton, I. (1687). *Philosophiæ Naturalis Principia Mathematica.* Book III, Prop. 20.

Nicholls, N., Tapp, R., Burrows, K., & Richards, D. (1996). Historical thermometer exposures in Australia. *International Journal of Climatology, 16*, 705–710.

Norris, R. P. (2016). Dawes Review 5: Australian Aboriginal Astronomy and Navigation. *Publications of the Astronomical Society of Australia, 33*, id. e039.

Page, S. (2014). Cited by Dow (2014).
Paish, L. (1989/2017). *William Dawes: The first white man on the Blue Mountains*. Blue Mountains local studies. http://bmlocalstudies.blogspot.com/2017/08/william-dawes-first-white-man-on-blue.html [accessed 10 September 2022].
Poll, M. (2019). Patyegarang's sky world. *News from Powerhouse*, 31 January 2019. https://www.maas.museum/magazine/2019/01/patyegarangs-sky-world/ [accessed 22 September 2022].
Pybus, C. (2009). 'Not Fit for Your Protection or an Honest Man's Company': A Transnational Perspective on the Saintly William Dawes. *History Australia, 6:1*, 12.1–12.7.
Russell, H. C. (1877). *Climate of New South Wales: Descriptive, Historical, and Tabular*. Sydney, Charles Potter.
Russell, H. C. (1882). *The Sydney Observatory. History and Progress*. Sydney, Thomas Richards, Govt. Printer.
Saunders, S. D. (1990). *Astronomy in New South Wales: 1788 to 1858*. Ph.D. Thesis, Department of General Philosophy, The University of Sydney.
Shelley, W. (1814). Letter dated 6 October 1814. Cited by Goodin (1848: 152).
Smith, K. V. (2008a). *Colebee*. Dictionary of Sydney. https://dictionaryofsydney.org/entry/colebee [accessed 13 July 2022].
Smith, K. V. (2008b). A Few Words from William Dawes and George Bass. *National Library of Australia News*, June 2008, 7–10. http://www.nla.gov.au/pub/nlanews/2008b/jun08/story-2.pdf [accessed 22 September 2022].
Steele, J. M. (2005). *The Aboriginal language of Sydney. A partial reconstruction of the Indigenous language of Sydney based on the notebooks of William Dawes of 1790–91, informed by other records of the Sydney and surrounding languages to c.1905*. MA (research) Thesis, College of Humanities and Social Sciences, Macquarie University.
*Sydney Morning Herald*. (14 July 2019). 'Something to remember': 18th century notebooks return to Sydney. https://www.smh.com.au/national/nsw/something-to-remember-18th-century-notebooks-return-to-sydney-20190712-p526ov.html [accessed 22 September 2022].
Tench. (1789a). *A Narrative of the Expedition to Botany Bay; with an Account of New South Wales, its Productions, Inhabitants, &c, to which is subjoined, A list of the Civil and Military Establishments at Port Jackson*. London, J. Debrett.
Tench, W. (1789b). *Sydney's First Four Years*, reprint of *A Narrative of the Expedition to Botany Bay* (1789b) and *A Complete Account of the Settlement at Port Jackson* (1793), introduction and annotations by L. F. Fitzhardinge. Sydney, Library of Australian History (1979, first published 1961).
Tench, W. (1793). *A Complete Account of the Settlement at Port Jackson Including An Accurate Description of the Situation of the Colony; of the Natives; and Of Its Natural Productions*. London, G. Nicol and J. Sewell.
Thomas, S. (2013). A Transnational Perspective on William Dawes' Treatment of Women. *History Australia, 10*(1), 187–204.
Torok, S., & Nicholls, N. (1996). A historical annual temperature dataset for Australia. *Australian Meteorological Magazine, 45*, 251–260.
Troy, J. (1992). The Sydney Language Notebooks and responses to language contact in early colonial NSW. *Australian Journal of Linguistics, 12*, 145–170.
Troy, J., 1993. *The Sydney Language*. Canberra, AIATSIS and the Australian Dictionaries Project.
Troy, J. (1994). *The Sydney language*. Aboriginal Studies Press.
Wales, W., & Bayly, W. (1777). *The Original Astronomical Observations Made In The Course Of A Voyage Towards The South Pole And Round The World*. London, W. and A. Strahan.
Wallace, J., & Hobbs, P. (2006). *Atmospheric Science, An Introductory Survey* (2nd ed.). Elsevier.
Welbergen, J. (2015). *The Flying Fox Heat Stress Forecaster*. Animal Ecology Laboratory, Hawkesbury Institute for the Environment, Western Sydney University. https://www.animalecologylab.org/ff-heat-stress-forecaster.html [accessed 4 September 2022].

Whitworth, R. P. (1866). *Bailliere's New South Wales gazetteer and road guide; containing the most recent and accurate information as to every place in the colony*. Sydney, F. F. Bailliere; cited by Barton (1889: 150).

Wilkins, D., Nash, D. (2008). *The European 'discovery' of a multilingual Australia: the linguistic and ethnographic successes of a failed expedition.* In: The history of research on Australian Aboriginal languages. McGregor, W. (ed.). *Pacific Linguistics, 591*. Pp. 485–507, Chap. 18.

Wood, H. W. (1951). Presidential address. *Journal and Proceedings of the Royal Society of New South Wales, 84*, 1–16.

Wood, H. W. (1958). Sydney Observatory 1858–1958. *Sydney Observatory Papers*, No. 31.

# Chapter 4
# Frustrations and Friction

## 4.1 Dawes' Grievances

As early as October 1788, Major Robert Ross (ca. 1740–1794), commanding officer of the small detachment of Marines in New South Wales, surveyed his men about their intentions upon the expiration of their three-year terms by late 1791. On 1 March 1790, Ross compiled a *List of such Officers, Non-commissioned Officers and Privates, as are desirous of remaining in this Country after the Time their Lordships intended to relieve the Detachment* (Phillip, 1789: 174; Cobley, 1963: 156–157). He notified the British Admiralty that

> I have likewise the honor [*sic*] of enclosing your lordship his returns of such officers as wish to be relieved at the expiration of the three years for which they were sent out, and of those who are desirous of remaining; as likewise copies of his letter and my answer, respecting the encouragement offered by Government to settlers. (Barton, 1889: 336)

Of the eleven officers in Ross's detachment, six expressed a desire to return to Britain at the end of their terms, or as soon afterwards as might be convenient. The other five were either not keen or as yet uncertain whether they wished to remain and become settlers. Two, including Dawes and Tench, responded that they wanted to continue to serve "… [a]s a soldier for one tour of three years more" (Barton, 1889: 201).

Until late 1791, Dawes remained intent on staying on for an additional three-year term, and possibly even settling, given

> ... the uncertainty of any other person being appointed to succeed him in the astronomical way, and the great loss, in his opinion, it would be to the science were the opportunity missed of making proper use of those valuable instruments. (Bladen et al., 1892: 423)



R. de Grijs and A. Jacob, *William Dawes*, Springer Biographies,
https://doi.org/10.1007/978-3-031-38774-6_4

Indeed, as early as 17 November 1788, Dawes had advised Maskelyne,

> My reason for wishing that a proper [astronomer] may come out with an established income before my six years are expired, is, that in that case the world will be much more benefited by his labours, than it is probable it ever will by mine, as I foresee that a great part of my time will be very differently employed; and not from any dislike or conscious inability to the business [of astronomy], as in the first place I could not be more happily situated, than to have that, and that alone to attend to, without any regard whatsoever to self interest ... (Dawes, 1788b)

In late 1790, Lord William Grenville (1759–1834), Secretary of State for the Home Department and in charge of the Colonial Office, informed Governor Phillip that the detachment of Marines would be replaced by a special New South Wales Corps (Bladen et al., 1892: 415). Marines already in the colony would be allowed to apply for appointments in the new hierarchy. Dawes considered applying, but he declared that he would only consider remaining in the colony if his worth to the settlement was adequately recognised by an appointment to a suitable new position.

In his missive to Governor Phillip, Lord Grenville had singled out Dawes, extolling his service to the colony. He authorised Phillip to retain Dawes as a lieutenant of engineers, in case the Marine had not yet been placed in charge of the auxiliary company of the New South Wales Corps. However, unbeknownst to Lord Grenville, Dawes' and Phillip's relationship had soured considerably, eventually culminating in a clear case of insubordination on Dawes' part.

Dawes' insubordination and the subsequent fall-out were merely the latest manifestations of a long list of minor and some major grievances that had affected the men's relationship over the years. As early as November 1787, Dawes had incurred Phillip's wrath when he had allowed their marine timekeeper, K1, to run down following the party's transfer from H.M.S. *Sirius* to H.M.A.T. *Supply*. One might even argue that Dawes had been inconvenienced to the point of obstruction by Phillip's refusal to allow the disembarkation of his telescope and astronomical equipment already at Santa Cruz (Canary Islands) early on during the First Fleet's voyage, and again when the convoy was moored in Table Bay in late 1787.

Moreover, Dawes was clearly annoyed by his initial duty-bound confinement on board the *Sirius*, upon arrival in Sydney Cove, which prevented him from getting on with the establishment of his Observatory as per Maskelyne's and the Board of Longitude's instructions. From Dawes' perspective, Phillip had added insult to injury by saddling the Marine with a range of additional tasks in service to the developing colony, including in his newly appointed roles as Engineer and Officer of Artillery. As we saw already, Dawes advised Maskelyne of the additional work he had been assigned:

> The extreme economy of Government, not allowing them to give salaries to proper persons to perform the necessary business of settlement has obliged the Gov.[r] to appoint me to the Duty of Engineer & Officer of Artillery, and the inability of the Surveyor [Augustus Alt] to perform his business has also thrown a great part of that on my hands. (Dawes, 1788a)

In a 'Memorial' written in 1826, Tench further clarified the situation that led Dawes to take on the tasks of the colony's surveyor and town planner:

> … the Surveyor General, being advanced in Years and not sufficiently acquainted with the practice of surveying and laying out Lands in a New Country; he [Dawes] was employed in laying out the Towns of Sydney and Paramata [*sic*], in surveying the Government Farm in the vicinity of the former and in various other similar Services. (Tench, 1926 [1826])

Augustus Alt was the closest person to a town planner Phillip had been provided with. Alt had served as an assistant Engineer at the Great Siege of Gibraltar in 1779 to Governor George Augustus Eliott (1717–1790). He was appointed Surveyor of Lands for the new colony in New South Wales in May 1787; in March 1788 Phillip appointed Dawes as Alt's assistant, given the young Marine's familiarity with building and engineering from his upbringing at the Portsmouth Naval dockyard.

Both men were tasked with laying out a plan for the new settlement according to Phillip's grand vision of the creation of a town for the residents to live in and take pride in as their home, as opposed to a prison to stop them escaping (Phillip, 1789). Phillip's vision for the settlement, which he proposed to call Albion, included 200 ft-wide streets to enable "… proper circulation of air …" (Phillip, 1789: 146) and standard allotment sizes of 60 by 150 ft:

> It is proposed by Governor Phillip that when houses are to be built here, the grants of land shall be made with such clauses as will prevent the building of more than one house on one allotment, which is to consist of sixty feet in front, and one hundred and fifty feet in depth. These regulations will preserve a kind of uniformity in the buildings, prevent narrow streets, and exclude many inconveniences which a rapid increase of inhabitants might otherwise occasion hereafter. (Phillip, 1789: 146)

Alt and Dawes presented Phillip with a plan for the new town in July 1788 (see Fig. 2.9). It was defined by a grid pattern centred around a town centre on the western hillside above Sydney Cove. Dawes proposed to call the ridge behind Sydney Cove Church Hill, an indication of their proposed site of Sydney's first church. The original major thoroughfare proposed by Alt and Dawes (labeled '4' on Fig. 2.9) would hence be Church Street. In the mean time, is has been renamed as Lang Street following a period as Wentworth Street (Yarrow, n.d.).

However, Phillips' grand plan proved somewhat impractical and none of the other streets proposed by Alt and Dawes ever materialised. Custom prevailed, and so a rough track from the Tank Stream to the hospital (on present-day Macquarie Street) became the settlement's main artery. It eventually developed into Bridge Street. The area Phillip envisioned for government offices, on the southern side of Bridge Street, is still occupied by Government buildings (Yarrow, n.d.).

Alt and Dawes also worked on planning the settlements of Parramatta and Tongabby (Toongabbie), in addition to surveying and recording early land grants. However, ill health forced Alt to retire on half pay in 1791, leaving Dawes as the colony's sole surveyor. His main legacy today is the rectangular, generous lay-out of Parramatta (see for details, Yarrow, n.d.; see Fig. 4.1 for an early town plan from 1814), which remains the only Australian town laid out by settlers from the First Fleet. On a visit to Parramatta in 1790, Tench commented,

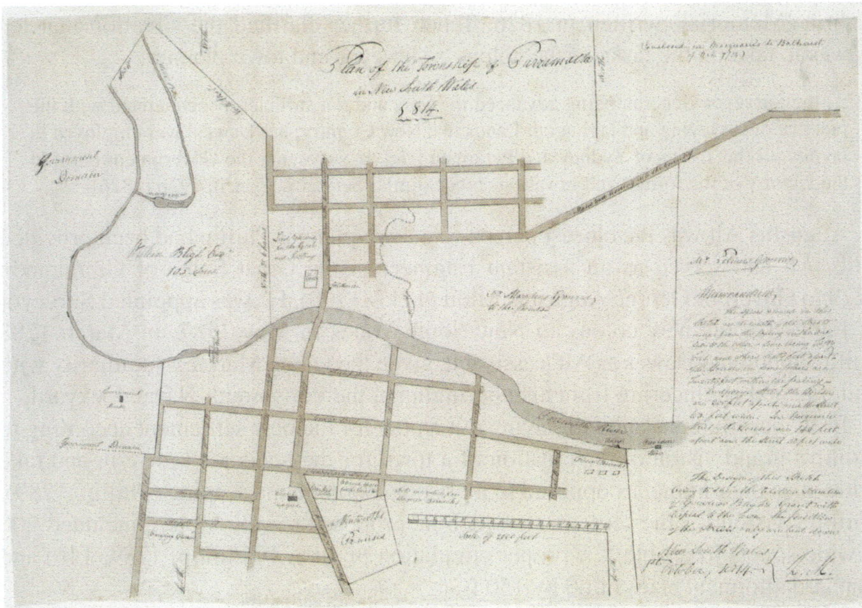


**Fig. 4.1** "Plan of the township of Parramatta in New South Wales 1814". (State Library of New South Wales, a1528520; out of copyright)

> The main street of the new town is already begun. It is to be a mile long, and of such breadth as will make Pall-Mall and Portland Place "hide their diminished heads." It contains at present 32 houses completed, of 24 ft by 12 each, on a ground floor only, built of wattles plaistered with clay, and thatched. Each house is divided into two rooms, in one of which is a fire place and a brick chimney. (Tench, 1793: Chap. X)

Despite Phillip's growing antipathy of Dawes, he clearly saw the benefit of assigning the Marine an increasing number of engineering tasks. The Governor was intent on fortifying the entrance to Sydney Cove, both in case of a convict uprising and to be able to handle any hostilities from enemy ships that might challenge British territorial claims. And so he instructed Dawes to build simple mud redoubts for the storage of explosives near his Observatory at Point Maskelyne and also across the cove at Cattle Point (present-day's Bennelong Point, site of the Sydney Opera House). In early October 1788, upon departure of the *Sirius* for Cape Town to secure much-needed provisions, Dawes oversaw the relocation of eight of her cannons to a small battery near the shore below his Observatory (see Fig. 2.17), so as to create more space onboard for the hoped-for supplies. This resulted in upgraded fire power for the colony, now consisting of two brass 6 pounders and four iron 12 pounders.

Yet, despite his dedication to the tasks he had been assigned, Dawes was clearly unhappy about the negative implications of his increased workload on the establishment of his Observatory at Point Maskelyne. Indeed, throughout his life, Dawes gained a reputation of engaging in direct communications and resisting interference with his responsibilities. This is clearly reflected in his letter to Maskelyne of 17

November 1788 (and we will see similar instances of his directness when dealing with opposing views in Sierra Leone and the West Indies; see Chaps. 5 and 6), where he also highlights the adverse consequences of his protestations:

> I have told you in former letters that the Governor had thought proper to order me to do the duty of Engineer & Officer of Artillery; when this was first proposed to me I did every thing in my power to prevent this taking place, as I was justly apprehensive that so much business would be found for me in consequence, that it would entirely prevent me from attending to the observatory so much as you would expect, or as I wish'd; but on its being mentioned sometime afterwards, and it then being said that the service required it; I thought it would have been highly improper to have declined it any longer, especially as the Gov.ʳ had frequently intimated that when I should be ordered to do those duties it would take me from the duty of the detachment which would give me good time for every thing, his words carrying an express meaning, that I should by that means get sufficient time to attend properly to the observatory: A very short time convinced me that my first fears were very justly founded, for I soon found that almost the whole of my time was to be employed in business, very different indeed from that of an Engineer or Officer of Artillery, but which I was ordered to do in consequence of doing those duties; in short the Gov.ʳ'ˢ expectations were so much greater than I could possibly fulfil [*sic*], that I was obliged very plainly tho' very respectfully to tell him my sentiments thereon; in consequence of which he was very highly offended, and severall [*sic*] letters past between us: this is however almost entirely worn off, and I now find myself rather less hurried than before; but I think by no means so much at liberty to fulfil your wishes as if I had been left to do my duty in the detachment without being ordered to do those of Engineer &c. and this is the situation I would willingly have remained in, had not the service required otherwise, which his Excell.ʸ has told me is absolutely the case: It is, I think, a very good maxim, that if a person cannot do as much as he would; he ought, nevertheless, to do as much as he can: this, I have adopted and am determined that no application shall be wanting in me, tho' I still find a great deal of my time taken up very differently from what either you or I had supposed it would be. (Dawes, 1788b)

It appears that Dawes was a reluctant engineer and artillery officer, whose real calling was as an astronomer. It also appears that he did not enjoy the survey, design and construction work he had been tasked with, regarding those duties as unwelcome, exhausting and demoralising but completed them competently and thoroughly. Yet, he found himself in an unenviable and politically fraught position where he had to display continued loyalty and please both his patron and his commander at the same time.

Dawes must have complained about the additional workload and the nature of his work to his father during this period, since a series of letters from 1789 to this effect from Benjamin Dawes to Maskelyne have survived (Dawes, 1789a, 1789b; Laurie, 1988). His father appeared most concerned with the young Dawes' wellbeing, as well as with his son's compensation for the additional duties. By virtue of the elder Dawes' appointment at the Office of Ordnance at Portsmouth and his professional network, Dawes Sr. was well-informed as to developments related to the Marine detachment in New South Wales. When he heard, some time in 1789, that the detachment would be withdrawn and replaced by a more permanent Marine Corps, he immediately asked Maskelyne to act on behalf of his son. Dawes Sr. wrote to Lord Grenville to secure an appointment for his son as artillery officer or engineer, or "… any other permanent post …" in New South Wales, remunerated at "… such sum as your Honor shall think proper" (Dawes, 1789c). As we saw already, by late 1790 Grenville supported this petition and requested that Phillip retain Dawes for another tour of three years, paid at a rank equal to his rank in the Marines.

On 26 July 1790, Dawes wrote to Maskelyne, once again complaining about his significantly increased workload. This now included laying out a city near Rose Hill (see Fig. 4.2), despite a persistent lack of supplies, but he felt that it would be "… as endless as fruitless …" (Dawes, 1790: f. 302v) to list all of the interfering impediments to his work at the Observatory, particularly given that—in Dawes' view—Phillip "… never had the business in the least degree, really at heart" (Dawes, 1790: f. 302r; original emphasis).

In the same letter, Dawes reflects on Phillip's drivers to withhold support for his astronomical endeavours, concluding that the Governor may have had ulterior motives:

> It was then to induce all of us to send home as favourable accounts as possible, of his care to render our situation comfortable and convenient; but that mask has long since failed of its effect. He is, however, never without a mask of some kind, but they have all happily, by much use, worn so thin as to be perfectly transparent. (Dawes, 1790: f. 302r)

Dawes had clearly lost faith in Phillip's goodwill and his support of the Board of Longitude's edicts, which by year-end led to a further, insurmountable deterioration in the men's relationship. Yet, although Dawes may have been among the most vocal officers in this regard, his fellow Marine officers had equally come to think that Phillip's main motivation in enforcing positive accounts about the colony was to retain his annual salary of £1500—which he would lose "… on the evacuation of this country" (Dawes, 1790: f. 303v).

## 4.2 Phillip's Perspective

The situation must have appeared quite differently from Phillip's perspective. The Governor clearly found himself in a position where his focus was necessarily much more comprehensive than that of Dawes. Phillip's unenviable task was to keep the fledgling colony functioning and its occupants fed in the face of severe external pressures and the growing risk of starvation. Dawes' evangelism and self-righteousness certainly did not help bridge the growing rift between both men. As such, their mutual antipathy may not, at least initially, have been a clash of personalities. It seems, however, that Dawes' attitude and behaviour did not contribute to a peaceful resolution of their many differences of opinion.

And so, despite Lord Grenville's express request to appoint Dawes as lieutenant of engineers, in November 1791 Phillip advised the British Home Office that he was unwilling to extend Dawes' term in the colony (Bladen et al., 1892: 543). Phillip explained to Lord Grenville that he had charged Dawes with three offences (Britton and Bladen, 1894: Sect. 13):

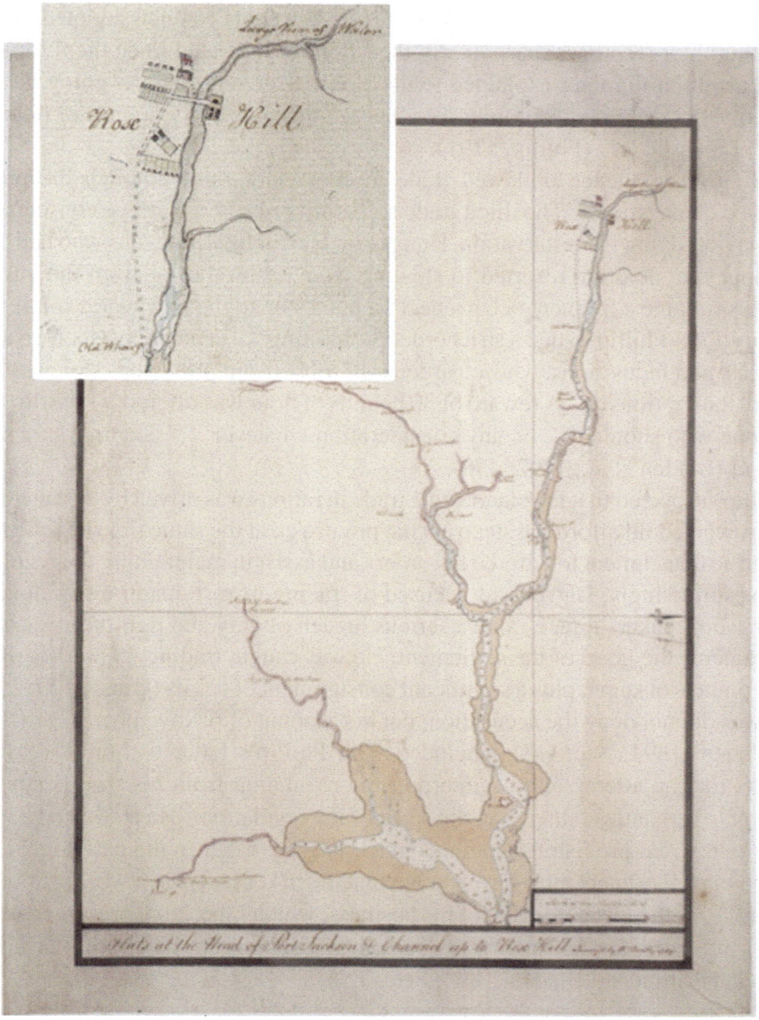


**Fig. 4.2** "Flats at the head of Port Jackson & channel up to Rose Hill, survey'd by Wm Bradley 1789". *Inset*: Rose Hill (detail). (Mitchell Library, State Library of New South Wales; Ref. SAFE/MT4 140/1792/1; out of copyright)

1. That Lieutenant Dawes had purchased rations from convicts contrary to the General Orders;
2. That he had refused to do duty;
3. That he had been guilty of unofficer-like behaviour to the Governor.

Conditions in the fledgling colony were dire in those early years, with provisions heavily rationed to avoid full-scale starvation. Each adult male was entitled to a weekly ration of four pounds of flour, which was barely enough—or, rather, insufficient—to survive on: Phillip is reported to have said, "… every man could eat his ration …" (Britton and Bladen, 1894: Sect. 13). Yet, some convicts adopted the habit of pooling their resources, particularly their flour rations, and trade them for spirits and tobacco. Phillip later explained to the Home Secretary that the going rate was, reportedly, "… ten pounds of flour for a bottle of rum or thirty pounds of flour for a pound of tobacco …" (Phillip, 1791).

The Governor hence outlawed trade in convict rations, although the practice continued clandestinely. The illicit trade sometimes resulted in tragic consequences, with convicts dying from starvation. Hungry for lack of food, convicts who had traded their flour rations often resorted to stealing from each other or from the garrison; robberies became a frequent occurrence. To put a stop to the downward spiral, on 11 February 1791 Phillip issued a strict order prohibiting "… a practice which distresses the honest and industrious, whose gardens are robbed and provisions stolen by those who sell their rations …" A reward of 30 pounds of flour was offered for the discovery of anyone who should "… on any consideration whatever …" acquire rations from a convict (Bladen et al., 1892: 450).

Phillip suspected that the clandestine trade in rations was driven by certain Marine officers, who could afford the steep asking price. Yet, at the same time he relied upon the Marine detachment to enforce his orders and assist in maintaining the regulation. Perhaps surprisingly, Dawes was accused of trading in vast quantities of flour with the garrison's baker, at face value a serious breach of duty and rightfully considered an act against the good of the settlement. He was caught trading 40 pounds of flour and 20 pounds of sugar, plus a significant consignment of spirits (Gibson, 2012: 173).

Dawes did not deny the accusation, but his account of 6 November 1791 (Britton and Bladen, 1894: Sect. 13)—included with Phillip's letter to Lord Grenville—suggests that he attempted to reinterpret the regulation from his own perspective, resorting to semantics rather than the spirit of the regulations. He readily admitted to having purchased provisions from a convict, the "… baker to the garrison". Dawes asserted that all officers must have known that the baker received a weekly allocation of flour "… as the just perquisite of his business, which I therefore presumed became his own property, and as such was deemed by everyone to be entirely at his own disposal" (Britton and Bladen, 1894: Sect. 13):

> It does not appear to me that I was guilty of any impropriety in purchasing goods from a convict, as I had every reason to suppose it was no part of his ration. (Barton, 1889: 543; Gibson, 2012: 176)

However, it transpired that this was not the first time Dawes had been reprimanded for illicit trade practices. As we learn from Phillip's letter to Lord Grenville of 17 November 1791, the Governor had "… some time before …" requested that Ross, his commanding officer, discipline the Marine for a similar infringement (Phillip, 1791). Phillip emphasised that he could not accept Dawes' denial of having purchased rations from a convict. Dawes' culpability—or, perhaps, his indiscretions—was plain for all to see at the time when he was summoned to appear before the magistrates charged with breaches of the illegal trade order. The convict acting as the colony's baker readily admitted to having given Dawes 40 pounds of flour and 20 pounds of sugar in exchange for "… spirits and other articles" (Britton and Bladen, 1894: Sect. 13).

In hindsight, Dawes' defence appears weak and self-centred. Although he may not have violated the law in the strictest sense, his actions certainly contributed to further pressure on the colony's starvation economy. The garrison's officers were expected to maintain adherence to and enforce the prevailing regulations, not trample on them in an attempt to change the moral boundaries and evade legal consequences. It is, therefore, no surprise that Governor Phillip took issue with Dawes' attempts at placing himself above the law. After all, Phillip was responsible for the colony's survival, and their joint provisions were key to that aim.

Yet, Dawes' insistence that he had merely traded in commercial goods in a free economy ignored Phillip's attempts to cater to the needs of all, at viable levels, in the fledgling colony. If Dawes had wanted to proceed with his purchase, he should have consulted with the Commissary, Andrew Miller, before making any purchase from a convict, even the colony's baker. The entire episode reflects poorly on Dawes' grasp of the bigger picture; it shows him as self-absorbed and self-righteous, and it highlights a character flaw of being unable to accept that he has breached any orders, conventions or requirements (Barton, 1889: 543–546).

Yet, Phillip had little to no recourse to any punishment. He was the colony's Governor and not the Marines' commanding officer. The colony's Judge-Advocate, David Collins (1756–1810), advised Phillip that he did not have the legal power to convene a court martial for Marines on land (Gibson, 2012: 173), and it was unclear whether the Marine Corps would agree with any disciplinary action against one of their own (Moore, 1987: 18–19). The Marine Corps' leadership insisted that the only authority to convene a court martial was the Lord High Admiral (Campbell, 1788). Dawes must have been well aware of these simmering tensions; he was certainly aware of the disdain in which the Marines held the Governor.

And so he took his time before responding to Phillip's accusation. It is perhaps telling that he took a full day, remaining confined to his Observatory, to prepare his defence, since it suggests that he did not have a convincing explanation available to explain his behaviour. It has been suggested (Gibson, 2012: 173) that this delay in Dawes' response may have been driven by the men's soured personal relationship, with Dawes taking the opportunity to spite Phillip in a tense standoff where the latter could not take any punitive actions owing to a lack of legal authority.

## 4.3 Insubordination

Once again, these developments are clear evidence that the ebb of the men's mutual respect was at an all time low. Their tense relationship had, however, suffered its worst setback almost a year before the flour scandal, in December 1790. For many months, indeed since a time shortly after the arrival of the First Fleet in Sydney Cove, the relationship between the British settlers and the Aboriginals in the Port

Jackson area had been steadily deteriorating (Cahill & Irving, 2010: Chap. 2). The Aboriginal population and the settlers were increasingly at loggerheads over issues like the competition for land and food. This was particularly pressing, since the invading force had appropriated Sydney's food-rich waterways. In turn, this led to famine among the Aboriginal clans in the winter of 1788. Second, the First Fleet's arrival had led to the spread of smallpox, resulting in an epidemic in 1789 that severely affected the local Aboriginals. In addition, Aboriginal cultural norms were routinely ignored by the British settlers, likely out of ignorance. For instance, the invaders regularly stole Indigenous weapons and artifacts for souvenirs. Finally, Phillip's approach to creating a dialogue between the two populations by kidnapping Aboriginals for use as cultural ambassadors backfired spectacularly.

Some clans resorted to attacking with spears any unarmed white person they would encounter in an act of revenge for injuries received from the convict population (see Fig. 4.3). Several convicts had been killed, whereas others were seriously wounded; at least 17 white men had been killed or seriously wounded since the arrival of the First Fleet (Barton and Britton, 1894: 9). The spearing to death of Phillip's convict 'gamekeeper', John McIntire (McEntire, McIntyre), on 9 December 1790 by an Aboriginal known as Pemulwuy (ca. 1750–1802), was Phillip's final straw. McIntire was one of just three convicts permitted to carry a musket. He was free to roam the bush in search of game.

Many of the men in the settlement suspected that the man had given cause for the Aboriginals' attack in retribution (Barton, 1889: 9). Captain Tench explained,

**Fig. 4.3** Port Jackson Painter, *Aborigines Attacking a Sailor*, also known as *The Hunted Rushcutter*; Port Jackson Coll. 44. (© The Trustees of the Natural History Museum, London; reproduced with permission)

> From the aversion uniformly shewn by all the natives to this unhappy man, he had long been suspected by us of having, in his excursions, shot and injured them. ... On being questioned with great seriousness, he, however, declared that he had never fired but once on a native, and then had not killed, but severely wounded him, and this in his own defence. Notwithstanding this death-bed confession, most people doubted the truth of the relation, from his general character, and other circumstances. (Tench, 1793: 91)

Whereas Phillip considered that most of the attacks by Aboriginals the invading party had endured were triggered by either misunderstandings—including the attack on himself at Manly in November 1789 (see Fig. 4.4)—or in retaliation for wrongs committed by the settlers, he considered the murder of McIntire as "... unprovoked, and the barbarity of their conduct admits of no extenuation" (Wood, 1924: 6). However, Phillip's opinion on this matter was not generally accepted among the colonists. Yet, as reprisal for McIntire's murder and in an effort to make an example of some of the offenders, on 13 December 1790 Phillip ordered a punitive expedition, meant as a deterrent. Dawes learnt of his appointment to the punitive party from Phillip's proclamation. Commanded by Tench, the party was

> ... to clear the peninsula between Georges River and Cooks River with orders to put ten natives to death, to cut off their heads and to bring them back in bags—which together with hatchets were provided. (Wood, 1924: 5)

**Fig. 4.4** Port Jackson Painter, *The Governor Making the Best of His Way to the Boat after Being Wounded with the Spear*. See insets for details. Watling Coll. 23. (© The Trustees of the Natural History Museum, London; reproduced with permission)

In his letter to Lord Grenville of 7 November 1791, Phillip summarised what happened next, clearly stating that it was ordinarily Dawes'

> … tour of duty … to go out with that party, [yet he] refused that duty by letter, … [even after Phillip had] … taken great pains to point out the consequences of his [Dawes] being put under an arrest. … Later in the evening, Lieutenant Dawes informed Captain [John] Campbell that the Rev. Mr. [Richard] Johnson thought he might obey the order, and that he was ready to go out with the party, which he did, but, after the service was over, informed the Governor that he was sorry he 'had been persuaded to comply with the order,' and very clearly showed that he would not obey a similar order in future. Lieutenant Dawes' expressions were such as would have subjected him to a court-martial had he been amenable to one. (Phillip, 1791)

The order clearly weighed heavily on Dawes' conscience. He immediately wrote a letter to Captain John Campbell (ca. 1758–1795), commanding officer of the New South Wales detachment (while Commander Ross was acting Lieutenant-Governor at Norfolk Island), outlining his conscientious objections and refusing to comply with the order. Dawes' letter is no longer available today. It is often said that this was most likely the first case of conscientious objection in the colony, while Dawes was the first European defender of Aboriginal rights.

Meanwhile, he turned to the Rev. Richard Johnson, the colony's Methodist chaplain, for moral and religious advice. Campbell forcefully remonstrated with the Marine, but he was initially unsuccessful in changing Dawes' mind. Campbell then showed Dawes' letter to Phillip, who informed Dawes that his refusal to serve was grounds for arrest:

> On this order appearing, Lieut. Dawes … refused that duty by letter to the senior officer of the detachment (Capt. Campbell), who, finding it impossible to persuade Lieut. Dawes to obey the order, brought the letter to the Governor, who likewise took great pains to point out the consequence of his (Lieut. Dawes) being put under an arrest. (Wood, 1924: 5–6; Cobley, 1963: 309)

It was the Rev. Johnson who eventually managed to convince Dawes to join the deterrent party:

> … late in the evening Lieutenant Dawes informed Captain Campbell [also by letter] that the Rev. Mr. Johnson thought he might obey the order, and that he was ready to go out with the party, which he did. (Britton and Bladen, 1894: Section 13)

The expedition was entirely unsuccessful—perhaps deliberately so. Any Aboriginals the party encountered disappeared as soon as the Marines were spotted. As such, after four unsuccessful days, on 17 December 1790 the party returned to the settlement at Sydney Cove, tired, demoralised and empty-handed (Saunders, 1990: 97)—much to the relief of both Dawes and Tench.

> At four o'clock on the morning of the 14th [December 1790] we marched: the detachment consisted, besides myself, of captain Hill of the New South Wales corps, lieutenants Poulder and Dawes, of the marines, Mr. Worgan and Mr. Lowes, surgeons, three serjeants, three corporals, and forty private soldiers, provided with three days provisions, ropes to bind our prisoners with, and hatchets and bags, to cut off and contain the heads of the slain. By nine o'clock this terrific procession reached the peninsula, at the head of Botany Bay; but after having walked in various directions until four o'clock in the afternoon, without seeing a native, we halted for the night.

> At daylight on the following morning our search recommenced. … at half past seven o'clock instead of finding ourselves on the south-west arm, we came suddenly upon the sea shore, at the head of the peninsula, ... Here we saw five Indians on the beach, whom we attempted to surround; but they penetrated our design, and before we could get near enough to effect our purpose, ran off. …
>
> The alarm being given, we were sensible that no hope of success remained, but by a rapid movement to a little village (if five huts deserve the name,) …, where possibly some one unapprized [*sic*] of our approach, might yet be found. Thither we hastened; but before we could reach it three canoes, filled with Indians, were seen paddling over in the utmost hurry and trepidation, to the opposite shore, where universal alarm prevailed. (Tench, 1793: Chapter XII)

It appears that Dawes had meanwhile been wrestling with his conscience, given that he clearly and publicly expressed regret that he had acquiesced to taking part in the expedition. And so, upon their return to the settlement and in the presence of Lieutenant John Long, the Marine Adjutant, he "… informed the Governor that 'he was sorry he had been persuaded to comply with the order', intimating at the same time that he would not obey a similar one in future" (Phillip, 1791; Britton and Bladen, 1894: Sect. 13). This represented a clear act of insubordination. Dawes' reasons for disobeying any such future orders remain unknown, given that neither Dawes himself nor Phillip provide any indications in their subsequent letters. Nevertheless, in his letter to Phillip of 6 November 1791, almost a year after the events, Dawes wrote, "… after so long a time having elapsed, and repeated reflections on the subject, I feel at this instant no reason to alter the sentiments I then entertained" (Britton and Bladen, 1894: Sect. 13).

Let us therefore briefly explore Dawes' possible reasons for his refusal to obey Phillip's orders, despite the potentially grave legal consequences. Both Tench and Dawes doubted that McIntire had been in the clear, given the man's brutal reputation. Both Marines had become familiar, even friendly, with a number of the local Aboriginals, and they saw that the Indigenous tribes were often the injured party in altercations with the British colonists.

Dawes refused to accept as universal principle, 'Our country, right or wrong', which hence made him the first conscientious objector in the colony, in December 1790. His motivations appear to have been twofold, both conscientious and religious. He had established cordial relationships with a number of Aboriginal people in the Port Jackson area, which had allowed him to acquire in-depth knowledge of their consciousness. This would become a red line throughout his life and career, and eventually prompted him to question the moral authority of colonialism (Coleman, 2005). Simultaneously, he was deeply engaged in learning the local Aboriginal language of the Gadigal people. Tench pointedly summarised Dawes' deep engagement after his own return to England:

> Of the language of New South Wales I once hoped to have subjoined to this work such an exposition, as should have attracted public notice; and have excited public esteem. But the abrupt departure of Mr. Dawes, who, stimulated equally by curiosity and philanthropy, had hardly set foot on his native country, when he again quitted it, to encounter new perils, in the service of the Sierra Leona company [Chap. 5], precludes me from executing this part of my original intention, in which he had promised to co-operate with me; and in which he had advanced his researches beyond the reach of competition. (Tench, 1793: 134)

Dawes had acquired a reputation as the most educated, conscientious and 'gentlemanly' of the colonists. Lieutenant Southwell, who had sailed on H.M.S. *Sirius* and was stationed at the South Head 'Outpost' in 1790, considered Dawes to be his "... most esteem'd friend and confidant ..." (Southwell, 1790: 719). He described his friend as "... a most amiable man ... truly religious, without an appearance of formal sanctity [and] ... kind to everyone" (Southwell, 1790: 711).

Given Dawes' conscientious objections, when a second party was sent out a few days after the return of the first (Tench, 1793: 98), again commanded by Tench, the Marine refused to join. Since Phillip did not, to his regret (Phillip, 1791), have sufficient authority to convene a court martial to try Dawes for insubordination, he charged the latter with "... unofficer-like behaviour ...", a charge he also notified Lord Grenville of in his letter of 17 November 1791. Dawes, in return, was "... exceedingly pained ..." that he was seen as having expressed "... anything either in word or manner in any degree improper or disrespectful" (Bladen et al., 1892: 545). However, Dawes' responses and behaviour were once again self-absorbed, rather than conducive to the colony's common good, and so it is no surprise that Phillip was frustrated by Dawes' public challenges to his authority as Governor.

The second punitive expedition came back empty-handed once again. In Tench's private journal we read,

> Our first expedition having so totally failed, the governor resolved to try the fate of a second; and the 'painful pre-eminence' again devolved on me. ... The orders under which I was commanded to act differing in no respect from the last, I resolved to try once more to surprise the village before-mentioned. And in order to deceive the natives, and prevent them from again frustrating our design by promulgating it, we feigned that our preparations were directed against Broken Bay; and that the man who had wounded the governor was the object of punishment. It was now also determined, being full moon, that our operations should be carried on in the night ...
>
> A little before sun-set on the evening of the $22^{d}$, we marched. Lieutenant Abbot, and ensign Prentice, of the New South Wales corps, were the two officers under my command, and with three serjeants [*sic*], three corporals, and thirty privates, completed the detachment. ... We rushed rapidly on, and nothing could succeed more exactly than the arrival of the several detachments. To our astonishment, however, we found not a single native at the huts; nor was a canoe to be seen on any part of the bay. ... Our final effort was made at half past one o'clock next morning; and after four hours toil, ended as those preceding it had done, in disappointment and vexation. At nine o'clock we returned to Sydney, to report our fruitless peregrination. (Tench, 1793: Chapter XII)

## 4.4 Moral Values

Dawes' moral convictions seem to have been complex and perhaps somewhat contradictory. Whereas he was disgusted by Phillip's order to capture and kill a number of Aboriginal men following McIntire's murder, at the same time he routinely served on the colony's court which dealt with transgressions by British settlers and convicts (Pybus, 2009). For instance, during one of his duty periods on the court, Thomas Orford (Halford), a desperately hungry black convict caught stealing potatoes from a

Marine officer's private garden, was sentenced to an inhumane 2000 lashes with the cat o' nine tails (Court of Criminal Jurisdiction, 1790)—a punishment that would almost certainly have led to the convict's death. Tench, usually also considered a man of laudable moral standards, was also on duty. He recorded the sentence as "… the melancholy lengths to which we were compelled to stretch our penal system …" (Flannery, 1996: 125) in the colony's starvation economy.

Phillip offered Dawes a means to show his goodwill in April 1791, when the Governor organised an exploration party that also included Tench, Collins, John White (1756–1832)—the First Fleet's principal surgeon—and Dawes (Clarke, 2015: 129), as well as the Aboriginal trackers Colebe and Balledery. They pushed into the wilderness east of Parramatta and into the Nepean–Hawkesbury confluence for several days (see Chap. 3). Despite valiant efforts, they became bogged down in an almost impassable landscape of escarpments and gorges. The colony's leading thinkers and decision makers quickly exhausted themselves and eventually returned without having achieved anything of note. Phillip himself never undertook another expedition into the wild side of the new colony.

Dawes' personality and his moral outlook thus caused him significant difficulties in the early colony of New South Wales, particularly when the time came for him to request an extension to his service. Governor Phillip appears to have been more lenient and open-minded than he was required to by the responsibilities of his authority. However, the relationship between Phillip and the Marine contingent was strained and tense. Phillip was thus forced to walk a fine line between insisting on discipline and causing a further escalation of the prevailing tensions. In his November 1791 letter to Lord Grenville, he explained that he had "… often found that the peculiar situation this colony has been in made it necessary to pass over improprieties which could not otherwise have passed unnoticed" (Clarke, 2015: 543).

And so, when he received Lord Grenville's direct request to retain Dawes as a member of the newly established New South Wales Corps, Phillip informed Dawes that his transgressions would be "… forgotten … provided he had seen his error …" and that he acknowledged his prior problematic conduct "… in such a manner as may leave no reason to suppose that anything similar will happen in future". Phillip was indeed keen to retain the Marine, as "… his services were wanted in surveying and marking out allotments of land for settlers" (Britton and Bladen, 1894: Sect. 13).

## 4.5 Homeward Bound

Nevertheless, a month later, on 18 December 1791, the first group of Marines, including Dawes, Tench, Commander Ross and the majority of the inaugural Marine detachment, embarked on H.M.S. *Gorgon* for the long voyage back to England (Bladen et al., 1892: 543–546). Dawes had, in fact, apologised for his behaviour and the implied disrepect, but he refused to admit that he was guilty as charged in relation to the other offences. In addition, Dawes was unwilling to accept an appointment as ensign, in essence a demotion, in the new hierarchy of the New South Wales Corps (Tench, 1926 [1826]).

The voyage home was reportedly uneventful. If Dawes wrote down any of his reminiscences, his writings have not survived. His last letter from this chapter of his life is dated 18 June 1792. It was written onboard the *Gorgon* and addressed to Maskelyne:

> Being at this time near Spithead, I write to inform you that the Instruments are all on board this Ship, having been order'd home by Governor Phillip in consequence of my being under the necessity of quitting the observatory on Point Maskelyne The Instruments which went out with us are not in my charge but in that of Capt.$^{n}$ Parker who commands this Ship. The Transit Instrument also which came out in the *Gorgon* is, at my request, included in the charge of Capt.$^{n}$ Parker … I should have excepted the Book & pocket Watch which remain with me as well as the new Sextant.
>
> I have made as many observations both on shore & on board as circumstances would permit tho' not so many as I wish'd. … I wish to have your opinion relative to the manner in which you would have me proceed. I think it would be proper to apply for leave to come to Greenwich as soon as circumstances will permit. Governor Phillip has thought proper to write against me, both to the Secretary of State and the Admiralty, so that I imagine it will take some little time to clear up that matter ...
>
> Altho' this first attempt has been so little successful, I do not yet despair of making some valuable observations in that (at present) dismal & distracted country, whenever a Chief Man may be appointed who is sincerely a lover & protector of scientific pursuits; had that been the case in the present instance, I most certainly ... should not have written to you, at this time, from so short a distance ... (Dawes, 1792; original emphasis)

Thus, most of the Board of Longitude-supplied instruments, with the exception of the new sextant and the pocket watch, had been placed in the care of Captain Parker. K1, the celebrated marine timekeeper, would be sent back later in 1792 on board H.M.A.T. *Supply* (Hawkins, 1979). In Chap. 1, we discussed the important role K1 played in the First Fleet's safe passage to New South Wales, first on H.M.S. *Sirius* and then on the *Supply*, when Dawes accidentally allowed it to run down. Following Dawes' careful recalibration at Sydney Cove, the timepiece made another circumnavigation of the globe on the *Sirius* during her voyage to Cape Town, via Cape Horn, from 2 October 1788 to 9 May 1789, which aimed at resupplying the settlement at Sydney Cove with desperately needed provisions (Howse, 1969: 197). It was still on board the *Sirius* when she shipwrecked off Norfolk Island in March 1790, but the timepiece was saved and transferred back to the *Supply*.

Dawes seems to have been torn by conflicting emotions on his return to England. On the one hand, he seemed to have been all too happy to draw a line under his trials at Sydney Cove. Soon after his return to Britain he sold the notebooks containing his annotations regarding the Gadigal language, drawn up during long and frequent sessions with his Aboriginal partner, Patyegarang, to William Marsden (Gibson, 2012: 273). Perhaps he needed the money, but the sale of this prized collection ensured that Dawes could not honour his agreement with Tench to jointly publish a full treatise on the Gadigal language.

On the other hand, from Dawes' final letter to Maskelyne it transpires that he was still hopeful to return to New South Wales at some point. He reportedly expressed that same desire on a visit to the Rev. John Newton (1725–1807) in July 1792 (Wood, 1924: 7–8). That wish never materialised, however, despite efforts to make it happen by a number of Dawes' supporters. As we will see in the next chapter, Dawes left for Sierra Leone by the end of 1792. Yet, in 1794, following Phillip's departure from Sydney Cove, William Wilberforce (1759–1833), the *de facto* leader of the movement to abolish the slave trade, recommended Dawes to the Home Secretary, Henry Dundas (1742–1811), as an eminently suitable first Superintendent of Education in New South Wales. Specifically, Wilberforce impressed on the Home Secretary that

> ... it would be highly desirable to send over some person to act as a general superintendent of all the schools which should be instituted. From past experience, I don't know where such an one could be found; but happily a man has fallen my way excellently qualified for what I reckon a very important trust; this is Mr Dawes. He went out to New South Wales with the original settlers, and was there between four and five years; he then came back to be married, meaning to return with his wife, and fix in New South Wales for the remainder of his life, employing himself in agriculture etc. (from purely benevolent motives), in the civilisation and instruction of the natives. We prevailed on him, however, to go out as Governor to Sierra Leone, where he has been till very lately, when the state of his health forced him away. It is impossible for me to speak in too high terms of his conduct in Africa. I don't believe there is in the world a more solid, honest, indefatigable man, more full of resources and common sense. Since he came back he called on me, and expressed his wish to put his former design into execution, as his constitution would not bear the weather of Sierra Leone. He meant to go out as a settler if he could meet with any encouragement ... I am confident that for the welfare of the settlement it would be well to give him all proper encouragement; at the same time he might act as the superintendent of schools. I mentioned the idea to him (not he to me), and he accorded, though not apparently very sanguine ab't it, and rather looking to the land which might be allowed him as his inducement for settling than to any appointment. Salary, I am sure, would be no object to him, but his being nominated to the office would [bind] him to attend to the object, and insure a consistency and a permanency of plans through any succession of Governors. Governor Hunter knows Mr Dawes well, and spoke of him to me in favourable terms ... I must add, and will then have done, that the expense of settling schools, Governor Hunter being directed accordingly, both for natives and for Europeans, and of establishing a superintendent, will, if you approve of it, be very trifling compared with the advantages which may follow—even the pecuniary advantages; for the more decently and orderly the colony will be maintained. (Bladen et al., 1892: 245–246; Wood, 1924: 8–10)

Wilberforce proceeded to write a similar letter to John King (1759–1830), Under-secretary at the Home Office. He added that he had asked Henry Thornton (1760–1815), Chairman of the Sierra Leone Company, to meet with King to discuss the matter of Dawes' return to New South Wales (Bladen et al., 1892: 245–247). It appears that Dawes did not receive any offer, or at least no offer that satisfied his family's needs and requirements, and so his wish to return to New South Wales remained just that, an unfulfilled desire. Wilberforce had indeed been correct in stating that then-Governor of New South Wales John Hunter had a favourable opinion of Dawes. In May 1798, Hunter advised William Henry Cavendish-Scott-Bentinck (1768–1854), Duke of Portland,

> I understand it was the intention of the Government to appoint an engineer, and that Lieutenant Dawes, then of the Marines, was the gentleman proposed on the occasion. If Mr Dawes

> could be found I shall be happy to have his assistance as an engineer, conceiving him to be eminently qualified; and your Grace may be satisfied, from the very extensive improvements in this colony, if such an appointment was deemed expedient in the early establishment, how far more necessary must it now be to take place. (Bladen et al., 1892: III, 391)

As far as we know, the Duke of Portland never responded, and so Dawes never returned to New South Wales. His service with the First Fleet had been his last engagement with the Marines, although he was still promoted to First Lieutenant on 18 April 1793. However, by that time he had already moved on to the next chapter in his life, as Governor of Sierra Leone.

Yet, his early return from New South Wales to England had left some loose ends. Maskelyne remained a staunch supporter of Dawes' work in the New South Wales colony. The Astronomer Royal eventually submitted a 'memorial' for consideration by the Board of Longitude at its meeting of 9 December 1799, seeking some level of compensation for Dawes' astronomical achievements (Board of Longitude, 1799). Dawes compiled a three-page manuscript listing more than 200 observations, sorted by date and type, starting from 19 February 1788 and continuing until shortly before his departure in late 1791 (Morrison & Barko, 2009). Dawes wrote to the Board of Longitude himself in December 1799, requesting additional compensation for his scientific work in New South Wales. He supported his request by pointing out that he had corrected errors in observations made on his voyage home on H.M.S. *Gorgon* that "… might have been attended with the most fatal consequences to the ship and all on board (Dawes, 1799; Board of Longitude, 1800a).

On 3 June 1795, the Board of Longitude had instructed Maskelyne to make five of Dawes' *folio manuscripts of Astronomical Observations made at Port Jackson and in the Voyage thither* available to William Wales, Master of the Royal Mathematical School at Christ's Hospital and Secretary of the Board of Longitude (Morrison & Barko, 2009). The Board of Longitude requested that Wales, in return for a generous stipend, would complete Dawes' calculations and prepare them for eventual publication. However, Wales' death precluded that task from being completed. None of Dawes' scientific observations were ever published, with the exception of his sixth journal, which covered his meteorological measurements.

Dawes eventually received his proper due for his scientific work:

> Received the 8 March 1800 of the Commissioners of Longitude by the hands of Mr [George] Gilpin [1754–1810; secretary of the Board] the sum of One Hundred Pounds, as an allowance for my Observations made in New South Wales. *William Dawes*. (Board of Longitude, 1800b)

In 1826, a second appeal, this time to the Colonial Office, for compensation for the surveying and engineering work Dawes had undertaken in New South Wales (Tench, 1926 [1826]), tasks that were well beyond his original job description, was rejected.

Dawes was clearly an interesting person, highly intellectual and cultured. His serious altercation with Governor Phillip halted what might have been a bright path forward as one of the fledgling colony's intellectual leaders. He apparently buried his astronomical interests with his departure from the colony (although his son, William Rutter Dawes, developed into a formidable astronomer in his own right; e.g., Denning, 1913).

## References


Barton, G. B. (1889). *History of New South Wales from the Records, vol. I: Governor Phillip.* Sydney, Charles Potter, pp. 1783–1789.

Bladen, F. M., Britton, A., & Cook, J. (1892). *Historical Records of New South Wales* (Vol. 2). Charles Potter.

Board of Longitude. (1799). *Confirmed minutes*. RGO 14/6, ff. 318–319.

Board of Longitude. (1800a). *Minutes*. RGO 14/6.

Board of Longitude. (1800b). *Accounts*. RGO 14/17, ff. 204r.

Britton, A., and Bladen, F.M. (ed.). (1894). *History of New South Wales from the Records. Vol. II: Phillip and Grose, 1789–1794.* Sydney, Charles Potter.

Cahill, R., & Irving, T. (2010). *Sydney: Places*. University of New South Wales Press.

Campbell, J. (1788). Letter to Robert Ross. Enclosing 'Officer's Objections'. *Historical Records of Australia, 1*, 92–94.

Clarke, R. M. (2015). *Vanguards of empire: the lives of William Dawes, Watkin Tench and George Worgan*. Ph.D. Thesis, The Australian National University.

Cobley, J. (1963). *Sydney Cove 1789–1790* (Vol. II). Angus and Robertson.

Coleman, D. (2005). *Romantic colonization and British anti-slavery*. Cambridge University Press.

Court of Criminal Jurisdiction. (1790). *Trial of Thomas Halford*. Minutes of the Court of Criminal Jurisdiction, 1147A. State Records of New South Wales.

Dawes, W. (1788a). Letter to Nevil Maskelyne, dated 30 April 1788. RGO 14/48, folio 281r.

Dawes, W. (1788b). Letter to Nevil Maskelyne, dated 17 November 1788. RGO 14/48, folio 291v–r.

Dawes, B. (1789a). Letter to Nevil Maskelyne, dated 30 May 1789. RGO 14/48, folios 294r–v.

Dawes, B. (1789b). Letter to Nevil Maskelyne, dated 9 June 1789. RGO 14/48, folios 295r–296r.

Dawes, B. (1789c). *Memorial*, dated 18 July 1789. In: Bladen et al. (1892: 423).

Dawes, W. (1790). Letter to Nevil Maskelyne, dated 26 July 1790. RGO 14/48, folios 302r–304r.

Dawes, W. (1792). Letter to Nevil Maskelyne, dated 18 June 1792. RGO 14/48, folio 307r.

Dawes, W. (1799). Letter to the Board of Longitude, dated 6 December 1799. RGO 14/11, folios 117–118.

Denning, W. F. (1913). The Rev. William Rutter Dawes. *The Observatory, 36*, 419–423.

Flannery, T. (Ed.). (1996). *1788: Comprising a Narrative of the Expedition to Botany Bay and a Complete Account of the Settlement at Port Jackson*. Text Publishing.

Gibson, R. (2012). *26 Views of the Starburst World. William Dawes at Sydney Cove 1788–91.* University of Western Australia Publishing.

Hawkins, J. B. (1979). Observatories in Australia 1788–1830. *The Australasian Antique Collector, 19*, 98–104.

Howse, D. (1969). Captain Cook's Marine Timekeepers. I: The Kendall Watches. In: D. Howse, & B. Hutchinson (Eds.). *The Clocks and Watches of Captain James Cook, 1769–1969.* Antiquarian Horological Society.

Hunter, J. (1793) *An historical journal of the transactions at Port Jackson and Norfolk Island, with the discoveries which have been made in New South Wales and in the Southern Ocean, since the publication of Phillip's voyage, compiled from the official papers; including the journals of Governors Phillip and King, and of Lieut. Ball; and the voyages from the first sailing of the Sirius in 1787, to the return of that ship's company to England in 1792.* London, John Stockdale.

Laurie, P. S. (1988). William Dawes, and Australia's first observatory. *Quarterly Journal of the Royal Astronomical Society, 29*, 469–482.

Moore, J. (1987). *The First Fleet Marines, 1786–1792.* University of Queensland Press.

Morrison, D., & Barko, I. (2009). Dagelet and Dawes: their meeting, their instruments and the first scientific experiments on Australian soil. *Historical Records of Australian Science, 20*, 1–40.

Phillip, A. (1789) *The voyage of Governor Phillip to Botany Bay; with an account of the establishment of the colonies of Botany Bay and Norfolk Island*. London, John Stockdale.

Phillip, A. (1791). Letter to Lord Grenville, dated 7 November 1791. *Historical Records of Australia, 1*, 290–294.

Pybus, C. (2009). 'Not fit for your protection or an honest man's company': A Transnational Perspective on the Saintly William Dawes. *History Australia, 6:1*, 12.1–12.7.

Saunders, S.D. (1990). *Astronomy in colonial New South Wales, 1788 to 1858*. Ph.D. Thesis, University of Sydney.

Southwell, D. (1790). Letter to Mrs Southwell, dated 27–30 July 1790. *Historical Records of New South Wales. 2*. Sydney, Charles Potter.

Tench, W. (1793). *A complete account of the settlement at Port Jackson. including an accurate description of the situation of the colony; of the natives; and of its natural productions*. In G. Nicol & J. Sewell (Eds.). London.

Tench, W. (1926). The memorial of Lieutenant William Dawes (Antigua, December 1, 1826). *Journal and Proceedings of the Royal Australian Historical Society*, *XII*, 227–228.

Wood, G. A. (1924). Lieutenant William Dawes and Captain Watkin Tench. *Journal of the Royal Australian Historical Society, 10*, 1–24.

Yarrow, S. (n.d.). *William Dawes. PocketOz: Pocket guide to Sydney*. http://www.visitsydneyaustralia.com.au/william-dawes.html. [Accessed 22 Sep 2022]

# Chapter 5
# Sierra Leone: Colonial Governance Against the Odds

## 5.1 The Sierra Leone Company

Upon his earlier-than-expected return to England in June 1792, Dawes does not appear to have reflected long on his aborted exploits in New South Wales. He returned armed with a letter of introduction from the Rev. Richard Johnson, the colony's Methodist chaplain, to the Rev. John Newton (1792). Dawes and Johnson had become close during their time in the settlement at Sydney Cove, with both professing a keen interest in 'improving' the local Aboriginal population.

Newton had meanwhile become an elder of the Evangelicals[1] in the Church of England. He maintained close links with other Evangelical clergy through the Eclectic Society, which had been established in 1783 "… for mutual religious intercourse and improvement, and for the investigation of religious truth" (Pratt, 1865: 1). Newton also acted as an important conduit for Dawes through his connections with like-minded leading laymen, including William Wilberforce—the parliamentary leader of the Abolitionist movement—and Henry Thornton (1760–1815). Wilberforce introduced Dawes to the Clapham Sect of Anglican reformers. He was keenly aware of Dawes' potential worth as Marine officer on half-pay who was "… the avowed friend of religion & good order in the midst of a dissolute and depraved Society" (Wilberforce, 1793), a clear reference to Dawes' conduct in New South Wales.

[1] Here, the capitalised spelling 'Evangelical' refers to the evangelical wing of the Church of England, and not to the general Protestant concept of seeking salvation through Bible studies.



R. de Grijs and A. Jacob, *William Dawes*, Springer Biographies,
https://doi.org/10.1007/978-3-031-38774-6_5

**Fig. 5.1** "Coast of Guinea and Bay of Sierra Liona, 1732." Attributed to Johannes 'Jan' Kip. (Public domain)

Granville Sharp (1735–1813), Wilberforce and their 'Abolitionist friends' had stood at the basis of the foundation of Sierra Leone[2] (Fig. 5.1) as the 'Province of Freedom', a home for freed black slaves from Britain's former North American colonies. A first attempt at establishing a free black settlement in Sierra Leone (see Fig. 5.2), fittingly named Granville Town, had occurred on 14 May 1787. The British Government, represented by proxy by the Committee for the Relief of the Black Poor, had intended to resettle the black poor of London in a new settlement founded on the principles of self-governance.

The Abolitionists' overarching aim was to prove that Africans could prosper and develop meaningful Christian lives, although still firmly guided by the colonial powers of the day. It was hoped that this 'experiment' would provide the Abolitionist movement with sufficient evidence to allow them to successfully argue for the abolition of the slave trade. They acted in the expectation that they would be able to show that former slaves could become hard-working Christians. However, the Sierra Leone enterprise initially came to nothing and was halted in 1789 (Peterson, 1969).

Around the time of his meeting with Newton, Dawes was approached by John (?) Lowes (Pearn, 1998), whom he knew as a former surgeon's mate on H.M.S. *Sirius*. Driven by religious fervour, Lowes had joined the fledgling Sierra Leone Company

[2] Note that the country name 'Sierra Leone' was formally adopted only on 5 July 1799. It replaced the designation 'Province of Freedom'.

**Fig. 5.2** "A view of the new settlement in the river at Sierra Leone", based on a drawing by the Abolitionist Carl Bernhard Wadstrom, ca. 1790. (British Library; public domain)

(Clarke, 2015: 52), "… a commercial [and] humanitarian undertaking …" (Thomas, 2013: 189). The Sierra Leone Company had been established on the foundations of Sharp's St. George's Bay Company. In June 1791, the British Parliament granted the Sierra Leone Company authority to establish a colony on the West African coast, with specific emphasis on trade and commerce—despite the dominance of Evangelicals among its leadership—specifically for

> … the Instruction of the Natives of Africa in religious and civil Knowledge; and, as a means to that Effect, they were desirous of establishing a fair Trade in the natural Produce of the Country, and, if possible, to put an end to the Slave Trade, which they understood was a great Obstacle to the Civilization of the Country. (House of Lords Sessional Papers, 1714–1805: 225; HLSP)

Several years later, in his testimony to the House of Lords on 17 June 1799, Dawes explained,

> The first Steps that were taken, I understand, to have been raising Subscriptions; procuring an Act of Incorporation; appointing a Governor and Council, with other Officers and Artificers; bringing between 1,100 and 1,200 Blacks from Nova Scotia, to settle on the Land which was procured for the Purpose of establishing the Colony. (HLSP: 226)

In essence, Sierra Leone would become a private colony incorporated by its own Act of Parliament, supported but not controlled by the British Government.

> The Quantity of Land purchased … [was] Twenty Miles square; namely, Twenty Miles directly South inland from the Coast, and Twenty Miles directly East nearly along the South Bank of the [Sierra Leone] River. (HLSP: 237)

As a privately funded settlement, the colony was meant to provide safe, secure and financially successful conditions for its investors, governors and colonists alike. Advice, although limited, lacking business acumen and incomplete (Clarke, 2015; Directors of the Sierra Leone Company, 1794: 15–16), had been obtained that the region was rich in arable land and provided good opportunities for trade (Clarke, 2015: 54; Directors of the Sierra Leone Company, 1791a: 2, 29; Matthews, 1788).

The Sierra Leone Company's directors made the case to the British Government for the mass resettlement of former slaves, runaways, mostly white loyalists (Fyfe, 1962: 31) and predominantly black soldiers who had supported the British cause during the American War of Independence (1775–1783). Most were Christians, and they were keen to own and develop their own parcels of land, two attributes that aligned well with the moral compass of the Sierra Leone Company. The British Government agreed to provide free passage to anyone who wanted to relocate from the former North American colonies to Sierra Leone. This was, at the same time, expected to relieve rising social tensions—mostly caused by competition for land grants—in the North American colonies. The first free black settlers, known as 'Nova Scotians', arrived in 1792 (see Figs. 5.3 and 5.4). They were former slaves and soldiers who had been displaced to Nova Scotia and New Brunswick (Canada).

The Company directors had taken note of Dawes' talents and achievements in Sydney Cove, prompting them to extend an invitation for Dawes to join the Sierra Leone Company. Dawes initially rebuffed the approach, explaining that he was too busy to travel from his barracks in Portsmouth to London. A week to ten days later,

**Fig. 5.3** Liberated slaves arriving in Sierra Leone. From *A System of School Geography Chiefly Derived from Malte-Brun* (1835), by Samuel Griswold Goodrich. (Public domain)

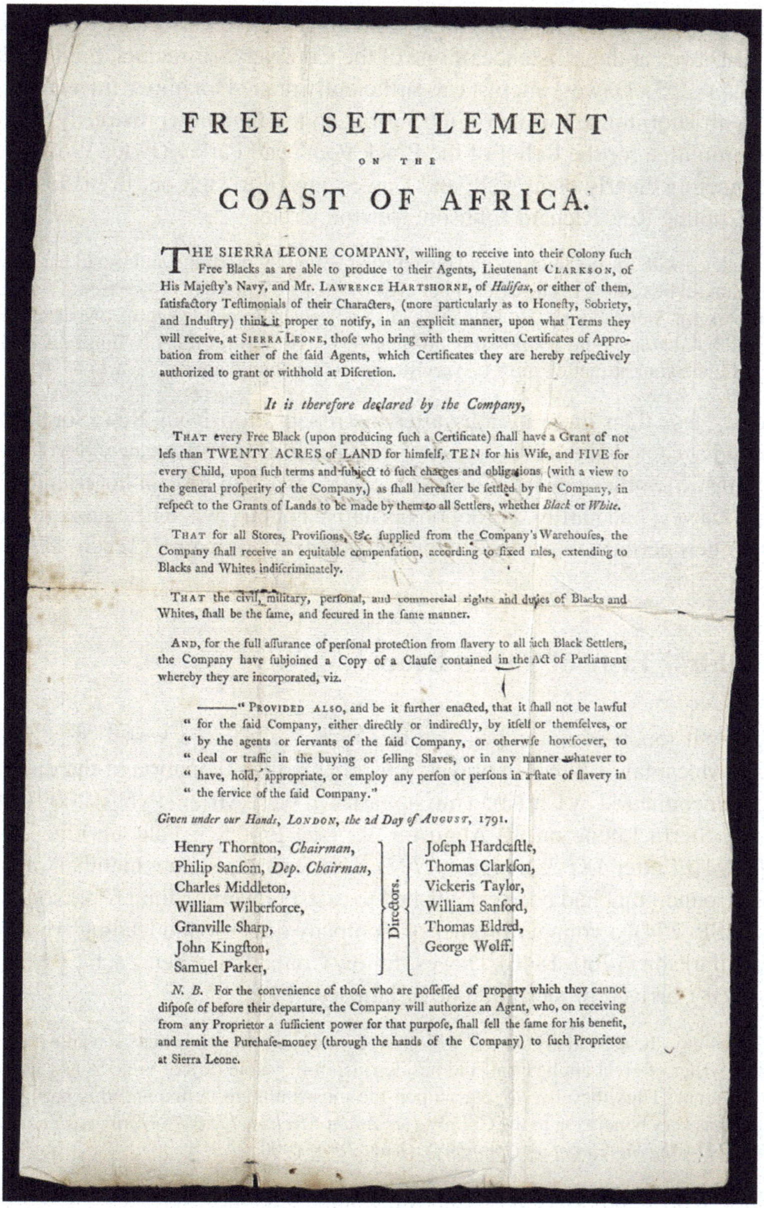

FREE SETTLEMENT

ON THE

COAST OF AFRICA.

THE SIERRA LEONE COMPANY, willing to receive into their Colony ſuch Free Blacks as are able to produce to their Agents, Lieutenant CLARKSON, of His Majeſty's Navy, and Mr. LAWRENCE HARTSHORNE, of *Halifax*, or either of them, ſatisfactory Teſtimonials of their Characters, (more particularly as to Honeſty, Sobriety, and Induſtry) think it proper to notify, in an explicit manner, upon what Terms they will receive, at SIERRA LEONE, thoſe who bring with them written Certificates of Approbation from either of the ſaid Agents, which Certificates they are hereby reſpectively authorized to grant or withhold at Diſcretion.

*It is therefore declared by the Company,*

THAT every Free Black (upon producing ſuch a Certificate) ſhall have a Grant of not leſs than TWENTY ACRES of LAND for himſelf, TEN for his Wife, and FIVE for every Child, upon ſuch terms and ſubject to ſuch charges and obligations, (with a view to the general proſperity of the Company,) as ſhall hereafter be ſettled by the Company, in reſpect to the Grants of Lands to be made by them to all Settlers, whether *Black* or *White*.

THAT for all Stores, Proviſions, &c. ſupplied from the Company's Warehouſes, the Company ſhall receive an equitable compenſation, according to fixed rules, extending to Blacks and Whites indiſcriminately.

THAT the civil, military, perſonal, and commercial rights and duties of Blacks and Whites, ſhall be the ſame, and ſecured in the ſame manner.

AND, for the full aſſurance of perſonal protection from ſlavery to all ſuch Black Settlers, the Company have ſubjoined a Copy of a Clauſe contained in the Act of Parliament whereby they are incorporated, viz.

——"PROVIDED ALSO, and be it further enacted, that it ſhall not be lawful
" for the ſaid Company, either directly or indirectly, by itſelf or themſelves, or
" by the agents or ſervants of the ſaid Company, or otherwiſe howſoever, to
" deal or traffic in the buying or ſelling Slaves, or in any manner whatever to
" have, hold, appropriate, or employ any perſon or perſons in a ſtate of ſlavery in
" the ſervice of the ſaid Company."

*Given under our Hands*, LONDON, *the 2d Day of* AUGUST, 1791.

| | | |
|---|---|---|
| Henry Thornton, *Chairman*,<br>Philip Sanſom, *Dep. Chairman*,<br>Charles Middleton,<br>William Wilberforce,<br>Granville Sharp,<br>John Kingſton,<br>Samuel Parker, | Directors. | Joſeph Hardcaſtle,<br>Thomas Clarkſon,<br>Vickeris Taylor,<br>William Sanford,<br>Thomas Eldred,<br>George Wolff. |

*N. B.* For the convenience of thoſe who are poſſeſſed of property which they cannot diſpoſe of before their departure, the Company will authorize an Agent, who, on receiving from any Proprietor a ſufficient power for that purpoſe, ſhall ſell the ſame for his benefit, and remit the Purchaſe-money (through the hands of the Company) to ſuch Proprietor at Sierra Leone.

**Fig. 5.4** Declaration of the Sierra Leone Company of their readiness to receive into their Colony certain 'Free Blacks', 2 August 1791. (Sleo.0001.0001.0006.0001, Sierra Leone Collection, Special Collections and University Archives, University of Illinois at Chicago; reproduced with permission) however, Lowes wrote to him once again, having travelled personally to Portsmouth to recruit Dawes at the insistence of one of the Company's directors, Samuel Parker (HLSP: 254–255). Dawes' interest was sufficiently piqued for him to travel to London to meet with Thornton, Chairman of the Sierra Leone Company (previously Chairman of the Committee for the Relief of the Black Poor) and Parker (Fyfe, 1962).

The meeting clearly went well, and Dawes agreed to sign on. In his letter of 17 July 1792 to the Rev. Richard Johnson, Newton wrote,

> He drank tea with me on Saturday and breakfasted yesterday morning, and I shall hardly see him again till he returns from Africa. I believe he will set off in a day or two from Plymouth to embark for Sierra Leone. He tells me he has not given up his design of revisiting New Holland but that he could not possibly go back yet ... However, I think if things go well at Sierra Leone your attraction must be very strong to draw him from it.

And so, less than three months after his repatriation from New South Wales, Dawes had accepted another overseas posting, keen to set off for new horizons. The speed with which these developments proceeded was not lost on his friend Watkin Tench: "[Dawes] had hardly set foot on his native country when he again quitted it to encounter new perils in the service of the Sierra Leone Company" (Tench, 1793: 201).

## 5.2 A First Taste of Sierra Leone

Having spent much of the voyage on the *York*, "... a large vessel of a thousand tons ..." (Macaulay, 1793a), in prayer, presumably in anticipation of the difficulties routinely encountered by Christian missionaries in West Africa (Pybus, 2009), Dawes arrived in Sierra Leone on 30 August 1792 (see Fig. 5.5; Falconbridge & Fyfe, 2000 [1794]: Letter IX, 25 August 1792).[3] He was the senior member of a two-member Council that had been established to assist Governor John Clarkson (1764–1828; HLSP: 254), a younger brother of Company director and leading Abolitionist Thomas Clarkson (1760–1846). Dawes' fellow Council member, Zachary Macaulay (1768–1838), arrived in Sierra Leone in January 1793:

> Mr. Macaulay, the second in Council, who arrived lately, is a very clever and sensible man, an elegant writer of great application, and besides just of the same turn of mind as Mr. Dawes, the Governor. Thus they live together upon the most intimate terms, and this harmony I consider as very beneficent to the Colony. (*Dr. Adam Afzelius, 1750–1837, Swedish botanist employed by the Sierra Leone Company*). (Knutsford, 1900: 22)

With Dawes' and Macaulay's arrival completed, the Sierra Leone governing council was reduced from its original seven-member composition to just three councilors. This reduction, Thornton had argued, was "... necessary for securing with tolerable certainty, that there shall be always one effective councilor" (Thornton, 1792).

[3] From his own recollection, Dawes tells the House of Lords Parliamentary Committee in June 1799 that he had arrived in the first week of September 1792 (HLSP: 225).

**Fig. 5.5** Engraved folding map in *An Account of the Colony of Sierra Leone, from its first establishment in 1793.* (London, James Philips; 1795). (Foreign, Commonwealth and Development Office Historical Collection, DT516 SIE; out of copyright)

Thornton had introduced Dawes to Clarkson, describing him as "… cool, correct and sensible …" (Coleman, 2005: 122, 185). The Company directors hoped that someone with Dawes' skills could help Clarkson overcome the difficulties the colony had found itself in lately and so help the Company retain control and, eventually, prosper (Clarke, 2015: 52). However, not everyone was convinced of the benefits of appointing a former Marine who had served in a penal colony to the role, given "… all the prejudices of [his] rigid military education …" (Falconbridge, 1794: 113).

Dawes arrived in a tense atmosphere. The Nova Scotians were frustrated by the lack of progress in getting their land allocations, a situation exacerbated by the apparent prioritisation of good-quality land grant allocations to the local Temne people, the largest ethnic group in Sierra Leone. On 21 May 1792, Clarkson warned his Council of the colonists' complaints about the lack of progress in 'Farm Lot' allocations. Yet, it took until 13 November of that year before land grants were allocated, initially to just 40 Nova Scotian families; more extensive action was not initiated until January 1793 (Public Record Office, PRO, 270/2: 21 May and 3 July 1792, 7 January 1793). Upon Dawes' arrival in August 1792, Clarkson had taken the blame for the colonial government's slow processing of land allocations: "I will take upon myself to scold and find fault, and his [Dawes'] business shall be to reward and praise" (de Hart, 1927: 31, 41, 43, 45).

To add insult to injury, the farm lots were by no means as generous as the colonists had been promised. In mid-1791, while still a naval lieutenant, Clarkson had travelled to Nova Scotia to recruit new settlers and organise their passage to Sierra Leone. On the Company's published terms, settlers would receive a fixed land grant of 20 acres for each man, ten acres for a wife and five for each child (Fyfe, 1962: 47). Importantly, the colonists would not have to pay anything, except for rates that would support the construction and maintenance of schools and provide for the sick and poor (Fyfe, 1962: 34). However, upon their arrival in Sierra Leone, Clarkson was only able to offer parcels of order one-fifth the size of those originally promised (Fyfe, 1962: 47), which further compounded the prevailing tensions. Dawes later testified to the House of Lords' Inquiry of June 1799,

> I believe that, on an Average, their Allotments do not exceed Six or Seven Acres; they have a Right to a greater Number; but, at their own Request, in consequence of the Difficulty and Time which it would have required to lay out the Whole of their Allotments at once, a Fifth Part only of the Proportion actually due to each has been yet laid out and granted to them. (HLSP: 302)

The colonists' agreement with the Sierra Leone Company was subject to burdensome requirements, which were spelt out in the fine print. They were given a 'Certificate of Character' that offered a land grant "... free of Expense" (Directors of the Sierra Leone Company, 1791b; see Fig. 5.6). However, the Company stipulated that the colonists were required to clear and cultivate the land for the Company's benefit:

> On the First Formation of the Colony almost the Whole of them were employed by the Company in clearing the Ground, erecting publick Buildings; and they employed themselves in erecting their own Habitations. In the more advanced State of the Colony they were employed in making Wharfs, improving the Landing Places, and other publick Services. (HLSP: 243)

A further stipulation imposed that their allotments "... shall be forfeited to the company ... [if] one-third of the said lots shall not be cleared and cultivated within two years ...". This was subsequently amended to require that two-thirds had to be cleared and cultivated by the third year. In addition, the contract set out that the colonists could access provisions "... on credit from the company's stores", but that credit had to be earned through produce sale to the Company (Directors of the Sierra Leone Company, 1791c).

When the first grants were allocated in November 1792, Clarkson immediately became the focal point of a barrage of complaints, for instance, that the lots were too

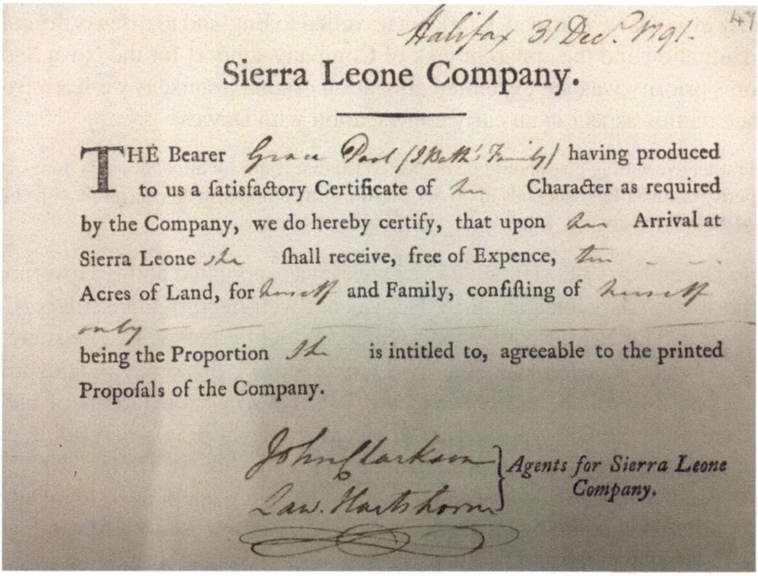

Halifax 31 Dec. 1791.

Sierra Leone Company.

THE Bearer Grace Pool having produced to us a satisfactory Certificate of her Character as required by the Company, we do hereby certify, that upon her Arrival at Sierra Leone she shall receive, free of Expence, ten Acres of Land, for herself and Family, consisting of herself only being the Proportion she is intitled to, agreeable to the printed Proposals of the Company.

John Clarkson
Law. Hartshorne
Agents for Sierra Leone Company.

**Fig. 5.6** Promises of land in Sierra Leone to Grace Pool. (British Library; Add. MS 41262 A, f.48; public domain, Creative Commons License)

rocky or too far away from the beneficiary's friends (see, for instance, Fig. 5.7 for a remote settlement allocated to the free colonists). Moreover, the Company directors intended to charge quit-rent, a land tax imposed to pay for services (Directors of the Sierra Leone Company, 1791a: 53).

**Fig. 5.7** Regent's Town, liberated settlement in Freetown, 1821. (Church Missionary Society; public domain)

Shortly after Dawes' arrival, Clarkson travelled to England to discuss the colony's strained situation and the apparent lack of Company support for the Nova Scotians. Clarkson's priority was the expedited allocation of land grants, as we learn from his insistence on this aspect in an early conversation with Dawes:

> I had rather everything else should be neglected till that [farm lot allocations] is done ... the prosperity of the colony depends upon our exertions in completing this desirable object. (de Hart, 1927: 16)

On 30 December 1792 (HLSP: 242), Dawes was appointed acting Governor with full authority, in Clarkson's absence. He was not well liked by the settlers (Clarke, 2015: 58–60). The Nova Scotians considered his public persona "… austere, [his] reserved conduct so reverse to the sweet manners of the other [Clarkson] they could not possibly relish, and consequently all hopes or expectations of the latter [Dawes] gaining popularity, proved abortive" (Macaulay, 1793a). They pointed out to Clarkson, "He may be a very good man, but he does not show it" (Pybus, 2009: 12.3).

Dawes, and his successor Macaulay, were much more adept at colonial administration than Clarkson had been. Dawes was reportedly pleased by Macaulay's "… firmness …", as opposed to what he perceived as Clarkson's "… harangues … upon almost all occasions …" (Kup, 1972: 220). Dawes implemented a regime where Council meetings were carefully minuted and department heads were required to provide regular progress reports. The Company directors hoped that the new regime might lead to a return on their investment in colonial affairs and that a range of trade articles they had shipped from England might finally be unpacked (Kup, 1972: 220).

Clarkson's paternalistic and caring attitude came to bite him in his dealings with the Company directors; they were concerned that Clarkson neglected the Company's interests by focusing on the colonists' welfare. Under Dawes' leadership, Macaulay recorded that the Europeans no longer associated with their African neighbours, noting with apparent approval that the local population had started to address him more respectfully as "My Father" (Macaulay, 1796: 6 October, 21 December). With Clarkson back in Britain, the Company directors decided that the time was ripe for Clarkson to be "… *dismissed* and … to be succeeded by Mr. Dawes" (original emphasis; Falconbridge & Fyfe, 2000 [1794]: Letter XII, 5 June 1793). Their decision was likely made easier by Dawes' reports that his firm leadership style yielded better results than Clarkson's approach of "… persuasion and promise" (Falconbridge & Fyfe, 2000 [1794]: 126).

Meanwhile, Macaulay had arrived in January 1793, following earlier employment "… in that most trying and dangerous of all occupations—the [slave] overseer of an estate in Jamaica" (Macaulay & Bedford, 1839: 5). The mere fact of his involvement in the slave trade led Anna Maria (Horwood) Falconbridge (1769–1835) to declare him unsuitable to assist in the governance of a colony "… founded on the principles of Freedom" (Falconbridge & Fyfe, 2000 [1794]: 108). Anna Maria Falconbridge had married Alexander Falconbridge (ca. 1760–1792), Abolitionist and former slave ship surgeon, at the young age of 19 in England. In 1791, Alexander Falconbridge was sent to Sierra Leone by the Anti-Slavery Society to reorganise the 'failed' settlement of freed slaves in Granville Town. Anna Maria twice accompanied her husband to the West African colony, where she became a correspondent of sorts, recording her experiences and impressions in a series of letters she eventually managed to have published. As a matter of fact, her vocal opposition to Macaulay's appointment was based on incorrect information. Macaulay had left Jamaica disgusted by his experiences there. His good intentions are reflected by the course of his later career. After his appointment in Sierra Leone, he became a leading member of the Church Missionary Society and long-time editor of the *Anti Slavery Monthly Reporter*. Anna Maria Falconbridge soon also became a thorn in Dawes' side, constantly criticising the Sierra Leone Company, those who worked for it and Dawes in particular, whom she called "... ridiculously despotic ..." (Falconbridge, 1794: Letter XII, 5 June 1793). She predicted that as soon as Dawes would take over the position of Governor, "... anarchy and discord ..." would ensue (Falconbridge, 1794: 116).

From June 1793, now formally appointed Governor of Sierra Leone, Dawes adjusted the colonial government's directives to better align with what he saw as the Sierra Leone Company's priorities: coastal commerce, the sale of wholesale goods to local traders, replacing the slave trade by the production of cash crops—cotton, sugar cane, tobacco, coffee, pineapples, pepper, coconuts and rice (Mouser, 1994)—and achieving self-sufficiency. Dawes elucidated on his objections to the slave trade in his later deposition to the House of Lords on 17 June 1799:

> I think it almost impossible that any considerable Progress should be made in communicating religious Knowledge, or in Civilization, while the Slave Trade prevails in that Neighbourhood; that I take to be the grand Obstacle to the Introduction of every Kind of useful Knowledge. ...
>
> The great Quantity of Rum which is usually given in Exchange for Slaves, and, as I have understood, in Presents to the native Chiefs, I take to be a great Obstacle to Civilization. The Natives in our Neighbourhood, who are mostly addicted to the Slave Trade, are in consequence of this Cause a very drunken, riotous, miserable Set of People. On our first arriving in the Country scarce a Day passed but several of them were seen drunk in the Town and Neighbourhood, in consequence of which frequent Quarrels and Affrays happened. ...
>
> I conceive it [cessation of the slave trade] would produce a very beneficial Effect, in that the Natives be excited to Industry, in order to procure those Articles of natural Produce, which it would be then necessary for them to raise, in order therewith to purchase those Articles of English Manufacture which they would want. (HLSP: 234–235)

Unfortunately, the region was not particularly wholesome to the European settlers. In June 1793, Macaulay accompanied Dawes on a visit to French-occupied Gambia Island,[4] some 30 km from Freetown, noting in his private journal,

> The French seem to live very wretchedly at present. We made them an offer of assistance; indeed, since the commencement of the disturbances [the Anglo–French war of 1793–1802], we have rather paid them more attention, and shown them more civilities, than before. The

[4] The Company considered purchasing Gambia Island, where a slave 'factory' was managed by Mr. C. Renaud (Renoux?), a Frenchman (referred to as "... one Rennieu ..." by Anna Maria Falconbridge). In early 1793, the Anglo–French war (a peripheral war of the French Revolutionary Wars) had broken out, and so Dawes was forced to tread carefully and diplomatically to maintain the Company's stance of neutrality (HLSP: 285ff; Crooks, 1903: 58; Fyfe, 1962: 59, 61; Mouser, 1994).

> soil is rich, but it is surrounded with swamps full of mangroves, consequently unhealthy. The Europeans there are very sickly. The buildings are mean. There is an open battery in front of them on which are mounted four four-pounders. (Crooks, 1903: 59)

Later, in his deposition to the House of Lords on 17 June 1799, Dawes expanded on the significant human loss they had suffered during the first year of the settlement,

> … many of the Company's Servants were long ill, and a considerable Proportion of them died during the first rainy Season; this was in a great Measure owing to the little Time the Settlers had to clear the Ground on which it was proposed to erect a Town; many of the Settlers also suffered severely from Sickness. … The chief Cause I take to have been the Rains coming on before sufficient Habitations could be provided. … many of them had died in the first Six Months which was previous to my Arrival; but, I believe, at least Two Thirds of the whole Number might have died, or have been obliged to return to Europe within the First Twelve Months. (HLSP: 228)

Farm lot allocations were delayed until a fort had been constructed at Freetown, a decision that followed from Dawes' previous experience in New South Wales. The Sierra Leone Company's surveyor, Richard Pepys (1762–1794; Armstrong, 2011), reversed a promise he had made under Clarkson's rule to begin surveying at the beginning of the New Year and instead worked with Dawes on the newly prioritised Freetown Fortress, "… plan[n]ed out upon a hill about a half a mile from the water side", which Anna Maria Falconbridge scathingly called "… the hobby-horse of Mr. Dawes …" (Falconbridge & Fyfe, 2000 [1794]: Letter X, 28 December 1792).[5]

In addition, Dawes and Macaulay redoubled their efforts to establish strong links with the local population to increase trade opportunities and acquire more extensive landholdings for agriculture (Clarke, 2015: 59). Although their expansionist tendencies had the potential to lead to conflicts with other stakeholders in the region, particularly those with vested interests on the Bullom Shore and islands in the Sierra Leone estuary, their efforts were clearly noticed and appreciated back in London. In 1794 the Company reported that Dawes and his Council had "… followed up the orders sent out on this subject with great spirit" (Directors of the Sierra Leone Company, 1794: 91), despite the significant loss incurred due to a fire on the store ship *York* on 30 November 1793 (Knutsford, 1900: 55):

> The transactions of this day I must communicate to you with no small regret, for it was marked by the most signal calamity that has yet befallen your Colony. Between nine and ten in the morning we were alarmed by the cry of fire, and on looking out beheld the *York* in a blaze. Before eleven o'clock the fire had completely ransacked her hold, and the loss is very great indeed. Upwards of four thousand pounds worth of African produce has been destroyed.

The Sierra Leone Company's focus on trade clashed with extant, long-term trade relationships in the region, particularly those involving the slave trade. Powerful and well-established slave trade operations had been in place in the region for decades prior to the Company's establishment of Sierra Leone as its main West African base.

[5] Isaac DuBois (b. ca. 1764), who would later marry Anna Maria Falconbridge and who maintained a personal disagreement with Pepys, called Dawes and Pepys "Fort Mad" (British Library, Clarkson Papers, MS Add. 41,263; Coleman, 2005: 164).

In addition, establishing a new colony on land that was already occupied by an Indigenous population was bound to result in clashing priorities and cultures and lead to competing interests. At the time of Dawes' arrival, the local Temne population was represented by King Naimbanna II (1720–1793; Falconbridge & Fyfe, 2000 [1794]: Letter IX, 25 August 1792). The King strongly defended the rights of his people (see also *AYV Newspaper*, 2020), claiming that they had tended the riverine jungles and coastal hinterlands for generations.

Dawes' refocused priorities did not go down well with the colonists either.[6] As a case in point, Anna Maria Falconbridge pointed out that Dawes had succeeded in "... displeasing the blacks and rendering them uneasy ..." (Falconbridge, 1794: 216; *Freetown Journal*, 1 January 1793). He nevertheless proceeded to reallocate the existing land grants, with any land alongside the waterfront reserved for Company use (see Fig. 5.8 for property boundaries in Freetown in 1791). This deprived the colonists from any means of communication and transport, which were all conducted on the water. They responded angrily, telling Dawes that

> ... when placed on the lots we at present occupy, we were informed, they were merely for our temporary accommodation, and we promised, when the plan of the town was fixed upon and surveyed we would remove, but we were assured no public or other buildings would be erected between our lots and the sea; now, in place of this, the sea shore is lined with buildings, therefore, your promise being broken, we consider ours cancelled, and will not remove unless the new lots are run from the water's edge, and we indiscriminately partake of them. Mr. Clarkson promised in Nova Scotia that no distinction should be made here between us and white men; we now claim this promise, we are free British subjects, and expect to be treated as such; we will not tamely submit to be trampled on any longer. Why are not our country allotments of land surveyed? Why are not all the Company's promises to us fulfilled? We have a high regard and respect for Mr. Clarkson, and firmly believe he would not have left us, without seeing every promise he made performed; if gentlemen here had not given him the strongest assurances they should be complied with immediately. (Macaulay, 1793a)

In response, Dawes' threatened to quit, hoping to get the colonists to comply out of a sense of solidarity. Clarkson had applied a similar strategy on a number of occasions, and successfully so. However, the colonists did not consider Dawes in the same light as Clarkson:

> Mr. Dawes met with a circumstance very galling to him this forenoon. He had in contemplation to palisade a piece of ground, for an immediate asylum, in case the natives should take it in their heads to attack us.
>
> The spot fixed upon, unfortunately took in part of a lot occupied by one of the Settlers, which, Mr. Dawes, conscious of his unpopularity, did not wish to encroach upon, without obtaining permission, although the Settlers only hold their present Town lots as a temporary accommodation, until their permanent ones are surveyed.
>
> He called on the tenant and took him out to explain what he wanted; many people in the neighbourhood having previously heard of Mr. Dawes's intentions, assembled about him, who declared they would not suffer an inch more ground to be enclosed, upon any pretence

[6] For instance, a letter to 'the Honourabel Company of Sirralone the Director', addressed to 'Mr. Granavil Sharp in England let this go with Care By the Beare', dated 10 May 1793, complains of Dawes' misuse of resources and accuses him of employing 'unfit' personnel (Kenneth Spencer Research Library, University of Kansas. MS P537).

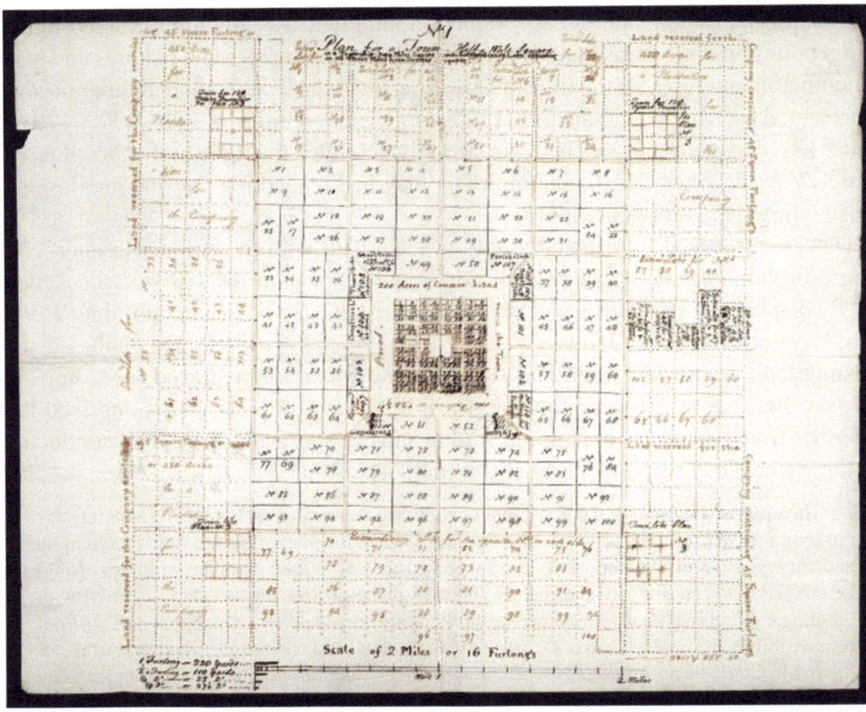


**Fig. 5.8** Plan of Freetown, 1791. (Sleo.0001.0001.0008.0001, Sierra Leone Collection, Special Collections and University Archives, University of Illinois at Chicago; reproduced with permission)

> whatever, before their town and country lots were given them, and most solemnly protested they would destroy every fence which might be erected till such time.
>
> Mr. Dawes endeavoured to persuade them by argument, what he wanted to do, was for their protection; but they were deaf to everything he said, and gave him language in return which he could not stomach: He told them if he had imagined they would have treated him with so much indignity, he should not have come among them: and is they continued to behave in the same way, he would certainly leave them as early as he could. (Macaulay, 1793a)

Their immediate response was, "Go! Go! Go! We do not want you here. We cannot get a worse [Governor] after you" (Falconbridge, 1794: 123). Others sought a confrontation, hinting that Dawes might suffer a similar fate as the French King Louis XVI, who had been executed by guillotine during the French Revolution (Fyfe, 1962: 51). Some even attempted a false raid to acquire guns Dawes suspected they might use against the colonial government (Pybus, 2006: 169).

Dawes never lived up to Clarkson's image in the eyes of the Nova Scotians. In comparison, the Nova Scotian Methodist minister Henry Beverhout called Dawes "Pharaoh"—a biblical reference to the oppression of Jewish slaves in ancient Egypt—to urge the locals to resist Dawes' leadership (Walker, 1992: 203). In contrast, Clarkson was like "… Mosis [*sic*] and Joshua was bringing the Children of Esaral [*sic*; Israel] to the promise[d] land" (Fyfe, 1991: 50). In a letter dated 19 October

1794, Freetown settler Luke Jordan (b. 1747) appealed to Clarkson on behalf of a number of other colonists to return to Sierra Leone, pointing out that "[i]n your Being here we wance [once] did call [the colony] Free Town, but since your absence we have a Raison to call it a Town of Slavery" (Fyfe, 1991: 43).

Under Dawes' rule as Governor of Sierra Leone, the colonists could only be employed by the Company, which in turn determined both the value of their labour and the price of any goods available in its stores. They considered the former too low and the latter artificially inflated. They also complained that the rum was watered down (Fyfe, 1962: 51), alleging that Dawes "… put thirty Gals. of water into a Peck of rum … & then [sold] it to us for a Shilling a Galln. more than we ever paid before" (Perkins & Anderson, 1793; Herrmann, 2019: 190). Anna Maria Falconbridge added that Dawes charged "… extortionate prices …" and exerted control over "… almost every kind of provisions in the neighbourhood" (Falconbridge, 1794: 105–106). Interestingly, a note added to this complaint adds deeper insights into Dawes' thinking:

> This is perfectly true, but upon investigation, it appeared to proceed from *religious* motives; Mr. Dawes said, he ordered a little water to be put into each puncheon, from a fear the consumers would neglect to dilute the spirit sufficiently. Had such a trick been played at a *Slave Factory*, how would it be construed? (Macaulay, 1793b; original emphasis)

The economic restrictions on the colonists imposed by Dawes, and previously by Clarkson, resulted in the most sustained pushback from the Nova Scotians. As early as December 1792 (PRO, 270/2, 31 December 1792: 66), Granville Sharp had proposed that the colonists be granted expanded political and legal rights. This would eventually allow them to elect representatives in a system that resembled colonial self-governance, thus creating a form of joint government. These black representatives were referred to as 'Hundred Men' or 'Hundredors' and 'Tythingmen': every ten householders formed a 'tithing' and every ten tithings formed a 'hundred'. Every ten freeholders were represented by a Tythingman; ten Tythingmen elected a Hundredor.

The Hundredors and Tythingmen jointly proposed new regulations, which were usually adopted by the colony's governing council (Fyfe, 1962: 16, 48). In response to Dawes' alleged price fixing, the Hundredors and Tythingmen asked the Council to set colony-wide prices for bread and meat, and to control alcohol distribution. In 1793, the Council approved standardised prices for beef, goat, pork and sheep mutton (PRO, 270/2, 4 July 1793: 77; Macaulay, 1794; PRO, 270/3, 23 August 1794: 5–6), while also legislating the ability of the local African population to sell meat to the colonists. The Council even commended the Hundredors and Tythingmen as "… highly proper & expedient …" when they proposed fining anyone convicted of selling liquor or wine without a license (PRO, 270/3, 12 October 1795: 230–233).

Nevertheless, and despite these small victories, the colonists claimed that their lives under Dawes were little better than bondage (Clarke, 2015: 59). In addition, Dawes fell out with a specific faction who felt that their (Christian) religious needs had been neglected (Clarke, 2015: 60). They proceeded by aggressively recruiting most of the recent immigrants. The situation in Sierra Leone further deteriorated with the arrival of two new cohorts of settlers in 1792. Yet, Dawes, and by extension the Company, could not be swayed to amend the strict rules governing the colonists' lives.

Relations with the Nova Scotians soon reached an all-time low. Dawes held firm to serving the Company's interests rather than making concessions to benefit the colonists' welfare. The only concession made was an offer to the colonists of a return passage aboard a Company ship in February or March 1794 (Falconbridge, 1794: 143–144). As a result of the tense stand-off, Dawes became ever more isolated, a situation that was further compounded by his poor health and the tedious administrative workload.

In November 1793 the colonists drew up a petition to Wilberforce and other Company directors to dismiss Dawes from his gubernatorial post and reinstate Clarkson:

> Mr. Dawes seems to wish to rule us just as bad as if we were all slaves which we cannot bear … [and] we are afraid concerning the happiness of our children for as we have not [had] justice shewn us we do not expect our children will after us. (Falconbridge, 1794: 143–144)

The petition was delivered, in person, to the Company's full Council by two of the colonists, Cato Perkins (d. 1805) and Isaac Anderson. The agitators' provocative action had unintended consequences. The Council concluded that "… the Nova Scotians may be said to furnish a less favourable specimen of the character to be expected in emancipated slaves, than may be commonly hoped for in other cases" (Sierra Leone Company Report, 1795: 88). They proceeded to deport most of the most vocal opponents of Dawes' governance from the colony (*Ibid.*: 31, 81).

Dawes' difficulties were worsened by the vocal opposition to the Company's policies and Dawes' governance by Anna Maria Falconbridge, particularly after her publication of a hefty tome full of complaints and criticism (Falconbridge, 1794). She clearly had an axe to grind with the Sierra Leone Company following the summary dismissal of her husband by the Company's governors owing to alcohol problems (Clarke, 2015: 58, 59), and his subsequent death. Despite being the widow of a deceased Company employee, Falconbridge was not awarded any compensation following her husband's death. Her second husband, Isaac DuBois (1764–1818), soon also ran into difficulties with the Sierra Leone Company.

Anna Maria Falconbridge did not hesitate to state her dislike of Dawes. She suggested that while in a penal colony, referring to his previous posting, "… no doubt it is necessary for gentlemen to observe an awful severity in their looks and actions …" (Falconbridge & Fyfe, 2000 [1794]: 100), she clearly did not appreciate Dawes' attitude in the free colony of Sierra Leone. Indeed, she declared, "We are in great tribulation about Mr. Clarkson's going away, for Mr. Dawes is almost universally disliked" (*Ibid.*: 104). However, the Sierra Leone Company continued to offer Dawes their full support. Dawes interpreted this support as confirmation that his governing style was on track to meet his brief. In a letter of 2 August 1794 to Henry Dundas (1742–1811), then Secretary of War, Wilberforce lavishly praised Dawes' achievements and conduct:

> It is impossible for me to speak in too high terms of his conduct in Africa. I don't believe there is in the world a more solid, honest, indefatigable man more full of resources and common-sense. (Wilberforce, 1794)

However, forced by intensifying bouts of fever, in late March 1794 Dawes returned to England on the Company's vessel *Harpy*, on leave of absence to regain his health and to take care of "... Private Affairs in England requiring my Presence" (HLSP: 174). Governing the colony was left to Macaulay, who had formally been appointed as 'Second in Council'. On 29 May, Dawes married Judith Rutter in Portsea, Hampshire, but further particulars of that period of Dawes' life in England are unknown. His first child, named Judith after his wife, was born in September 1795, by which time he had returned to Sierra Leone.

## 5.3 A Second Chance to Make a Difference

Although he had been promoted *in absentia* to First Lieutenant of the Marines,[7] on 18 April 1793, Dawes was placed on the Half Pay List on 13 December 1794. He found life difficult on his meagre Marine Corps salary, and so he sought Wilberforce's assistance in his efforts to return to New South Wales. As we saw at the end of the previous chapter, despite the goodwill of a number of his ardent supporters, these efforts did not result in a new opportunity to resume his former life in New South Wales.

Instead, Dawes returned to Sierra Leone on 6 May 1795, leaving the wife he had married just a year earlier behind in England. Upon his arrival on 31 May,

> I was taken ill of a Fever on the last Day of May, immediately upon my Arrival at Sierra Leone the Second Time, which proved to be a severe Fit; I recovered from that, and during the rest of my Residence in Africa I enjoyed remarkable good Health. (HLSP: 265)

His second stint as Governor of Sierra Leone would only last until March 1796.[8] Upon his return to the West African colony, he arrived at a settlement that had been almost completely destroyed, in September 1794, during an unprovoked raid by the French—a "... ragged, lawless set of rascals ..." (Coleman, 2005: 127), according to Lowes.

A report by the Company directors to its owners clearly laid out the damage, with the contents of all Company buildings and the Freetown church completely devastated. Added to the destruction, a Company vessel that had arrived from England during this time was also taken by the French, complete with its cargo of trade goods and supplies. The Company directors estimated that the total damage amounted to

---

[7] According to the official records, "... the Adjutant-General, Royal Marines, London, under date of September 27, 1926, writes: ... promoted 1st Lieutenant, R.M., 18th April, 1793; ..." (Wright, 1927). Note, however, that the Marines did not become 'Royal Marines' (R.M.) until 1802.

[8] Dawes stated that he was in government "[f]rom June 1793, having acted as Governor from the 30$^{th}$ of December previous to that Time, to my final Departure from it in April 1796, including the Year I was absent on Leave" (HLSP: 242).

losses of £40,000 (Directors of the Sierra Leone Company, 1795: 27). To add insult to injury, a significant fraction of the settlers interpreted the French raid, opportunistically supported by two American slave traders, as retribution for the allegedly poor treatment they had received from the Sierra Leone Company (Clarke, 2015: 61).

Upon hearing of a "… tumult in the colony …", in essence a brief uprising by the Nova Scotians in response to the French raid, Dawes volunteered to return to Sierra Leone, motivated by "… his zeal in the cause" (*Ibid.*).

> The Directors wished me to return there in consequence of Tumults having taken place in the Colony; in consequence of which they wished the Hands of the Government there to be strengthened; and I understood, that having been there before, they deemed it more proper that I should return, than that they should engage with any Person who had not been there before.
>
> I saw the Remains of I believe every Publick Building which had been erected, and which I understood the French had destroyed by Fire, including the Company's Store House, which had cost a very considerable Sum in erecting, and which at the Time of its Destruction, or rather immediately preceding the Time of its Destruction, had contained Goods to a very considerable Amount; which Goods I understood to have been taken by the French, and converted to their own Use, or at least the much greater Part of them. (HLSP: 233)

By the time Dawes had returned to the colony, the uprising had subsided and Dawes decided that it was not worth his time and effort to find out the source of the colonists' dissatisfaction (HLSP: 175–176). This clearly reflects his single-mindedness of character. However, given his apparent disdain for the Nova Scotians, it is perhaps not surprising that he felt that no further investigation was warranted once the matter of contention had apparently been settled.

The French raid also led to the replacement of the colony's original silver currency of dollars, half-dollars, twenty and ten-cent pieces, and a copper currency of cents, by a paper currency of dollars, half-dollars, shillings and sixpence to pay for goods and services and exchange for labour. Nevertheless, the availability of any currency remained an issue of importance to Dawes, as is obvious from the following extract from an appeal made to the Company directors on 13 July 1795:

> We would not recommend your sending any more Dollars while the risk of sending them here and preserving them afterwards remains so great. The wear of our paper dollars is very slow, and of course the making a number as occasion requires to keep a sufficient sum in circulation is not attended with much loss of time or other inconveniences:– but, it is far otherwise respecting shillings and sixpences. Governor Dawes has often with extreme reluctance been absolutely obliged to give up one, two, or more days to the making of shillings and sixpences, when business of importance has rendered it very desirable that his time might have been employed in a different way. We therefore deem it extremely expedient that a number of sixpences or ten cent pieces to the value of about 40 or 50£ should be sent by the first opportunity—and these may be either silver or paper as the Court of Directors may think proper—only in the latter case two blanks for the [illegible] (one on each side) and room for two signatures at the bottom must be left, and it will require a much larger sum than above mentioned to allow for wear. (Vice, 1983: 28)

Meanwhile, the devastation caused by the French attack, combined with his deteriorating health, forced Macaulay to return to England. Dawes returned to a colony in need of rebuilding that was even more hostile to him than before he had left in March of the previous year. A number of factors contributed to the increased tension. First, the colonists interpreted the French attack as retribution on the Company for their allegedly poor treatment. Second, a number of American slave traders had seen the French attack as an opportunity to seek revenge for the Company's negative impact on their business (Clarke, 2015: 61).

In response to the colonists' hardships, Dawes adjusted his style of governance. He endeavoured to improve their food supply, streamline agricultural practices and he introduced monetary incentives ('premiums') for improvements in land clearing and cultivation,[9] and in cattle breeding (Sierra Leone Council, 19 May 1795). In addition, the Sierra Leone Company provided stores and equipment in support of the reconstruction effort.

From a practical perspective, particularly to source local supplies, Dawes was forced to walk a fine line to maintain a working relationship with the large slave trading company at Bance (Bunce) Island, a short distance (~30 km) upriver from Freetown (see Fig. 5.9). At times, this required a liberal interpretation of the Sierra Leone Company Act. As a case in point, at a time when the colony was short of supplies, the captain of a French ship was prepared to provide supplies, but he would only accept payment in slaves. Dawes attempted to circumvent the prohibition of trading in slaves imposed on the Freetown settlement by channeling his purchase through the Bance Island 'factor' (agent), John Tilley (ca. 1768–1813). Obviously, this upset the colonists, adding fuel to Anna Maria Falconbridge's incessant criticism of Dawes' governance (Falconbridge, 1794: 106). She was particularly scathing about Dawes' 'dubious morality', accusing him of covertly engaging in the slave trade, thereby pointing at his dealings with Tilley (Falconbridge, 1794: 113, 116; Pybus, 2009: 12.3).

The nigh impossible demands of the job, once again combined with his deteriorating health, eventually led Dawes to resign as Governor in March 1796 and return to England on the Bristol-based trading vessel *Mohawk* (HLSP: 271). The gubernatorial authority fell to Macaulay once again. From Macaulay's letter of 17 May 1796 to his fiancée Selena (Selina) Anne Mills (1767–1831), who remained in England, we gain further first-hand insights into Dawes' character and personality:

> Dawes and [Dr. Thomas Masterman] Winterbottom [1766–1859; physician to the Sierra Leone colony] will come to see you as soon as they arrive in England, and our secret will be safe with them. Dawes is one of the excellent of the earth. With great sweetness of disposition and self-command, he possesses the most unbending principles. For upwards of three years have we acted together, and in that time many difficult cases arose for our decision; yet I am not sure that in the perplexities of consultation or the warmth of discussion, we either uttered an unkind word, or cast an unkind look at each other.
>
> You will be surprised, after saying so much, to hear me declare that I believe I love Winterbottom still more. He certainly is far from having attained the distinguished eminence of Dawes. But there is so much warmth of affection in him, and the expression of it so often bursts from him involuntarily, that one is constrained to love him. Besides, I was accustomed

[9] Dawes estimated that "An exceeding small Proportion, perhaps not more than the One Hundredth Part" of the coastal area immediately adjacent to the colony's settlement had been cultivated by the local population prior to the arrival of the Sierra Leone Company (HLSP: 230).

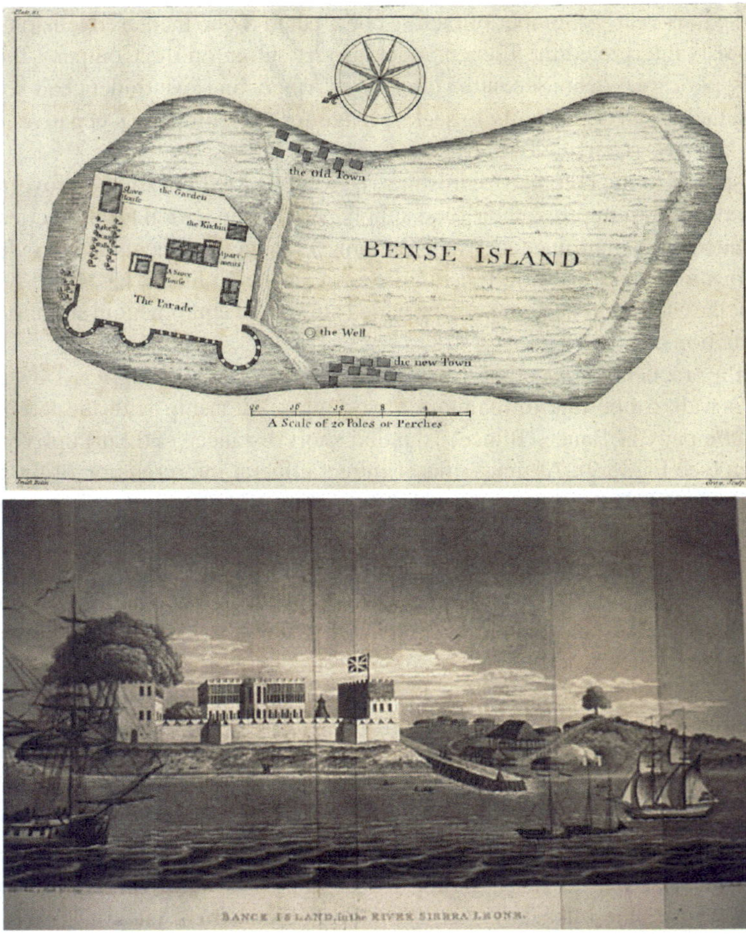


**Fig. 5.9** (*Top*) Map of Bunce Island, ca. 1727. (Public domain); (*Bottom*) Bunce Island in 1805, during the period the slave factory was run by John and Alexander Anderson. (Public domain)

to look up to Dawes for advice and assistance, while Winterbottom, on the contrary, has been in the use of looking to me for counsel, and you know how naturally a dependence of that sort begets in a degree all the tender feelings of parents to children. He is a man of general science and great professional skill. He is also a serious inquirer after truth; and though his religious principles are by no means so fixed as those of Dawes, he may be fairly reckoned a disciple of Christ. (Macalay [*sic*], 1796)

## 5.4 Master of the Royal Mathematical School

Back in England, Dawes initially lived on half-pay. On 25 January 1799, at a meeting of the Council of Governors of Christ's Hospital, he was appointed Master of the Royal Mathematical School (established in 1673; see Fig. 5.10; Plumley, 1976), as successor of the previous, recently deceased Master, William Wales (d. 29 December 1798). Although there were four candidates for the post, Dawes was the overwhelming favourite. This may have been facilitated by the support and recommendations he appears to have received from a number of political and scientific heavyweights, including Henry and Samuel Thornton, Nevil Maskelyne and Joseph Banks (Council of Governors of Christ's Hospital, 1799).

**Fig. 5.10** The Royal Mathematical School at Christ's Hospital, from a view published by N. Smith in 1793. Image extracted from *Old and New London, Illustrated* (1873; Vol. 1, p. 973), by Walter Thornbury. (British Library, HMNTS 010349.l.1; public domain)

Shortly afterwards, on 19 March 1799, his son, William Rutter Dawes, was born (Royal Astronomical Society, 1868).[10] Dawes Sr. appears to have boarded in the Wales' household. This must have been uncomfortable for all involved, although it allowed him access to his old journals, from his posting in New South Wales. Dawes reported that William Wales' widow was rather cool and unwelcoming to him, possibly because her late husband had not been paid for his work. In any case, Dawes appears to have been less than impressed by Wales' contributions to the reanalysis and completion of his astronomical observations from 1787–1791 (Dawes, 1799). He also informed Maskelyne, in a letter dated 5 December 1799, that he wanted his old manuscripts to be collected by George Gilpin (1754–1810), clerk of the Royal Society of London, who would soon be appointed as Secretary of the Board of Longitude (Morrison & Barko, 2009: 33, note 125). For unknown reasons, Maskelyne declined Dawes' request.

Dawes' new and distinguished role was intense, but he was well-placed as Master given the wealth of practical experience he brought to the appointment. The School had been established to prepare young boys for a life at sea in both the Royal Navy and the country's merchant marine, given "... the publick importance it appeared to be to the Crowne, that a Nursery might on this occasion be Erected of Children to be educated in Mathematicks for the particular Use and Service of Navigacon [*sic*]" (Plumley, 1976: 52). There would be up to 40 pupils at any given time, who would be taught "... the art of Arithmatique and Navigacon ..." until age 16. Any boy admitted to the School should "... have attained a competent skill in Grammar and common Arithmatique as far as the Rule of Three" (a writing technique based on the principle that any ideas, thoughts, events, characters or sentences that are presented in threes are most effective). This represented a significant hurdle. Each year, Dawes was required to make ten boys available for service at sea, once they had been duly examined at Trinity House and passed their examination (Kennerley & Seymour, 2000).

In addition, the Royal Mathematical School was dedicated to teaching mathematics to boys from the Hospital's Grammar School, to help them prepare for university. The School's mathematics curriculum included arithmetic; navigation, including Euclidian geometry and trigonometry; calculations of the times of tides; longitude calculations using the lunar distance method, observations of the Sun and by means of marine timepieces; as well the use of navigational instruments.[11] Dawes was initially paid an annual salary of £170 to meet these requirements, which was increased to £200 in January 1800 (Council of Governors of Christ's Hospital, 1800).

[10] Note that Clarke (2015) suggests that William Rutter Dawes was born in December 1797, whereas Dawes' Wikipedia page includes an explicit date of 17 December 1797, further stating that he was baptised in Portsea, Hampshire, on 11 February 1798; see https://en.wikipedia.org/wiki/William_Dawes_(British_Marines_officer) and the Parish registers for St. Mary's Church, Portsea (1666–1875). However, Wright (1927) suggested a later birth date: "... Miss Ella Jones, a great granddaughter of Lieutenant William Dawes, ... dated November 17 last [1926], ... states:— Mr. Dawes married Miss Rutter (my grandmother) some time before 1799, as in that year his only son was born".

[11] Royal Mathematical School Regulations for Management. Christ's Hospital archives, MS 12878/002.

Despite his generous remuneration and distinguished appointment, Dawes resigned after just two years as Master of the Royal Mathematical School. In his resignation letter of 6 November 1800 to the School's Board of Governors he only referred to "... a very peculiar and pressing occurrence ...", which had brought about "... very uncommon circumstances ..." that required him to resign (Dawes, 1800). It is possible that he alluded to the death of his wife Judith three months earlier, just two weeks after their third child, John Macaulay Dawes, had been born, in July 1800 (Clarke, 2015: 104). John Macaulay died in infancy, in February 1801. Dawes had meanwhile returned to Sierra Leone for his third term as the colony's Governor.

## 5.5 Witness to a House of Lords' Inquiry

Meanwhile, in June 1799 Dawes had been called away from his ailing wife and his duties as Master of the Royal Mathematical School to appear before the Slavery Committee. This committee had been convened as part of a House of Lords' Inquiry into the proposed Slave Trade Limitation Bill, which aimed at abolishing the slave trade along part of the West African coast. Its proponents, most notably Wilberforce and his fellow Abolitionists, hoped that by focusing on a relatively small region, they might sway the Lords to approve an interim restriction and encounter only modest opposition. They were fully aware that the wholesale abolition of slavery was as yet unachievable.

Both Dawes and Macaulay had been summoned to appear before the Lords. They discussed their strategy over dinner with Wilberforce prior to their formal evidence sessions (Wilberforce, 1799a). Dawes was the first to appear. He was cross-examined for three full days from 17 June 1799 (House of Lords Papers, 28 May 1799). The Bill's opponents questioned Dawes relentlessly for extended periods, firing probing questions at him about most aspects of the colony's trading environment. Nevertheless, Dawes firmly maintained that abolition of the slave trade in the region would lead to further trading opportunities for the Sierra Leone Company (HLSP: 151). Macaulay's keen observations of the proceedings are reflected in his letter of 20 June to his fiancée, Selena Mills:

> My examination comes on Monday [24 June 1799]; and I can perceive from the mode of proceeding with respect to Dawes that the Duke of Clarence [later King William IV] is disposed to make it as harassing as he can. Dawes' examination occupied three days, and his extreme want of recollection gave our opponents some advantage over him. But his simplicity, integrity and coolness were very striking and I think must have impressed the House as forming a strong contrast to the loose, rash, hasty manner in which the witnesses on the other side delivered themselves. It seems to be a main object ... to prove us visionaries, Dissenters, and Democrats. (Macaulay, 1799)

The House of Lords' proceedings also provided clear evidence of Dawes' lack of political awareness and an unwillingness to compromise, similar characteristics we already saw in the context of his squabbles with Governor Phillip in New South

Wales. Yet, Wilberforce recorded "... Dawes's evidence middling, but the lawyers charmed with his honesty ..." (Wilberforce, 1799b). Macaulay's letter to his fiancée offers some additional insights into Dawes' qualities and shortcomings (Gibson, 2012: 7):

- his coolness;
- his extreme focus on details in the present moment and his habitual self-absorption;
- his tenacity;
- his vagueness or complete lack of concern around issues that do not fall within his own definition of things worth considering;
- his inability to read the politics of a situation and to finesse the complexities for his own quick benefit.

Macaulay appeared before the House of Lords a week later. His precise and courteous evidence provided more insights into the colony's dealings than Dawes had offered while hiding behind his characteristically evasive, brief and uninterested demeanour. The opposing Counsel had become rather irritated by Dawes' apparent lack of attention to pertinent detail or, perhaps, his deliberately evasive answers. In particular, Dawes could not recall the population of the colony, nor the proportion of Nova Scotian colonists to locals; he could not quantify the size of land being cultivated by the Company, nor the quantity of rice produced during his period as the colony's Governor, nor even whether the value of the Company's exports to Britain covered the expenditure incurred (HLSP: 161–163). Dawes considered such issues the remit of the Company's commercial agent.

> The Governor and Council established a Plantation on the Bulam [Bullom] Side, on the North Side of the River. I think, in the Season immediately preceding my last Return to England, there were Sixteen Acres in Cultivation there, on the North Side of the River. The Company did not make any considerable Attempts at Cultivation on their own Account; it was our Plan rather to stimulate the Settlers to the Cultivation of their Farms. (*Ibid*.: 291)

The only aspect about which Dawes expressed a forceful opinion was his disdain for the Nova Scotian settlers (Clarke, 2015: 66–67). Dawes' attitude was provoked by the colonists' perceived lack of work ethic. They had fallen short of their expected contribution to repay the Company for its investment in their resettlement. Dawes stated that while he was aware of the colonists' complaints, these resulted from "... their ignorance and misapprehension of the measures which were taken with them ..." by the Sierra Leone Company (HLSP: 159, 176).

Given the strength of the opposition to the proposed Bill, led by Lord William Wyndham Grenville (1759–1834), leader of the House of Lords and Foreign Secretary, it is no surprise that it failed when put to a vote in early July 1799. In a two-hour speech, the Duke of Clarence (Prince William, third son of King George III) convinced the majority of the Lords of the potential negative economic impact of the *total* abolition of the slave trade on Britain's treasury income, with most of the lost proceeds—including £3 million in trade with the West Indies—going to other countries (*The Times*, 6 July 1799).

**Fig. 5.11** "A view of Freetown, 1803". Published in *An account of the native Africans in the neighbourhood of Sierra Leone, to which is added an account of the present state of medicine among them* by Thomas Masterman Winterbottom (London, C. Whittingham; 1803; public domain)

Ultimately, the Lords' Inquiry into the Slave Trade was decided by the politics of factions, ideologies and the interests of business groups. Although some members of the House of Lords were genuinely concerned about the inhumane conditions that prevailed in the slave trade, others were more worried about losing the competitive (trade) edge with respect to rival nations. Personal interests in wealth accumulation and religious motivations to save African souls also contributed to the volatile political mix in the House of Lords (Gibson, 2012).

## 5.6 Third Time Not So Lucky?

Following his resignation in November 1800 as Master of the Royal Mathematical School, Dawes' financial position deteriorated rapidly, and so he leant once again on Wilberforce (Pybus, 2009: 12.4). The latter managed to get him re-appointed as Governor of Sierra Leone, where he returned on 4 January 1801, accompanied by his nephew, whom we only know as "Mstr. Wm. Dawes".[12] Dawes had sent his son to stay with his father in Portsmouth (Laurie, 1988). This time, Dawes remained in the colony until February 1803 (Fig. 5.11 is a representation of Freetown in 1803). In the mean time, the Chief Administrator post had been filled first by Macaulay and, briefly, by Thomas Ludlam (1775–1810; see also Leach, 2018).

Ludlam governed from May 1799 until after 10 December 1800 (Sierra Leone Council, 1800), when he resigned following the suppression of a full-scale rebellion

[12] Note that the Sierra Leone Council minutes of 6 June 1801 suggest that Dawes and his nephew arrived instead on 6 January 1801.

**Fig. 5.12** Trelawney town, residence of the Maroons. (Public domain)

by the arrival, in September 1800, of 550 'Maroons' and a detachment of soldiers. This second wave of settlers, known as Maroons, came from Jamaica—where they had been engaged in a fierce guerilla war with the British colonial government—by way of Nova Scotia (Turner, 1997). They were more disciplined, contented and industrious than the earlier colonists, whom Wilberforce described as "… the worst possible subjects, as thorough Jacobins as if they had been trained and educated in Paris" (Fyfe, 1962: 87). Figure 5.12 shows the residential area allocated to the Maroons.

Nevertheless, traumatised by overseeing the court martial and banishment of 31 rebels and the execution of two of their leaders, Ludlam offered his resignation in the rebellion's aftermath. Dawes was asked to step in and take up the reins once again. Yet, despite the Maroons' arrival, Dawes' time in the colony remained troublesome. With the Nova Scotians somewhat subdued, tensions with the local Temne people about the Company's land rights and the construction of Dawes' fortifications soon boiled over. They launched an attack on Freetown and its Fort in November 1801, killing a number of colonists and wounding many more, including Dawes. Dawes was wounded on "… shoulder and breast …", but he was reported to be "… already recovered or likely to recover" (Committee on the Petition of the Court of Directors of the Sierra Leone Company, 1800–1807: f. 16). Meanwhile,

> ... Miss Ella Jones, a great grand-daughter of Lieutenant William Dawes, ... [on] November 17 last [1926], … states:—
>
> During Mr. Dawes' tenure of office [at Sierra Leone] he strongly opposed the Slave Trade, which aroused the ill-will of the traders and the uncivilised tribes. A skirmish was the result, in which Mr. Dawes was wounded in the leg, and incapacitated for many a long day. This prevented his acceptance [in 1802] of administering the Government of the Seychelles. (Wright, 1927: 64)

But luck was on the side of the colonists, as a passing British war ship was about to quell the unrest with the help of a group of newly arrived settlers (Clarke, 2015: 69–70; Fyfe, 1962: 90).

It appears that Dawes attempted to restore cordial relations with the local Indigenous population. Figure 5.13 is a letter of credentials in both Arabic (in West African handwriting) and English, in Dawes' handwriting, for one Surakhata of Sendugo to "... the Head man of the Scarcies Rivers". The original document is contained in the collection of the National Library of Australia.

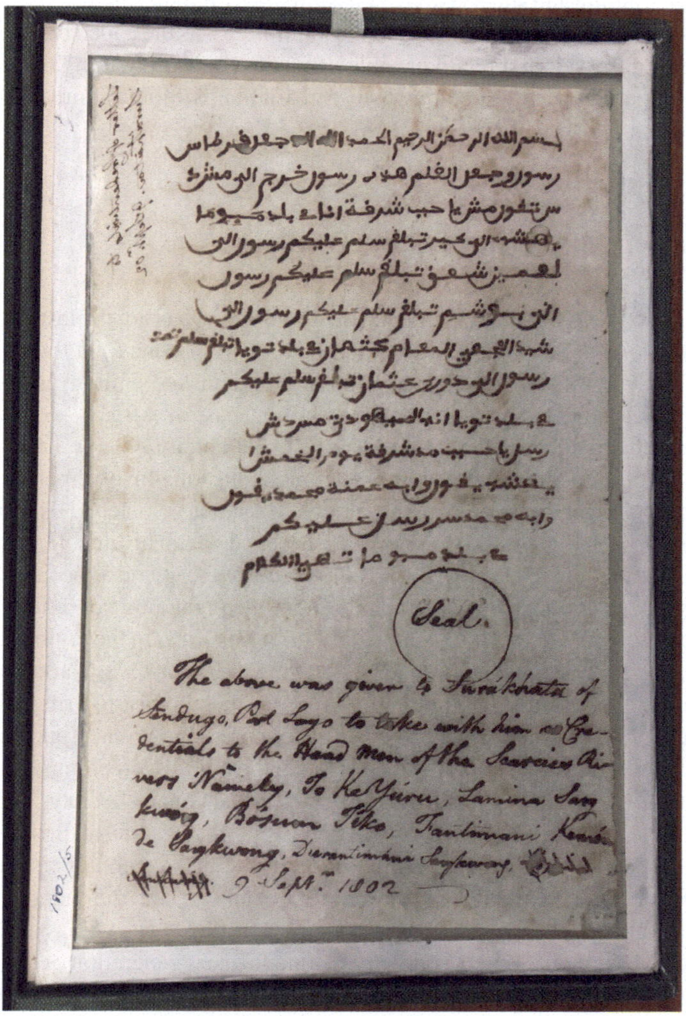
Seal.

The above was given to Surákhata of Sendugo, Port Logo to take with him as Credentials to the Head man of the Scarcies Rivers Namely, To Ke Yúru, Lamina Sangkwóng, Bósman Tiko, Fantimani Kemé de Sangkwong, Dinantimáni Sangkwong.

9 Septr. 1802

**Fig. 5.13** Dawes' letter of credentials to one Surakhata of Sendugo to "... the Head man of the Scarcies Rivers". (Photo: Author supplied; National Library of Australia. MS 4067; reproduced with permission)

Dawes stayed in Sierra Leone only until 15 February 1803, when he returned to England and handed over the reins to Post Captain William Day, R.N. (d. 1805/6?).[13] Day left Freetown for England for war service just six months into his tenure. The gubernatorial role reverted to Ludlam, whose duties as Governor were somewhat challenged by the active presence of an informal Committee of Correspondence composed of four representatives that Dawes had established in 1802 at the request of the Church Missionary Society. Dawes was one of the inaugural members (Church Missionary Society, 1802). The Sierra Leone Committee was meant to resemble the Society's London-based Committee of Correspondence, which was tasked with the recruitment, training and supervision of missionaries. Ludlam's main concern about a Freetown-based counterpart of the London Committee was the added burden it would place on him to recruit and retain capable officials who were willing to tolerate the poor living conditions in the colony. Most of Ludlam's officials already occupied multiple roles in the colony (Leach, 2018).

## 5.7 Church Missionary Society Matters

Although Dawes was bound for England once again, this was not the last we will hear about his exploits in Sierra Leone. He returned to Sierra Leone for a fourth and final time on 21 July 1808, but this time as Commissioner tasked with investigating 'the State of the Settlements and Governments on the Coast of Africa'—a plush senior appointment that paid a princely salary of £1500 per annum (Clarke, 2015: 75).

Dawes' appointment was the natural outcome of his long involvement in colonial governance and his association with the Church Missionary Society. His insights into the missionary work undertaken in West Africa were deemed invaluable, and so when he returned to England in December 1804, he was asked to attend a special meeting of the Society's Committee to provide details about the prevailing conditions in Sierra Leone. He observed that the "… Mahometans …" (Muslims) in the region comprised "… the zealous …" and, more commonly, "… the indifferent …" (Clarke, 2015: 71). Dawes recommended specifically that missionaries be allowed to bring their wives along, both for moral support and to encourage more local women and children to attend church services (Church Missionary Society, 1804). His recommendation was adopted at the Committee's next meeting (Church Missionary Society, 1805).

Given his aptitude in languages—including "… the Arabic of Susoo …" (the language spoken by the local population in Sierra Leone), Arabic, Persian, the "… vulgar language of the Hindus …" (Dawes, 1806c) and "… the rudiments …" of five European languages (Dawes, 1806d)—in October 1806 Dawes was asked to head a new Church Missionary Society seminary to train the next generation of missionaries at Bledlow, some 20–25 km east of Oxford. He would teach his students Arabic and Susu (for destinations in Africa) or Persian and Hindi (to those aiming for India).

[13] Some sources (e.g., Crooks, 1903: 61; Currer-Jones, 1930: 66) suggest that Dawes stayed in the colony for a short time to focus on laying out and allocating land grants.

He would also teach them tropical medicine, astronomy, mathematics, 'mechanical arts', surveying and 'building good, plain, comfortable houses'.

Dawes had been recommended for the role by the Rev. Thomas Scott (1747–1821), who considered his own accommodation too small to host a class of trainee missionaries. Scott knew that Dawes, his son, William Rutter, and his nephew, William, had meanwhile moved from his London address in South Lambeth (Mander-Jones, 1966) into an old mansion at Bledlow, Buckinghamshire, that was owned by Robert Smith (1752–1838), first Baron Carrington and former director of the Sierra Leone Company. Scott suggested that Dawes' experience in Sierra Leone, combined with his "... exceedingly frugal and hardy habits ..." made him an excellent choice for the Society's purposes (Scott, 1806).

Dawes was keen to take on this new role (Dawes, 1806a, 1806b), partly to educate his "... little boy and my nephew ..." (Dawes, 1806d) to prepare them for appointments in the church and, perhaps, in part driven by his unenviable financial situation. However, he expressed concern (Dawes, 1806a) that Lord Carrington was unable to guarantee a long-term lease of the type needed for a seminary to thrive and become sustainable. Figure 5.14 is an example of the type of instruction materials provided to trainee missionaries destined for West Africa.

Dawes considered Bledlow a suitable site for a distraction-free seminary given its remote location, and he wrote to Pratt that he would accept any terms considered reasonable by Scott and the Rev. Nathaniel Gilbert (1761–1807) (Dawes, 1806c). Gilbert had been the first clergyman appointed in Sierra Leone, prior to Dawes' first arrival in the colony. He resided in Bledlow and would become an important ally of Dawes, introducing him to his later life in Antigua (see Chap. 6).

Despite Lord Carrington's noncommittal attitude to the seminary's lease arrangements, Dawes readily agreed to the terms negotiated for the Society by Scott and Gilbert (Dawes, 1806e). Lord Carrington eventually resolved the uncertainty surrounding the lease in 1807, as we learn from Dawes' letter to the Society's Committee announcing that he would most likely be required to leave the premises by Michaelmas (late September) of 1808 (Dawes, 1807). However, the numbers

HOUSES AT SIERRA-LEONE.

**Fig. 5.14** "Houses at Sierra-Leone, from The Wesleyan Juvenile Offering: A Miscellany of Missionary Information for Young Persons (1844–1878)". (Public domain)

of candidate missionaries at the Bledlow seminary fell far short of expectations. Combined with delays in converting his remaining assets in Sierra Leone into cash, this left Dawes in significant financial strife (*Ibid.*).

By the time Dawes returned to Sierra Leone as newly appointed Commissioner, the seminary had all but died a quiet death, and its remaining trainees were transferred to nearby Aston Sandford under Scott's guidance. William Rutter Dawes was also transferred into the care of Scott. This represented yet another upheaval in the boy's young life,[14] following a period at his paternal grandfather's in Newton, Portsmouth, after his mother's death, while his father was based in Sierra Leone. This time, William Rutter was left in the care of Scott, since his grandfather had meanwhile passed away. It is not known what happened to Dawes' oldest daughter, Judith, during this period; we first find out about her family involvement when she travels with the Daweses to Antigua in 1813 (see Chap. 6).

## 5.8 Sustained Persecution

On 1 January 1808 the Sierra Leone Company handed over the financially struggling colony's full control to the British Crown. Wilberforce was, once again, instrumental in securing Dawes' lucrative venture as Commissioner to investigate the state of the British colonies on the West African coast. He was offered the appointment in April 1808 and arrived in the colony shortly afterwards, on 21 July. To understand the context of what happened next, it is important to highlight that on 25 March 1807, the 'Act for the Abolition of the Slave Trade'—introduced by Wilberforce and Lord Grenville—had received Royal assent, thereby formally terminating Britain's involvement in the slave trade.

Dawes set sail on the same ship as Thomas Perronet Thompson (1783–1869), the British Crown's first Governor. At the young age of just 26, in August 1808 Thompson took over the colony's reins from Ludlam, his interim predecessor who had governed since 1805. Thompson had been recommended to wait half a year and orient himself prior to assuming his new role, but Dawes and Ludlam convinced him to take charge almost immediately upon his arrival in the colony (Clarke, 2015: 76; Thomas, 2013: 190). One of Ludlam's final official decisions was to appoint Dawes as the colony's engineer, just one day before ceding his authority to Thompson (*Sierra Leone Gazette*, 18 August 1808).

However, already in December 1808, the decision was made to recall Thompson because he had made a number of "... very strongly worded ..." and unsupported allegations to the Foreign Secretary, Robert Stewart (1769–1822), second Marquess of Londonderry (better known as Lord Castlereagh), against "... agents of the

[14] Despite these upheavals, Dawes' decision to leave William Rutter in England allowed his son to have a full education, including a period at Charterhouse School. William Rutter chose to study medicine at St. Bartholomew's Hospital, although he eventually pursued a career as an astronomer (Clerke, 1885–1900).

Sierra Leone Company …", *viz.* Ludlam, Dawes and the directors of the Company (Turner, 1997: 335). Thompson's termination became official in mid-August 1809. In February 1810, he passed the baton to Edward Henry Columbine (1763–1811; Turner, 1997: 326, 327, 329), a Governor supported by the African Institution.

Dawes' relationship with Thompson had already soured in 1808, on the voyage from England to Sierra Leone, but it turned truly toxic shortly after their arrival. During the seven-week voyage, Thompson questioned Dawes about the use of so-called 'apprenticeships' in Sierra Leone, which he had heard about from Macaulay prior to his departure for the colony. He was greatly concerned that these apprenticeships were nothing but another term for forced labour in the form of 'domestic slaves', indentured servants or involuntary military recruits in the Royal African Corps (Stewart, 2013: 141–142), while some "… women or girls were selected for the basest of purposes …" (Thorpe, 1815a), as concubines or sex slaves.

This prompted Thompson to express his alarm in a letter to his fiancée shortly after his arrival in Freetown: "… you may have heard me speak of Macaulay's apprenticeships; and it is as I suspected, that these apprenticeships have … introduced actual slavery" (Thompson, 1808b; see also Turner, 1997: 335). It transpired that during their governorships, both Dawes and Ludlam had purchased African workers, whom they worked without proper pay, hired out and pursued when they tried to escape (Thorpe, 1815b: 26–29). Thompson soon concluded that Dawes' reference to apprenticeships was really a euphemism for continued enslavement, a practice that continued in the colony even after the 1807 Act for the Abolition of the Slave Trade had become law across the British Empire.

Thompson uncovered evidence that not a single apprentice was ever set free (Pybus, 2009). As a consequence, he accused Dawes of continuing the slave trade under false pretenses, warning Wilberforce that Dawes was "… not fit for your protection or an honest man's company" (Thompson, 1808c). This was a clear escalation from the note on Dawes' character he had made in his journal just two days after their arrival from England: "… shall keep a sharp eye on Mr. D" (Thompson, 1808a). However, from Thompson's perspective, the conspiracy ran much deeper, as we learn from the letter of 23 July 1808 he wrote to his fiancée: "Messers Wilberforce Thornton and Co. have at last become slave traders with a vengeance in their old age" (Thompson, 1808b). Thompson appears to have been consumed by the need to expose what he considered hidden slavery that was apparently openly condoned by the Sierra Leone Company.

Thompson would later claim that Dawes had said, "I have always thought slavery necessary in the colony, I think as still" (Thompson, 1808a). However, Dawes maintained that he strongly objected to that characterisation (Dawes, 1810). Whether or not Dawes made the statement, perhaps in a context that has meanwhile been lost, his attitudes and behaviour throughout his life support his anti-slavery views. Nevertheless, Thompson made it clearly known that he was convinced that Ludlam and Dawes "… endeavour[ed] with all reasonable dispatch to establish a monopoly of the slave-trade on the very foundation of the Abolition Act" (*The African Herald*, 26 August 1809).

Yet, theory and practice do not always get along easily. Where Britain had formally abolished the slave trade, other countries stepped into the void and expanded their trading networks. Most slaves captured from foreign ships were relegated to apprenticeships in the colony (Fyfe, 1962: 115). Thompson considered this an excuse by the Sierra Leone Company to continue the slave trade and an abuse of the anti-slavery act.

Dawes' relationship with Thompson quickly deteriorated even further. An early confrontation between the men, in August 1808, involved a number of slaves imprisoned on a captured Swedish vessel. Ludlam, Dawes and three other Company employees attempted to convince the slaves to escape from captivity. The Sierra Leone Court of Vice-Admiralty, solely represented by Thompson, found that 18 of the indentured men had been purchased from a trader on the River Pongas, in British territory (*The African Herald*, 26 August 1809). These 18 slaves were confiscated, but the remainder of the enslaved cargo was given the 'all clear' to proceed to their next destination, to the great consternation of Dawes and his Abolitionist allies.

The Abolitionists requested a writ of *habeas corpus*—at the time their only recourse to challenge illegal detention in court—to allow the remaining slaves to be removed from the ship. Thompson was incensed by Dawes' "... daring and fraudulent attempt to defeat the decision and to injure a neutral ... [rather than] do justice upon neutrals and not to plunder them", accusing unnamed Government officials—that is, Ludlam and Dawes—of "... conduct so disgraceful to a civilised nation and so contrary to the good understanding which exists between our Sovereigns" (Thompson, 1809b; *The African Herald*, 26 August 1809). Although Dawes would have acted illegally if he had managed to remove the remaining slaves from the Swedish vessel, his actions underscore his determination to act on his beliefs, irrespective of the political consequences.

The new Governor became meanwhile horrified even more upon learning that Dawes may have been involved in the fairly common practice of infanticide of children born from illicit encounters between British settlers and local women (Clarke, 2015: 79). Around the same time, Anna Maria Falconbridge's public exposure of Dawes' alleged 'moral turpitude' became more widely known (Pybus, 2009). Despite her accusation, Dawes' behaviour was more likely morally ambivalent. He is known to have made good-faith efforts to curb morally reprehensible excesses in the colony, including sexual promiscuity, alcohol abuse, corrupt practices and complicity in the slave trade, although these efforts mostly targeted the population of black colonists, leaving the white officials relatively free rein (Leach, 2018: 87).

Dawes' and Macaulay's world views were shaped by their Christian faith, and so they interpreted the prevailing colonial excesses in those terms. Given this context, they formally allowed colonists to divorce if adultery could be proven, they dismissed Company employees who behaved morally reprehensibly, they only recognised marriages that had been sanctioned by an ordained priest or by the Governor himself (black preachers were known to encourage bigamy) and they even tried to enforce rules that required fathers to take responsibility for their illegitimate children (PRO, 270/2, 1 May 1794 [ff. 154–158], 8 July 1796 [ff. 44–45], 1 May 1806 [f. 44]).

Nevertheless, Thompson's opinion of Dawes as a supporter of infanticide was cemented by an event involving a single local woman, Anne Edmonds, which had occurred in the colony when Dawes was still in England. Edmonds had a strict father, David Edmonds. She was accused of having committed infanticide following the birth of a child resulting from an intimate encounter with Thomas Robson, formerly a surgeon employed by the Sierra Leone Company (Thomas, 2013: 194; Clarke, 2015: 85). However, she was acquitted for lack of evidence. The jury found that while she "… has born a child and has concealed its birth … it is not proved that the child found [in the garden of a Mrs Hannah Bennett] is hers or that she murdered it" (Thomas, 2013: 192, 195; Thompson, 1809a).

Yet, Thompson (1809a) alleged that Ludlam, Dawes and others—including the Freetown Mayor and storekeeper, Alexander Smith, a former Company employee—had conspired "… to white-wash the character of this unfortunate woman and to bury the recollection of the murder of the illegitimate offspring of a European for ever". In other words, "… the men who ought to have discovered the guilty were the first to join in an endeavour to conceal them". He described Anne as "… a reputed murderess & a convicted whore …".

Thompson became obsessed with the need to expose what he saw as questionable and immoral behaviours practised by the Sierra Leone Company, in particular by Dawes and Ludlam. He hence decided to re-try Anne Edmonds, who had meanwhile married John Morgan on 16 February 1809 and taken her husband's surname to become Anne Morgan. She was charged under an ancient and obscure law, in part of being "… moved and seduced by the instigation of the devil …" (cited by Thomas, 2013: 194) and for concealing the death of the child herself or with the assistance of others. The trial was held on 29 March 1809 and presided over by Thompson himself, in his role as the colony's magistrate. Anne was convicted and sentenced to death, although her sentence was commuted on 3 April 1809, on the verge of execution, to expulsion from British territory.

During Thompson's stint as Governor, he came across the Edmonds family once again. Anne Morgan's mother, also called Anne Edmonds, was found to give shelter to an approximately 12-year-old runaway female apprentice known as Fenda. Under the prevailing property rights, Thompson advised Anne Edmonds, whom he clearly disliked,[15] that the girl had to be returned to her rightful owner, a Mr. Botefeur. However, he soon learnt that Dawes had come to the family's support (Thompson, 1809c). In a deposition to Governor Columbine at a public hearing in April 1810, Dawes clarified that he was resolved "… to prevent the poor girl from being unjustly and cruelly reduced a second time to a state of slavery …" (Columbine, 1810b)—thus suggesting that apprenticeships were, in fact, equivalent to 'a state of slavery'. In fact, he intended to take the matter public, suggesting that

> … the most effective method to put a stop to so disgraceful & iniquitous a transaction, …, would be, to take some step, which would infallibly force the whole transaction into public notice, and by strongly attracting the eyes of the whole Colony towards the helpless object of these base designs; to place an insurmountable obstacle in the way of their accomplishment. (Columbine, 1810b)

---

[15] Thompson accused her of being "… a woman of infamous character and reputation …', allegedly involved in the infanticide of some of her daughters' children by "… different Europeans in the service of the Sierra Leone Company" (Clarke, 2015: 79–80).

Thompson saw this episode as further confirmation of his opinion of Dawes as morally bankrupt, provocative and intent on undermining his authority: "… you & others are … continually abetting & encouraging every kind of disaffection & insubordination in the Colony" (Dawes, 1809). Thompson had had enough of Dawes and his interference, declaring him the

> … evil maintainer of a Woman of infamous character and reputation being by common report hold [*sic*] and supposed to have been concerned in privately causing the deaths of three of the children of her own daughters, begotten by different Europeans in the service of the Sierra Leone Company & being otherwise a woman of exceeding[ly] ill fame. (Sierra Leone Council, 1809)

Understandably, Dawes strongly objected to Thompson's accusations, insisting that he should be judged on "… the purity of his views and of his past life and actions" (*Ibid.*).[16] Recent scholarship has re-ignited a heated debate as regards Dawes' character. His evangelical worldview and "… undeviating rectitude and unbending firmness of principle …", in Macaulay's words (Knutsford, 1900: 135, 188), combined with his self-confessed "… well meant plainness which I am accustomed to use …" (Dawes, 1796), reflected his rather austere personality. However, some of the most serious allegations, accusations of him having committed unthinkable crimes (Pybus, 2009; Thompson, 1810), not just in New South Wales but throughout his life, have been thoroughly refuted (Atkins, 2009; Clarke, 2015: 84–85; Thomas, 2013).

Nevertheless, Thompson felt compelled to document all of the colony's perceived ills, moral shortcomings and instances of dereliction of duty, particularly those committed by "… agents of the Sierra Leone Company" (Thompson, 1808d). He sent a lengthy and detailed report to Lord Castlereagh in late 1808, as well as a series of letters that became more strongly worded and alarmed as time went by (Wilberforce, 1808). Thompson's accusations included a claim that he had unearthed evidenced that 168 local Africans had been "… illegally sold or disposed … [with the] consent and active cooperation [of the Sierra Leone Company] … and of the Directors of that Company in England" (Thompson, 1808d). Although it is clear that Thompson aimed this latter accusation at Dawes, who was in England at the time the alleged offences were committed, he clearly overreached his remit, angering powerful political opponents in the British establishment by his accusations. Whereas many of his accusations may have been valid (Sierra Leone Council, 1809), most were, in retrospect, based on hearsay and suspicion, rumour, gossip and innuendo rather than hard evidence (Clarke, 2015: 80).

Thompson relied, in part, on letters from Macaulay, Secretary of the Sierra Leone Company, to Governor Ludlam. In those letters, Macaulay wrote that he had "… always been of the opinion [that a] likely means of promoting civilisation in the colony would be by indenting the natives for 7 years or [until] age 21, … [whose duties should be] well defined and rigidly supervised" (Macaulay, 1807). Thompson considered this type of indenture equivalent to slavery, although Henry Thornton

[16] Transcript of a public meeting held on 8 April 1810 to hear the various accusations against Dawes and Ludlam. Kew, UK National Archives. (Colonial Office) 267/25.

made a clear distinction, with the proviso that those in authority must be careful in preventing slave-like conditions (Thornton, 1808).

In addition, Thompson levelled unfounded allegations and "... charges sent home to be laid before ..." the Privy Council

> ... that infant murders and the procuring of abortion among the women of Colour with whom the European Servants of the Sierra Leone Company were connected, were systematically encouraged by the highest servants of the Company for that purpose as we imagine of maintaining uninjured the religious reputation of their Colony that to this day, the encouragement and maintenance of the persons concerned in these horrible transactions is the watchword of those who chuse [*sic*] to distinguish themselves in the support of the illegal practices of the Sierra Leone Company in opposition to the Government of the Colony, among the foremost of whom we regret to be obliged to class the persons whom we have understood to have been appointed His Majesty's Commissioners whom we have on several occasions found ourselves obliged in the execution of our duty as Magistrates for the preservation of some tolerable degree of public morality and sense of decency to put to such personal and public shame, as we apprehend in England, would be perfectly equivalent to an exposure in the Pillory. (Sierra Leone Council, 1808)

Thompson's allegations and unfounded allegations exasperated Wilberforce and his powerful political allies, prompting them to urge Lord Castlereagh to dismiss Thompson as the colony's Governor and recall him. Lord Castlereagh advised Thompson that his charges "... do not appear in themselves sufficiently satisfactory without further explanation ..." (Castlereagh, 1809). He ordered Thompson to provide full evidence on his return to England. Having been dismissed from his gubernatorial role, Thompson was henceforth effectively shut out by the British establishment.

## 5.9 Commissioner

Following Thompson's departure and the arrival of Columbine as the colony's new Governor, perhaps unsurprisingly, a public meeting held in April 1810 dismissed the former Governor's allegations and accusations of both Ludlam and Dawes as unfounded (Columbine, 1810b). Columbine was under special instructions to rekindle the work of the Commissioners of Inquiry. In a letter of 8 April 1810, Columbine urged the new Colonial Secretary, Robert Jenkinson (1770–1828), second Earl of Liverpool, that "... [i]f any tolerable proof or even fair ground of suspicion were brought forward ... [it was] his duty ... immediately to suspend Mr. Ludlam and Dawes ... and send them to England" (Columbine, 1810b). No such grounds were identified and Columbine proceeded to appoint Ludlam and Dawes to his governing council, unceremoniously dismissing and expelling the lone opposing member from the colony (Sierra Leone Council, 1810).

Yet, despite his appointment as Commissioner tasked with investigating 'the State of the Settlements and Governments on the Coast of Africa' since April 1808, little progress had been made. In 1810, Dawes estimated that only 448 acres of arable land were under cultivation in the colony, "... about one-half of which had been cleared within these last thirteen months" (Dawes, 1810). This was a significant reduction since 1801, when Ludlam had estimated that between 600 and 700 acres had been under active agricultural management (Madden, 1841).

Nevertheless, the survey proper had yet to commence. Thompson had rightfully questioned Dawes' and Ludlam's generous salaries given the absence of any effort in return. The survey's late start was, in part, owing to the lack of a clear timetable and deliverables, the deferred appointment of a third Commissioner and the loss of their survey vessel, which had caused a 15-month delay (Directors of the African Institution, 1815: 83). In 1809, Lord Liverpool proceeded more proactively than his predecessor, Lord Castlereagh, by appointing Columbine as the third Commissioner, with instructions to commence their survey as soon as conditions allowed (Liverpool, 1809).

Lord Liverpool was aware of Columbine's duties as the colony's new Governor, and so he suggested that the practical survey might be carried out by Dawes and Ludlam on their own, whereas their final report to the British Government could then be signed off by all three Commissioners. Their brief included instructions to travel along the Gold Coast (Ghana), to visit and assess the conditions and longer-term viability of all forts and settlements to gain "... full and accurate ... [information] on the State of Africa ...", both as regards the state of coastal agriculture and "... the social & civil conditions of the inhabitants ...". In addition, they were instructed to liaise with the African chiefs with a view to abolish the slave trade on the Gold Coast, and to advise as to how the Portuguese could be persuaded to do the same, "... with a view to the extinction of the slave trade on the Windward Coast" (Liverpool, 1809), approximately covering the present-days coastal regions of Liberia and Ivory Coast. Figure 5.15 provides an overview of (top) the major slave trading regions along the West African coast and (bottom) the main trading routes.

Despite Lord Liverpool's encouragement and the provision of adequate resources, the survey team was plagued by challenging weather conditions and significant delays owing to sickness. Ludlam died during the survey, while Columbine would be dead by the time the final report was sent to the British Parliament in 1812.[17] Dawes, meanwhile, suffered from severe bouts of fever, similar to his predicaments during his earlier stints in the colony. This delayed the completion of the final report until 1811.[18] Extracts were eventually published by the House of Commons in October 1812 (House of Commons, 1812). Dawes published his own, somewhat dissenting report a month later, on 1 November 1812 (Dawes, 1812), to be read alongside the "... Report of Captain *Columbine* ..." (Dawes' emphasis).

The main conclusion of the survey was that the southern American states had significantly increased their slave trading activities, through vessels sailing under the Spanish flag, following an initial slowdown after the British and American

[17] In his supplementary report to that released by the three Commissioners, Dawes (1812) writes, "Had Captain *Columbine* lived to become acquainted with these material changes, ..." (original emphasis).

[18] In fact, Dawes first compiled a *Report on the Gold Coast* (12 January 1811), followed by his *Report on the Slave Trade* (27 August 1811).

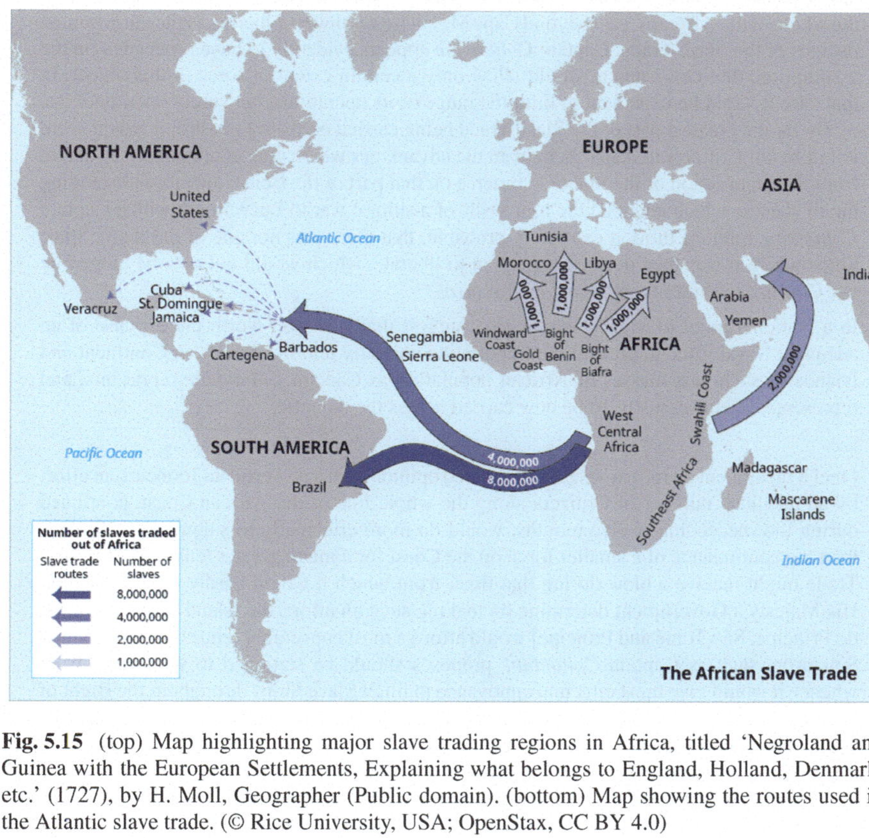


**Fig. 5.15** (top) Map highlighting major slave trading regions in Africa, titled 'Negroland and Guinea with the European Settlements, Explaining what belongs to England, Holland, Denmark, etc.' (1727), by H. Moll, Geographer (Public domain). (bottom) Map showing the routes used in the Atlantic slave trade. (© Rice University, USA; OpenStax, CC BY 4.0)

Governments had abolished slavery. Although the arrival of a British naval fleet in early 1810 had put a temporary halt to these practices by capturing a significant number of Spanish-flagged traders along the Sierra Leone coast, freeing "... about 2,800 Africans ...", most of the slave trade occurred further south—along the coasts of Benin, Gaboon (Gabon), Congo and Angola. In his follow-up report, Dawes concurred with the main report's findings, but he suggested that a significant increase in the British naval presence covering a much larger stretch of the West African coast was required. He appears to have alienated his fellow Commissioners by demanding very expensive and uncompromising adjustments to Britain's policing of the slave trade (Gibson, 2012: 82), *viz.*,

> It is admitted by Captain *Columbine*, that the chief part of the Slave Trade is carried on from the Gold Coast, Whidá [Benin], the Bight of Benin and Angola; and it cannot be denied, that if the same vigorous means were resorted to on these parts of the Coast [Sierra Leone], which he proposes should be adopted to windward, the same beneficial effects, which he so confidently anticipates in the one case, would not fail equally to follow in the other. The Naval force which would be required in this case, would be much larger than that proposed by Captain *Columbine*. It would require to be at least doubled; but that if the force were adequate, its effects would prove equally efficacious to Leeward [the coast northwest of Sierra Leone, covered by modern Guinea] as they have been to Windward, there can be no doubt.
>
> While Great Britain possesses her present Naval means, it would seem hardly justifiable in any point of view, to adopt the policy which Captain *Columbine* has proposed, of abandoning the whole of the African Coast beyond Cape Mount [in northwest Liberia], to the unrestrained ravages of the Slave Trade. Captain *Columbine* appears evidently to have proceeded on the assumption, that Government would allow only a certain extent of force to this object. In that case it might be expedient to limit the range of its operations, but surely not otherwise. ... By far the greatest part of the Slave Trade being carried on from Leeward, it would seem unfair to our Cruizers to debar them from the advantages which they could not fail to derive from the suppression of the illicit Commerce on that part of the Coast; and notwithstanding the ill success which attended the four visits of a ship of war to Leeward, to which Captain *Columbine* refers, I think it extremely probable, that there was not one of the many Slave Ships which it was then deemed expedient to liberate, which would not now be judged by our Courts of Admiralty good and lawful prize.
>
> In a National point of view, I would also suggest that it is well worth the expense of an adequate Naval force to prevent the importation annually into the American Continent and Islands of such vast masses of African population as Captain *Columbine* represents, and represents, I believe, justly, to be now carried across the Atlantic.
>
> ...
>
> I feel it incumbent on me to express my decided opinion, that a powerful and concurrent effort by an adequate number of Cruizers along the whole line of the African Coast, continued during the succeeding twelve months, would do more effectually to suppress this Traffic, than the continuance of a smaller force on the Coast for a much greater length of time. The Trade might receive a blow during that time, from which it would hardly recover. Should His Majesty's Government determine on making such an effort, the Island of Princes [Ilha do Príncipe; São Tomé and Príncipe] would afford a most convenient rendezvous for a small Squadron, such as Captain *Columbine* proposes should be stationed to windward, from whence it would give most effectual annoyance to illicit Slave Ships throughout the Bight of Benin, and even occasionally as far as Cape Three Points [Ghana], westward, and Angola, southward. (Dawes, 1812)

In addition, he fully agreed with the recommendation in their joint report that Bance Island should be nationalised by the British Government for reasons of national security:

> I concur entirely in the recommendation which is contained in the Report with respect to this Island, that it should be taken into the hands of Government It would serve greatly to strengthen Sierra Leone, and to increase the British influence in that part of Africa. The present Proprietors would, I apprehend, be willing to part with it on very reasonable terms. They formerly offered to dispose of it for £2,000. and it is now, as it appears to me, not worth to them above half that sum. (Dawes, 1812)

Having concluded his appointment as Commissioner, Dawes returned to England one final time. The aftermath of his time in Sierra Leone was not pretty. The slow start of the coastal survey and the generous salary he had drawn for a significant period before producing results were raised once again by Robert Thorpe (1773–1836), former Chief Justice of Sierra Leone, in 1815. Thorpe contended that Dawes "… had a little knowledge of land surveying, yet he was perfectly inadequate to such an undertaking as surveying the coast …" (Thorpe, 1815c).

In response, the Directors of the African Institution highlighted Dawes' "… proficiency in the science of nautical surveying", as demonstrated already on the First Fleet's voyage. Moreover, they pointed out that the survey was not intended to chart the West African coast, but to obtain an inventory of the prevailing conditions. Ironically, they also countered that Thorpe had drawn a generous salary as Chief Justice of Sierra Leone for the three-and-a-half years prior to his arrival in the colony (Directors of the African Institution, 1815).

## References


Armstrong, J. (2011). Richard Pepys—a pioneer in Sierra Leone. *Cameo—The Journal of the West Africa Study Circle, 12*, 105–106. Retrieved August 09, 2022, from https://www.wasc.org.uk/Cameo%20digital/Cameo%2083%20complete.pdf

Atkins, G. (2009). *Wilberforce and his milieux: The worlds of Anglican Evangelicalism, c. 1780–1830*. Ph.D. thesis, University of Cambridge.

*AYV Newspaper*. (2020). *The death of John Frederic Naimbana: Did the British Kill A Sierra Leonean Prince?* 10 January. Retrieved August 07, 2022, from https://ayvnews.com/the-death-of-john-frederic-naimbana-did-the-british-kill-a-sierra-leonean-prince/

Castlereagh, L. (1809). Letter to T. P. Thompson, dated 3 April 1809. Colonial Office 268/6. Kew, UK National Archives.

Church Missionary Society. (4 January 1802). *Minutes*. Church Missionary Society Archives, G/C1/1.

Church Missionary Society. (21 December 1804). *Minutes*. Church Missionary Society Archives, G/C1/1.

Church Missionary Society. (7 January 1805). *Minutes*. Church Missionary Society Archives, G/C1/1.

Clarke, R. M. (2015). *Vanguards of empire: The lives of William Dawes, Watkin Tench and George Worgan*. Ph.D. Thesis, The Australian National University.

Clerke, A. M. (1885–1900). Dawes, William Rutter (1799–1868). *Dictionary of National Biography*, 14. Retrieved September 01, 2022, from https://en.wikisource.org/wiki/Dictionary_of_National_Biography,_1885-1900/Dawes,_William_Rutter

Coleman, D. (ed.) (1999). *Maiden voyages and infant colonies. Two women's travel narratives of the 1790s*. Leicester University Press.

Coleman, D. (2005). *Romantic colonization and British anti-slavery*. Cambridge University Press.

Columbine, E. (1810a). Letter to Lord Liverpool, dated 8 April 1810. Colonial Office 267/27. Kew, UK National Archives.

Columbine, E. (1810b). Notice of a public meeting on 13 April 1810, dated 8 April 1810. For the full transcript, see UK National Archives entry, Colonial Office 267/27.

Committee on the Petition of the Court of Directors of the Sierra Leone Company. (1800–1807). War Office 1/352. Kew, UK National Archives.

Council of Governors of Christ's Hospital. (25 January 1799). *Minutes*. Christ's Hospital archives. MS 12806/13.

Council of Governors of Christ's Hospital. (8 January 1800). *Minutes*. Christ's Hospital archives. MS 12806/13.

Crooks, J. J. (1903). *A history of the colony of Sierra Leone, Western Africa, with Maps and Appendices*. Kent and Co., Ltd. Retrieved August 07, 2022, from https://archive.org/stream/in.ernet.dli.2015.506192/2015.506192.history-of_djvu.txt

Currer-Jones, A. (1930). *William Dawes, R.M., 1762–1836*. Torquay, Smith.

Dawes, W. (1796). Letter to John Clarkson, dated 9 December 1796. British Library, Clarkson papers. Add. MSS 41263 (Vol. III, pp. 138–139).

Dawes, W. (1799). Letter to Nevil Maskelyne, dated 5 December 1799. Australian Joint Copying Project. Microfilm reel No. 1759 (PRO 1759).

Dawes, W. (1800). Letter to the Council of Governors, dated 6 November 1800. Reproduced in *Minutes of the Council of Governors of Christ's Hospital*, 7 November 1800. Christ's Hospital Archives. MS 12806/13.

Dawes, W. (1806a). Letter to Josiah Pratt, Secretary of the Church Missionary Society, dated 1 October 1806. Church Missionary Society Correspondence, G/AC 3/3/32.

Dawes, W. (1806b). Letter to Zachary Macaulay, dated 1 October 1806. Church Missionary Society Correspondence, G/AC 3/3/33.

Dawes, W. (1806c). Letter to the Rev. Josiah Pratt, dated 4 October 1806. Church Missionary Society Correspondence, G/AC 3/3/34.

Dawes, W. (1806d). Letter to Zachary Macaulay, dated 4 October 1806. Church Missionary Society Correspondence, G/AC 3/3/35.

Dawes, W. (1806e). Letter to Josiah Pratt, dated 31 October 1806. Church Missionary Society Correspondence, G/AC 3/3/39.

Dawes, W. (1807). Letter to Josiah Pratt, dated 25 September 1807. Church Missionary Society Correspondence, G/AC 3/3/73.

Dawes, W. (1809). Statement respecting Mr. Dawes accompanying Mrs. Edmonds to Fort Thornton, 25 October 1809; enclosed in Columbine (1810a).

Dawes, W. (1810). Deposition enclosed in Columbine (1810a). Letter to Lord Liverpool, dated 8 April 1810. Colonial Office 267/29. Kew, UK National Archives.

Dawes, W. (1812). *Further Paper Relating to the Slave Trade: viz. Observations by William Dawes, Esquire, one of the Commissioners of African Enquiry*, printed 22 April 1812. House of Commons Parliamentary Papers Online.

de Hart, J. (ed.). (1927). *Diary of Lieutenant J. Clarkson, R.N* (p. 8). Sierra Leone Studies, O.S.

Directors of the African Institution. (1815). *Special Report made at the Annual General Meeting on 12 April 1815* (pp. 81–83). London.

Directors of the Sierra Leone Company. (1791a). *Substance of the Report Delivered by the Court of Directors of the Sierra Leone Company to the General Court, 19 October 1791*.

Directors of the Sierra Leone Company. (1791b). *Certificate of Character*. University of Illinois at Chicago (Sierra Leone Collection). Cited by Stewart (2013).

Directors of the Sierra Leone Company. (1791c). *Terms of Service*. University of Illinois at Chicago (Sierra Leone Collection). Cited by Stewart (2013).

Directors of the Sierra Leone Company. (1794). *Substance of the Report Delivered by the Court of Directors of the Sierra Leone Company to the General Court, 1794*.

Directors of the Sierra Leone Company. (1795). *Substance of the Report Delivered by the Court of Directors of the Sierra Leone Company to the General Court, 26 February 1795*.

Falconbridge, A. M. (1794). *Two Voyages to Sierra Leone*. In Coleman (1999).

Falconbridge, A. M., & Fyfe, C. (ed.). (2000 [1794]). *Narrative of two voyages to the river Sierra Leone during the years 1791–1792–1793*. Liverpool University Press.

Fyfe, C. (1962). *A history of Sierra Leone*. Oxford University Press.

Fyfe, C. (ed.). (1991). *"Our Children Free and Happy": Letters from black settlers in Africa in the 1790s*. Edinburgh University Press.

Gibson, R. (2012). *26 Views of the Starburst World: William Dawes at Sydney Cove 1788–91*. University of Western Australia Publishing.

Herrmann, R. B. (2019). *No useless mouth: Waging war and fighting hunger in the American Revolution*. University Press Scholarship.

House of Commons. (1812). *Extracts from the Report of the Commissioners appointed for investigating the State of the Settlements and Governments on the Coast of Africa*, printed 9 March 1812. House of Commons Parliamentary Papers Online.

House of Lords Papers. (28 May 1799). Minutes of evidence taken on the second reading of the Bill, '*An Act to prohibit the Trading for slaves on the coast of Africa within certain limits*'. House of Lords.

House of Lords Sessional Papers [HLSP]. (1714–1805). Evidence of William Dawes. Minutes of Evidence taken on the Second Reading of the Bill, '*An Act to prohibit the Trading for Slaves on the Coast of Africa, within certain Limits*' (Vol. 3).

Kennerley, A., & Seymour, P. (2000). Aids to the teaching of nautical astronomy and its history from 1600. *Paedagogica Historica, 36*, 151–175.

Knutsford, V. (1900). *Life and letters of Zachary Macaulay*. Edward Arnold.

Kup, A. P. (1972). John Clarkson and the Sierra Leone Company. *International Journal of African Historical Studies, 5*, 203–220.

Laurie, P. S. (1988). William Dawes, and Australia's first observatory. *Quarterly Journal of the Royal Astronomical Society, 29*, 469–482.

Leach, F. (2018). *Reclaiming the women of Britain's first mission to West Africa: Three lives lost and found* Chap. 3 (pp. 80–102). Leiden, Brill.

Liverpool, Lord. (1809). Letter to Edward Columbine, dated 27 December 1809. Colonial Office Out Letters, Sierra Leone. Colonial Office 268/6. Kew, UK National Archives.

Macalay [*sic*], Z. (1796). Letter to Selena Mills, dated 17 May 1796. In Knutsford, Viscountess, 1900. *Life and letters of Zachary Macaulay* (pp. 135–136). Edward Arnold.

Macaulay, Z. (1793a). *Journal*. In Falconbridge, 1794. Part 6. Retrieved July 20, 2022, from https://www.natinpasadvantage.com/essays/two_voyages_to_Sierra_Leone,part_6.htm

Macaulay, Z. (1793b). *Journal*. In Falconbridge, 1794. Part 7. Retrieved July 30, 2022, from https://www.natinpasadvantage.com/essays/two_voyages_to_Sierra_Leone,part_7.htm

Macaulay, Z. (21 August 1794). *Journal*. Macaulay Papers, MY 418 (3). San Marino CA, Huntingdon Library.

Macaulay, Z. (6 October and 21 December 1796). *Journal*.

Macaulay, Z. (1799). Letter to Selena Mills, dated 20 June 1799. In Knutsford, Viscountess, 1900. *Life and Letters of Zachary Macaulay* (p. 226). Edward Arnold.

Macaulay, Z. (1807). Letter to Thomas Ludlam, dated 1 May 1807. In Report of the Sierra Leone Council to Lord Castlereagh, November 1808. Colonial Office 267/27. Kew, UK National Archives.

Macaulay, Z., & Bedford, P. (ed.). (1839). *A brief sketch of the life of the Late Zachary Macaulay, Esq., F.R.S: As connected with the subjects of the abolition of the slave trade and slavery*. J. Hatchard and Son.

Madden, R. R. (1841). *Report of Her Majesty's Commissioner of Inquiry on the State of the British Settlements on the Gold Coast, at Sierra Leone, and the Gambia, with some Observations on the Foreign Slave Trading Factories along the Western Coast of Africa, in the Year 1841.* Appendix to 'Report from the Select Committee on the West Coast of Africa', II, 9ff.

Mander-Jones, P. (1966). Dawes, William (1762–1836). *Australian Dictionary of Biography,* 1. Retrieved July 30, 2022, from http://adb.anu.edu.au/biography/dawes-william-1968

Matthews, Lt. J. (1788). *A voyage to the river Sierra Leone.* London.

Morrison, D., & Barko, I. (2009). Dagelet and Dawes: Their meeting, their instruments and the first scientific experiments on Australian soil. *Historical Records of Australian Science, 20,* 1–40.

Mouser, B. L. (1994). *Journal of James Watt: Expedition to Timbo, Capital of the Fula Empire in 1794.* Introduction. Retrieved August 07, 2022, from https://www.webfuuta.site/bibliotheque/watt/intro.html

Newton, J. (1792). Letter to Richard Johnson, dated 19 July 1792. *Historical Records of New South Wales,* 2, 473.

Parish Registers for St Mary's Church, Portsea. (1666–1875). Church of England. St. Mary's Church (Portsea, Hampshire).

Pearn, J. (1998). Surgeon's Mate Lowes of H. M. S. 'Sirius' and the First Fleet. *Health and History, 1,* 65–71.

Perkins, C., & Anderson, I. (1793). Letter from 'Cato Pirkins and Isaac Anderson to the Hble the Chairman & Court of Directors of the Sierra Leone Company London, sent to John Clarkson 30 October 1793', dated 13 October 1793. British Library, Add. MS 41263, folios 98–99.

Peterson, J. (1969). *Province of Freedom. A History of Sierra Leone, 1787–1870.* Faber and Faber.

Plumley, N. (1976). The Royal Mathematical School within Christ's Hospital. The early years. Its aims and achievements. *Vistas in Astronomy, 20,* 51–56.

Pratt, J. H. (Ed.). (1865). *Eclectic notes; or, Notes of discussions on religious topics at the meetings of the Eclectic Society, London during the years 1798–1814* (p. 1). Nisbet.

Public Record Office, Colonial Office (Sessional Papers, Sierra Leone), 270/2. *Minutes.* Kew, UK National Archives.

Public Record Office, Colonial Office (Sessional Papers, Sierra Leone), 270/3. *Resolutions of Council.* Kew, UK National Archives.

Pybus, C. (2006). *Epic journeys of freedom: Runaway slaves of the American Revolution and their global quest for liberty.* Beacon Press.

Pybus, C. (2009). 'Not fit for your protection or an Honest Man's Company': A transnational perspective on the Saintly William Dawes. *History Australia, 6*(1), 12.1–12.7.

Royal Astronomical Society. (1868). Report of Council. *Monthly Notices of the Royal Astronomical Society, 29,* 116.

Scott, T. (1806). Letter to the Rev. Josiah Pratt, Secretary of the Church Missionary Society, dated 17 June 1806. Church Missionary Society Correspondence, G/AC 3/3/23.

Sierra Leone Company Report. (1795). Cited by Stewart (2013).

Sierra Leone Council. (19 May 1795). *Minutes.* Colonial Office 270/3. Kew, UK National Archives.

Sierra Leone Council. (10 December 1800). *Minutes.* Colonial Office 270/5 (p. 75). Kew, UK National Archives.

Sierra Leone Council. (6 June 1801). *Minutes.* Colonial Office 267/6, f. 7. Kew, UK National Archives.

Sierra Leone Council. (November 1808). Report to Lord Castlereagh. Colonial Office 267/27. Kew, UK National Archives.

Sierra Leone Council. (25 October 1809). *Minutes.* Colonial Office 270/11. Kew, UK National Archives; see also the T. P. Thompson papers, University of Hull. DTH/1/45, DTH/1/57.

Sierra Leone Council. (March 1810). *Minutes.* Colonial Office 217/63. Kew, UK National Archives.

*Sierra Leone Gazette.* (18 August 1808). Colonial Office 267/24. Kew, UK National Archives.

Stewart, I. D. (2013). *The tenacity of bondage: An anthropological history of slavery and unfree labor in Sierra Leone.* Ph.D. Thesis, University of Michigan.

Tench, W. (1793). *A complete account of the settlement at Port Jackson. Including an accurate description of the situation of the colony; of the natives; and of its natural productions*. In G. Nicol & J. Sewell (eds.).

*The African Herald*. (26 August 1809). Court of Vice-Admiralty. P. 2. Colonial Office 267/25. Kew, UK National Archives.

Thomas, S. (2013). A transnational perspective on William Dawes' treatment of women. *History Australia, 10*, 187–204.

Thompson, T. P. (1808a). *Rough Journal*, 22 July 1808. T. P. Thompson papers, University of Hull. DTH/1/21.

Thompson, T. P. (1808b). Letter to Anne Elizabeth (Nancy) Barker, dated 23 July 1808. T. P. Thompson papers, University of Hull. DTH 4/1, 455–461.

Thompson, T. P. (1808c). Letter to William Wilberforce, dated 23 August 1808. T. P. Thompson papers, University of Hull. DTH 1/61; see also DTH 1/23 (undated draft statement).

Thompson, T. P. (1808d). Colonial Office papers, Secretary of State. Original Correspondence, Sierra Leone. Colonial Office 267/27. Kew, UK National Archives; see also the minutes of the Sierra Leone Council. Colonial Office 270/11. Kew, UK National Archives.

Thompson, T. P. (1809a). Letter to Lord Castlereagh, dated 17 February 1809. Colonial Office 267/25. Kew, UK National Archives.

Thompson, T. P. (1809b). Letter to the Swedish Governor at St. Bartholomew, dated 12 August 1809. Colonial Office 267/25. Kew, UK National Archives.

Thompson, T. P. (1809c). Letter to Lord Castlereagh, dated 25 October 1809. Colonial Office 217/63. Kew, UK National Archives.

Thompson, T. P. (1810). Letter to Edward Columbine, dated 1 March 1810. Colonial Office 217/63. Kew, UK National Archives.

Thornton, H. (1792). Letter to John Clarkson, dated 14 September 1792. Clarkson Papers. British Library, Add. MSS 41262a.

Thornton, H. (1808). Letter to T. P. Thompson, dated 20 October 1808. In Report of the Sierra Leone Council to Lord Castlereagh, November 1808. Colonial Office 267/27. Kew, UK National Archives.

Thorpe, R. (1815a). Letter to William Wilberforce, dated 1 February 1815. William L. Clements Library, University of Michigan. Folios 12, 23.

Thorpe, R. (1815b). *Point by Point Reply to the Special Report of the Directors of the African Institution*. Rivington.

Thorpe, R. (1815c). *A letter to William Wilberforce, Esq., M.P. Vice President of the African Institution, &c. &c. &c., containing remarks on the reports of the Sierra Leone Company, and African Institution: With hints respecting the means by which an universal abolition of the slave trade might be carried into effect*. London, Law and Gilbert.

Turner, M. J. (1997). The Limits of Abolition: Government, Saints and the 'African Question', c. 1780–1820. *The English Historical Review, 112*, 319–357.

Vice, D. (1983). *The Coinage of British West Africa & St. Helena 1684–1958*. Birmingham, Peter Ireland (Format) Ltd.

Walker, J. W. St G. (1992). *The black loyalists: The search for a promised land in Nova Scotia and Sierra Leone*. University of Toronto Press.

Wilberforce, R. I., & Wilberforce, S. (1838). *The life of William Wilberforce*, 1. J. Murray.

Wilberforce, W. (1793). Letter to John Clarkson, dated September 1793. British Library, Clarkson Papers. Add. MSS 41262A.

Wilberforce, W. (1794). Letter to Henry Dundas, dated 2 August 1794. *Historical Records of New South Wales*, 2, 246.

Wilberforce, W. (1799a). Diary entry, 31 May 1799. Quoted in Wilberforce and Wilberforce (1838: 337).

Wilberforce, W. (1799b). Diary entry, 24 June 1799. Quoted in Wilberforce and Wilberforce (1838: 339).

Wilberforce, W. (1808). Letter to Thompson, dated 19 October 1808. T. P. Thompson papers, University of Hull. DTH 1/61.

Wright, H. (1927). Further note on Lieutenant William Dawes. *Journal of the Royal Australian Historical Society, 13*, 63–64.

# Chapter 6
# Pride and Prejudice in Antigua

## 6.1 A Last Overseas Posting

Back in England once more, on 25 May 1811 Dawes married his second wife, Grace Gilbert (ca. 1776–1844), at St Pancras Old Church in London (Parish Registers of St Pancras Old Church, 1660–1916; London Metropolitan Archives, 1811). Grace, a white woman of mixed European and Caribbean descent, was a cousin of the Rev. Nathaniel Gilbert (1761–1807) with whom Dawes had worked closely at his Bledlow seminary. In 1760, Gilbert's father, also Nathaniel Gilbert (1721–1774), had established the first overseas Methodist Society, in Antigua (Thompson, 1960: 22; Glen, 2001). Dawes and Gilbert the Younger had first met in Sierra Leone.

Grace was born in Antigua into a family of 'planters', plantation managers. After the death of their mother in ca. 1776, Grace and her brother George were initially raised by their great-aunt, Grace Bryant. Upon her death in St. Vincent in ca. 1787, Bryant's will (Oliver, 1896: 14) stipulated that Grace be cared for[1] by Mary Horne, mother of another of Grace Gilbert's cousins, the Rev. Melville Horne (ca. 1761–1841), who had accompanied Dawes on his 1792 voyage on the *York* to Sierra Leone (Currer-Jones, 1930).

The Daweses' financial situation was precarious, however, and so William turned to William Wilberforce, his old patron, for assistance. Although Dawes had been cleared by a local inquiry from Thompson's allegations of impropriety in Sierra Leone, his reputation was irreparably damaged, and even Wilberforce could not easily provide him with gainful employment in the Home Counties (Pybus, 2009). On Wilberforce's recommendation and driven by his own religious convictions, on 19 May 1812 Dawes joined the Committee of the Church Missionary Society during its twelfth anniversary meeting (Hole, 1896: 220; see Fig. 6.1).

[1] Grace Gilbert was orphaned around 1780 (Thomas, 2011a; see also Gilbert, 1835: 12).

 
R. de Grijs and A. Jacob, *William Dawes*, Springer Biographies,
https://doi.org/10.1007/978-3-031-38774-6_6

**Fig. 6.1** Founding meeting of the Church Missionary Society at Aldersgate Street in the City of London on 12 April 1799. (Wikimedia Commons; out of copyright)

Meanwhile, he reverted to his former occupation, training aspiring missionaries prior to their deployment overseas. However, never content to sit still, by April 1813 he advised the Church Missionary Society of his intention to resign and relocate to the British slave colony of Antigua (see Fig. 6.2), in the West Indies (Hole, 1896: 242). He indicated that he would be willing to represent the Society's interests by being "… accredited as a gratuitous catechist and correspondent of the Society …" as he "… should have the superintendence of a considerable number of Negroes" (Church Missionary Society, 1813a).

He offered to represent the Society in both Antigua "… and any of the neighbouring islands where [he] may find opportunity" (Church Missionary Society, 1813b). This offer was accepted at the Society's meeting of 10 May 1813, which was also Dawes' last appearance at a London Committee meeting. In addition, the Church Missionary Society tasked him with establishing Sunday schools for slaves in the West Indies, thus revealing its true desire: safeguarding slave souls (Pybus, 2007).

William and Grace left England shortly after that final Committee meeting, accompanied by Dawes' daughter Judith. They settled on the Gilberts' estate with Grace's sister, Martha Gilbert, and some 200 "… choice Books in Divinity" (Dawes, 1823a). Dawes' sickly son from his first marriage, William Rutter, remained in the care of Thomas Scott at Aston Sandford (Wright, 1927: 64).

Although Dawes was officially on half-pay as First Lieutenant of the Royal Marines, for the first few years of his residence in Antigua his salary had been

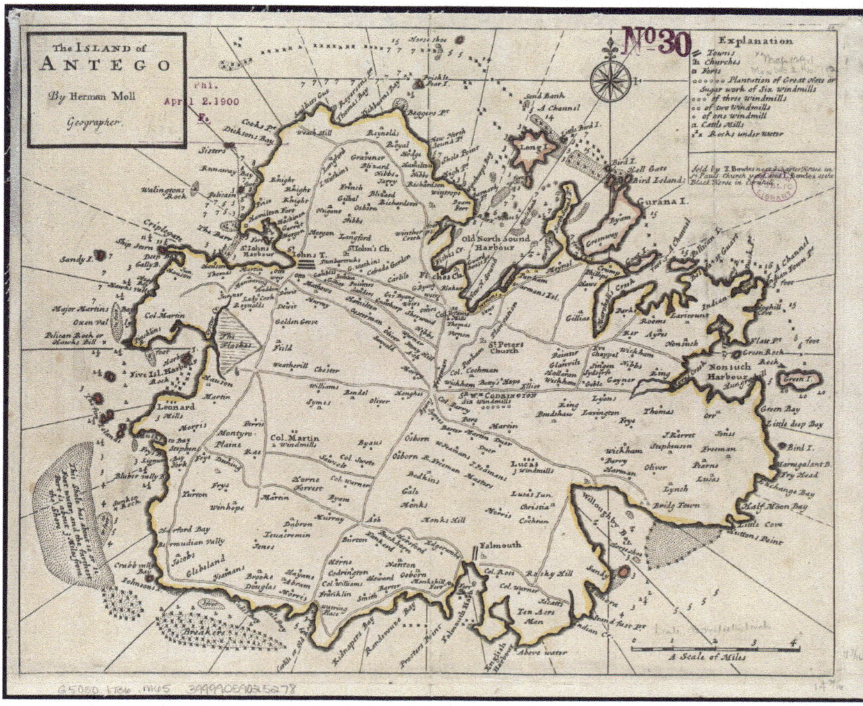


Fig. 6.2 Topographic map of Antigua by Herman Moll (d. 1732). Published by T. and J. Bowles (London, ca. 1736). (Courtesy of the Norman B. Leventhal Map Center, Boston Public Library; Call #: G5050 1736.M65; no known copyright restrictions)

suspended. It took until 1821 before he received his outstanding pay claim, in arrears, dating back as far as 1810. In a letter of 17 December 1821, he notified the Church Missionary Society that he had been assured that his half-pay would continue going forward (Dawes, 1821b).

Meanwhile, Dawes drew a small commission on account of his appointment as an Attorney[2] for Nathaniel Gilbert the Younger on the Gilberts' family estate (see Fig. 6.4), a "… middling-size sugar plantation of 300 acres that had between 150 and 200 slaves" (cited by Thomas, 2011a). In addition, the couple received two small annuities of £20 each and interest from two bequests ('legacies') to the value of £500 each (Gilbert, 1822), "… charged contingenting upon the Estates of the Reverend Nathaniel Gilbert [the Elder] in Antigua" (cited by Thomas, 2011a). The Daweses released their annuities and both legacies on 31 December 1822 (see Fig. 6.3).

Following an altercation with the then-owner of the Gilberts estate in February 1815, the Daweses moved in with Grace's brother John and his wife Anne (née Hart) (Thomas, 2011a) at English Harbour. John was a Methodist lay preacher and

[2] A number of accounts mistakenly suggest (cf. Thomas, 2011a) that Dawes relocated to Antigua to support Wilberforce's efforts to abolish the slave trade, specifically to "… strive to assist to erase the blot from the British nation" (Wright, 1927: 64).

**Fig. 6.3** Release from Miss Martha Gilbert and Mr William Dawes and Grace his Wife of two annuities of £20 and two legacies of £500 charged contingenting upon the Estates of The Reverend Nathaniel Gilbert in Antigua. (Courtesy Hamilton College/Central New York Library Resources Council; no known copyright restrictions)

Storekeeper at the local naval dockyard. Because of this direct access to the navy complex, Dawes witnessed first-hand the dangers of a potential "… life of unchastity …" to which many non-white, "… free coloured …" girls and young women were exposed. Distressed, he set out to improve the conditions of women subjected to a life "… to which they have constant solicitations after the age of fourteen or fifteen years …" (Dawes, 1816b: 141). Charitable support of exploited women was severely lacking in Antigua, since non-white women, irrespective of how destitute they were, did not qualify for parochial relief of the poor (Thomas, 2011b, 2013b).

He resolved to assist his sister-in-law, Anne Gilbert (1768–1834), in raising funds for the fledgling Female Refuge Society. Anne had established the Society in 1815 to operate around the English Harbour naval base (see Fig. 6.5 for a late-nineteenth-century view of the English Harbour dockyard), encouraged by the prospects of transatlantic fund raising mooted by Dawes (Thomas, 2011a). The Female Refuge Society attempted to address the societal problems caused by the prevailing power imbalance between the white residents of the naval base and the poor coloured girls and young women, many of whom were abandoned with their mothers or left destitute by their white fathers.

**Fig. 6.4** (*top*) Planting Sugar Cane on Bodkins Estate, Antigua, West Indies, 1823; (*bottom*) Sugar Cane Harvest on Delap Estate, Antigua, West Indies, 1823. Ladies' Society for Promoting the Early Education of Negro Children (London, ca. 1833). (Courtesy Slavery Images: A Visual Record of the African Slave Trade and Slave Life in the Early African Diaspora; Creative Commons Attribution-Non-commercial 4.0 International license)

**Fig. 6.5** English Harbour dockyard, ca. 1880. (Courtesy UK National Archives; CO 1069/411; no known copyright restrictions)

## 6.2 The English Harbour Sunday School

A few years earlier, in September 1809, Anne had been inspired by Beilby Porteus (1731–1809), Bishop of London, to establish the English Harbour Sunday School (Thomas, 2011a), jointly with her sister Elizabeth and her sister's husband, Charles Thwaites (1780–ca. 1860). The first such school in the West Indies, the English Harbour Sunday School had attracted as its patron the "Hon. Mrs. [Lady] Grey" (Hole, 1896: 540), whose husband, Charles Grey, 2$^{nd}$ Earl Grey, would become British Prime Minister in 1830. Bishop Porteus explicitly prohibited the teaching of writing to slaves, to

> … always preserve a proper distinction and subordination between them and their superiors, and present an insurmountable barrier against their approaching to anything like an equality with their masters. (Porteus, 1808: f.22n)

Anne Gilbert's driving force was, instead, to teach the children to read and write.

The School blossomed under the patronage of Lady Grey at the time, in September 1814, when Dawes became President of the English Harbour Sunday School Society—"… one of whose rules was 'That the Schools should be open to all &c.'" (Dawes, 1825k). This role was entirely separate from his work for the Church Missionary Society, although the Society supported Dawes' efforts in providing educational opportunities to the Antiguan slaves and their children (Clarke, 2015: Chap. 4)—or, as has been suggested, "[t]o instil Christian doctrine, together with obedience and respect …" (Pybus, 2009: 12.5). Arguments in support of the latter interpretation come from Dawes himself: he complained that some of the senior boys portrayed "… a disposition to ape their superiors in rank and station … truly painful to behold" (Dawes, 1825a).

The Daweses expanded their fund-raising efforts to also benefit the Distressed Females' Friend Society, a society with similar aims that operated around St John's, on the other side of the island. They successfully leveraged their networks by writing letters of appeal to well-placed contacts in England. William attained a success rate of some 20–25 per cent in his fund-raising efforts (Thomas, 2011a). The Female Refuge Society and the Distressed Females' Friend Society both aimed to provide support to young, non-white women in the face of their frequent, often sexual exploitation as 'concubines'. Dawes considered these impoverished freed women and coloured children to be "… doomed … to a life of prostitution" (Dawes, 1825f), which often resulted in a vicious cycle of poverty and malaise.

Dawes was clearly inspired by a number of nondenominational Protestant social and cultural reform projects initiated by Anne, with whom he also worked to establish the Antigua Auxiliary Bible Society—to significant opposition from Church of England clergy (Thomas, 2011a). Jointly with the Thwaites, Anne and her husband John were driven to support and sustain local communities, irrespective of gender, religious conviction or ethnic background, a sphere of 'Creole benevolence' that counteracted the excesses of the plantocratic elite (Thomas, 2011a), that is, the plantation elite.

The Gilberts and Thwaites wanted to foster "… a race of creditable, though humble, and industrious young people … in these Colonies …", using methods such as 'othermothering' (James, 1993: 44–54)—a form of community care and education—that were culturally comfortable for members of the African diaspora. With the consent of their mothers or guardians, the Female Refuge Society and the Distressed Females' Friend Society both placed destitute children into foster care with pious women, who would take care of the children's educational needs and prepare them to earn their own living. The Distressed Females' Friend Society also financially assisted a number of women who were no longer able to support themselves, either through illness or old age. Both Societies had all-female and multi-racial management committees; they accepted subscriptions from women and donations from anyone, irrespective of their gender (Thomas, 2011a).

The first mentions of the Female Refuge Society and the Distressed Females' Friend Society appeared in print around 1816 in the *Missionary Register*, the journal of the Church Missionary Society. Dawes routinely forwarded the Societies' annual reports to the Church Missionary Society. In times of financial difficulty, he also made available correspondence from Anne Gilbert (Dawes, 1825b, 1825c), acting Agent of the Female Refuge Society, which often evoked a sense of distress and helplessness (Thomas, 2013b). Dawes' wife Grace was the Female Refuge Society's Treasurer.

In a letter reproduced in the *Missionary Register* of 1 April 1816, Dawes "… painfully reminded …" the Church Missionary Society of the "… actual … STATE OF BRITISH COLONIAL SOCIETY, IN RESPECT OF MORALS AND RELIGION" (capitalised in the original; Dawes, 1816a: 144), exposing the societal ills that usually remained hidden from public view, by means of an "… affecting appeal …". Dawes' purpose in publishing his letter was "… to shew that there is a loud call at present upon Christians to deny themselves superfluities; and, as they have received of God, to help the distressed ..." (Dawes, 1816a: 142). He points out that the "... distressed ..." children around English Harbour are, on the whole, the "... illegitimate offspring of white men, principally in the Navy and Army" (Dawes, 1816a: 141). He writes that their mothers' status as concubines "... almost universally terminate[s] with the decay of personal appearance, or the incumbrance of children". It has been suggested that Dawes' reference to a 'decay of personal appearance' may be a euphemism for having contracted venereal disease (Thomas, 2013b).

Dawes also offers a number of 'affective anecdotes'—"... [s]ecret, private, or hitherto unpublished narratives or details of history ..."—in support of his characterisation of the "... wretched ..." consequences of the British naval presence in Antigua. One such anecdote refers to "... the natural daughter of the late Lord Falkland, ... still a slave ...", with a mother "... sunk in sin, poverty, and wretchedness, ... daily witness[ing] scenes, the most improper and shocking that can well be conceived, ... [and at risk of] being reduced to the dreadful alternative[s] of either famine or prostitution". Dawes' aim was to point out that orphaned local girls were at risk of being "... dispose[d] of ... as concubines to men who are able to provide them with food and raiment, and to put it in their power to relieve their distressed relations" (Thomas, 2013b: 29–30).

The commentary on Dawes' exposé in that same issue of the *Missionary Register* states that "... it is as impolitic as it is criminal, to thwart or retard the influence of genuine Christianity on the population of our Foreign Possessions ...", whereas the Church Missionary Society declared the situation in Antigua a "... melancholy subject ...", that is, a deplorable environment. In response, a letter referred to in the next issue of the Society's journal quotes an 'Anonymous Lady' offering 'trinkets'[3] for sale, with the proceeds meant "... towards the education of those pitiable girls in Antigua, who, without some affectionate aid, are doomed to the wretchedness of vice, or the accumulated distress of poverty and slavery ..." (Church Missionary Society, 1816b).

## 6.3 Church Missionary Society Representative

The Church Missionary Society was keen to capitalise on Dawes' services, skills and experience. And so, when its Committee discussed the Society's Sunday schools in Sierra Leone at a special meeting in 1816, Dawes' name naturally rose to the top of a short list of suitable candidates for an appointment as General Superintendant of Church Missionary Society operations in Africa, "... at a competent salary ...". At the same time, 'King Henry' of 'Hayti' (Haiti) had asked Wilberforce and the British and Foreign Bible Society for assistance in setting up local schools.

The Committee agreed to offer both appointments to Dawes, although mindful of the possibility that he might want to stay with his family in Antigua. If he decided to stay put, the Society would offer to furnish him with an assistant and appoint

[3] Dawes referred to the items as "... superfluities ...".

him as Superintendent and Catechist of the English Harbour Schools, a position that attracted remuneration at a level sufficient "… to make up his present means of living" (Church Missionary Society, 1816a). In response to the Society's offer for him to return to Sierra Leone, in January 1817 Dawes declined the opportunity "… chiefly from a conviction of the great importance of improving his opportunities for good in Antigua" (Church Missionary Society, 1817). Meanwhile, the Haiti post quietly disappeared from the collective consciousness, most likely because of the demise of King Henry's reign (Clarke, 2015: 88).

Nevertheless, and despite Dawes' rejection of the Sierra Leone role, the Society was in dire need of authoritative information about the prevailing conditions on the West African coast. They particularly were keen to find a way to circumvent the high levels of sickness, and even death, among their missionaries in Africa. Dawes took this request for information seriously, composing a seven-page letter in which he carefully described a number of case studies, treatments and remedies he had administered himself while acting as the West African colony's medical practitioner around 1801 (Dawes, 1820c).

Combined with much work undertaken by the Gilberts and the Thwaites,[4] by 1819 Dawes' energy, determination and resourcefulness had contributed significantly to an expansion of the Antiguan Sunday schools network. Dawes had taken an active approach to representing the Church Missionary Society, appointing teachers as required and sending updates on his progress in establishing and operating the schools. Society schools had been established in three towns—English Harbour, Bethesda and Falmouth—as well as one on the Codrington plantation, all in south-eastern Antigua. The school at English Harbour alone educated more than one thousand children (Church Missionary Society, 1819a, 1819b). As the network grew, it became harder for Dawes to exercise his role as the schools' Superintendant, and so the Society agreed to purchase and maintain a horse for his use (Church Missionary Society, 1819c).

Meanwhile, the Church Missionary Society had either financially supported or condoned Dawes in setting up three different businesses to provide him with the level of income he needed to take care of his family. As we saw on several occasions throughout his career, money was often tight and Dawes and his family were forced to live off a meagre income. To supplement this income, in 1816 he had opened a chemist's dispensary (Church Missionary Society, 1816c), based on experience from his time in Sierra Leone and gained at home in England. He also edited *The Antigua Journal*, a local newspaper, and he operated a rum-improvement business. The Daweses continued to live in accommodation provided to them free of charge by their supporters, which somewhat eased their financial concerns. Nevertheless, on one occasion in early 1823, as Dawes was absent for a period of time, Grace was forced to take out an advance of her husband's salary to make ends meet (Dawes, 1823b).

In 1820, Dawes wrote to the Church Missionary Society that the Distressed Females' Friend Society convened a meeting with a number of children in its care,

[4] Dawes had recruited Charles Thwaites as Superintendent of the Church Missionary Society Sunday Schools in Antigua in 1817 (Thomas, 2011a).

Committee members and foster parents. The aim of the meeting was such that the girls' "... improvement in reading, writing, working and behaviour might be enquired into ..." (Dawes, 1820a), thus suggesting that the children were merely experimental specimens (Van Sant, 1993).

In view of the successful expansion of the Church Missionary Society's schools network in Antigua, the Society was keen to cover more of the West Indies—despite quiet opposition from the Established Church[5] (because of the schools' evangelical and ecumenical curriculum, which included non-conformist, Methodist and Moravian or 'United Brethren' teachings) and plantation owners (Clarke, 2015: 90–91). In March 1820, Zachary Macaulay convinced the Church Missionary Society that Dawes was "... a most eligible person ... from his known prudence, economy and integrity, and from his intimate acquaintance with the state of things in the West Indies ...". Macaulay "... enthusiastically encouraged ..." (Midgely, 1992: 48) female anti-slavery societies, thus showing his close alignment with Dawes' efforts in the West Indies. Macaulay's recommendation resulted in an offer for Dawes to become full-time Superintendent of Sunday Schools "... now formed or to be hereafter formed, in connection with the [Church Missionary] Society, throughout the West Indies", at an annual salary of £300 plus travel expenses (Church Missionary Society, 1820a).

Dawes was pleased to accept the Society's offer and continued to hold the position until 1829. He gracefully responded, "I should deem it an honour to live and die in such a cause ..." (Thomas, 2011a). Dawes' appointment as Superintendent of Sunday Schools in the West Indies represented his first proper, salaried position since his departure from England. The Church Missionary Society stipulated that he was required to relinquish his independent businesses. Trying to avoid "... a ruinous loss ..." by selling his stock at overly discounted prices, he advised the Society that the winding up of his chemist's dispensary would take somewhat longer than expected, but he committed to have the process completed by late 1821 (Dawes, 1821b).

In addition to the patronage of Lady Grey, the English Harbour Sunday Schools also received the backing of Sir Benjamin D'Urban (1777–1849), Governor of Antigua (1820–1824), and his wife, Lady Anna D'Urban née Wilcocks (d. 1843) (Church Missionary Society, 1822; Dawes, 1822c).[6] The Governor was impressed by Dawes' work and his dedication to the Church Missionary Society, calling him a "... Most Active and Zealous ..." agent in a letter to the politician Edward Smith-Stanley (1799–1869), Lord Stanley, 14th Earl of Derby, in 1824 (D'Urban, 1824).

Lady D'Urban participated actively in charitable activities on behalf of the English Harbour Sunday School Society. On one of her visits to the English Harbour Sunday School in the early 1820s, she was introduced to a number of girls and their residential hosts supported by the Female Refuge Society and its sister charities. Financial contributions to the Society were mostly used to support destitute girls with accommodation and meals, clothing and shoes (Thomas, 2011a). On that particular visit, Lady D'Urban witnessed a number of performances which appeared to contain

---

[5] This refers to the High Church movement within the Church of England.

[6] Governor D'Urban had also agreed to act as patron of the Antigua Auxiliary Bible Society.

hidden messages. For instance, one of the girls recited from Isaac Watts' (1674–1748) hymn *Praise to God for Learning to Read*,

I.
THE Praises of my Tongue
I offer to the LORD,
That I was taught, and learnt so young
To read his holy Word.
II.
That I am brought to know
The Danger I was in,
By Nature and by Practice too,
A wretched Slave to Sin. (Watts, 1715)

This was followed by the congregation singing *Rule, Brittania*, another case of 'sly civility' (Bhabha, 1994: 93), given that in this case the children of slaves proclaimed that "Britons never, never, never shall be slaves"...

Anne and her sister Elizabeth had established the English Harbour Sunday School Society before the Church Missionary Society had become involved. From its inception, they were committed "... not to annex the Schools to any particular denomination of Christians whatever" (English Harbour Sunday School Society, 1822a)—Methodist, Wesleyan, Moravian or Anglican. Figure 6.6 shows a few representative missions in Antigua. By the early 1820s, that approach had started to attract increasingly negative commentary from the Church of England, and Dawes suddenly found himself in the midst of yet another controversy.

The English Harbour Sunday School Society board felt that it had to clarify its position, passing a number of resolutions in support of cross-denominational education. Dawes passed these on for information to the Church Missionary Society. The School Society resolved that they would not ostracise "... any body of Christians ...". However, further involvement from the Church Missionary Society was felt to jeopardise that position, and so they eventually resolved to abandon all their schools except for those at English Harbour. Despite this apparent setback, Dawes was pleased with the clear guidance and so he told the Church Missionary Society that he was "... relieved by the turn which the matter has taken" (Church Missionary Society, 1822).

Already in September 1820, at the first Church Missionary Society Committee meeting following his full-time appointment, Dawes filed a report regarding his visit to St. Bartholomew, where he had been "... kindly received by the Governor" (Church Missionary Society, 1820b). By 1822, the English Harbour Sunday School Society had expanded its remit to the management of six Country Schools. That year, the Church Missionary Society took over the management of the Country Schools at Bethesda, Cook's, Golden Grove, Hope, Old Road, Popeshead, Sion Hill and Union (Thomas, 2011a). The Sunday School at English Harbour taught 122 boys and 163 girls; of the girls, eight were reported as white, 100 as free (black and coloured) and 55 as children of slaves (English Harbour Sunday School Society, 1822b). In keeping

**Fig. 6.6** (*top*) Moravian mission at St. John's. (Johann Heinrich Stobwasser, Ansichten von Missions-Niederlassungen der evangelischen Brüder-Gemeinde/Vues des establishments missionaries fondés par la Communauté évangélique des Fréres-Unis; Basel: Birmann, 1830). (Courtesy Slavery Images: A Visual Record of the African Slave Trade and Slave Life in the Early African Diaspora; Creative Commons Attribution-Non-commercial 4.0 International license) (*bottom*) Roman Catholic church at St. John's, ca. 1910 (postcard). (Courtesy Robin McDonald via Facebook; no known copyright restrictions)

with their parents' wishes, the Society did not require any of the children to attend church or be of any given denomination. In 1823, Dawes explained to Governor D'Urban that the English Harbour Sunday School Society

> … is a Catholic Institution, to every office in which, pious persons of all denominations of Protestant Christians are admissible. The Church Catechism, and Watts's Catechisms [Watts, 1825] are taught. Children of all classes under 14 years of age are received and continued in

> the school until they are 16; at which time they are dismissed unless circumstances render it conducive to their best interests that they should continue in which case they are appointed to the office of teachers or continued as scholars. (D'Urban, 1824)

It thus appears that most parents were predominantly concerned about their children's education and not necessarily about their salvation.

Dawes was apparently uninterested in politics; there is no evidence that he was involved in any political activity in Antigua (Thomas, 2011a). In mid-1823, he warned the Church Missionary Society that teachers "... sho$^{d}$ be especially guarded against meddling with the politics or prevailing system of the country [slavery], and even of expressing an opinion on these subjects ..." (Dawes, 1823a), advice he carefully heeded himself. An apolitical attitude was particularly important given that the plantation owners and their families resolutely opposed evangelical missionary activity and influence, opposition that became increasingly fierce in response to the slave rebellions in Barbados in 1816 and Demerara (Guyana) in 1823. In this context, his involvement with the Antigua Auxiliary Bible Society, which he had founded and of which he remained a Vice-President, was seen as supporting an organisation whose "... object is not of a religious but [a] political nature ..." by members considered "... auxiliaries of the African Institution" (Dawes, 1824a).[7]

In early 1823, Dawes visited Dominica on behalf of the Church Missionary Society, where he received full support from the Governor and his local officials. As a result of this visit, he requested that the Society send two teachers to Dominica, adding the proviso that they ought to carry a certificate from the Bishop of London, who wielded religious authority over the West Indies (Church Missionary Society, 1823a). His other suggestion proved contentious: he requested that the Society would urge absentee plantation owners to support his schools. At the Society's London Committee meeting of 8 September 1823 Macaulay pointed out that plantation owners were generally opposed to educating slaves and their children, and that the Church Missionary Society should therefore concentrate on educating "... free people of colour ..." (Church Missionary Society, 1823b).

## 6.4 In Trouble

Dawes soon found himself in hot water back in Antigua. In a report for Governor D'Urban on the state of the schools on the island, he complained that he regretted a "... want of sound preaching ..." among the island's Anglican clergy, that is, a lack of the evangelical approach he had come to favour. A copy of his report was sent to the Church Missionary Society, where it caused significant embarrassment. Adding insult to injury, at least from a London-based perspective (Clarke, 2015: 94),

---

[7] In Barbados a broadsheet described 'Methodist Missionaries' as "... agents to the villainous African Society ..." (Lambert, 2005: 153).

he openly admitted to having cooperated with the local Methodist missionaries,[8] which he portrayed as essential for the success of the schools, however:

> In this part of the world real Christians are unavoidably associated in some respects, however they may differ in name. I have not been so unmindful of the necessity of a distinction in our public acts however, ..., for I have more than once declined taking the Chair at the Wesleyan Auxiliary Missionary Society, though ... I did not think it necessary to refrain from being present, or making a motion on such occasions. (Dawes, 1824b; emphasis in original)

> When therefore I had to select Assistants in the important work of instruction, I did not make leaving the Methodists a condition with those who were paid, nor interdict uniting with them ... My aim was to make Bible Christians, and to that end, while the inculcation of the principles of the Christian Faith was not neglected, teaching our Pupils to read the Scriptures for themselves was our grand object. (Dawes, 1826c)

He also reported to having used Methodist chapels for his adult and night schools; the Methodists had, in return, used the English Harbour school building for their own educational purposes, given the "... scarcity of proper places for instruction ..." (Dawes, 1824b). The Society was aghast and greatly concerned that publication of Dawes' letter would damage their reputation. Dawes defended his approach, on account of having to navigate a complex local situation:

> It gives me considerable uneasiness to find that any part of my letter to Sir B[enjamin]. D'Urban has incurred the censure of the Committee. I certainly entertained no apprehension of any such effect, I beg leave to offer a few explanatory remarks. The Governor's request was clear & comprehensive. He desired to be furnished "with any observations as the existing state & deficiency of Religious Instruction throughout the Island generally, so far as it may be within my own knowledge." I thought it also important & desirable that the friends of Religion in England should be acquainted with the real state of it here. I considered myself as called upon, therefore, in the course of God's Providence, to give such evidence as, had it transpired here, would have exposed me to reproach & hostility; nevertheless, having long mourned over the exceedingly corrupt state of the Clergy generally in this part of the world, I considered it as the fruitful source of incalculable evils, it appeared an impervious Duty (particularly at a time when an increase of the ecclesiastical establishment was in contemplation) to say what I did, & I conceived, had I not done so, I should have incurr'd the guilt of disobedience to the Governor, & of dereliction of Duty to God, to the cause of Religion generally & to that Church especially of which I esteem it a happiness & privilege to be a member. (Dawes, 1824b; emphasis in original)

A Church Missionary Society letter requesting that the offending sections would be erased from any report to Parliament was forwarded to Robert Wilmot-Horton (1784–1841), Under Secretary of State for War and the Colonies, but the damage had already been done (Church Missionary Society, 1824a, 1824b, 1824c). Dawes' outspoken criticism of the Anglican clergy, combined with his open association with the Methodists, caused him significant difficulties when an uncensored copy of the letter was made available to the new Bishop, William Hart Coleridge (1789–1849), shortly after his arrival in the West Indies on 29 January 1825, to take charge of the newly established Diocese of Barbados.

---

[8] As a case in point, John Gilbert had been appointed as Vice-President of the English Harbour Sunday School Society (Thomas, 2011a).

This was not the first time that Dawes had openly courted involvement with other denominations. As the new Superintendent of Sunday Schools in the West Indies, he had asked the Church Missionary Society for clarification as to whether schools on neighbouring Caribbean islands might request assistance from other denominations. He pointed out the "... great zeal ..." with which the Methodists and Moravians pursued Sunday School education, and so he considered their assistance essential. The Society acquiesced to Dawes' request. Although they had initially passed a resolution that schools in the West Indies could employ teaching staff "... without respect to the denomination to which they belong ...", this was clearly contentious, given that their public response included the proviso that the schools' curriculum had to follow the Society's regulations (Church Missionary Society, 1820b, 1821).

Yet, although the Society's leadership was sympathetic to the pragmatic need to cooperate with other denominations and 'non-conformists', from a London perspective such collaboration was fraught with difficulties in the face of strong opposition from the Evangelical wing of the Church of England. In response, in December 1824, Dawes pointed out the "... exceedingly corrupt state of the Clergy generally in this part of the World". To underscore the "... awful results ... [for] the slave population and free white and coloured people ..." (Dawes, 1824d), in September of that year he had highlighted his scorn for the local Anglican clergy:

> It is not from the slaves I apprehend danger; it is not from any political measures adopted by the Colonists, that I assign evil consequences, nor yet from any of the enactments of His Majesty's Government; but from a lukewarm Clergy whose lives disgrace even their inadequate views of Xtian [Christian] Theology.

In his letter of 15 November 1825, he expanded on the Clergy's moral deficiencies:

> These dignitaries [Bishop Coleridge and Archdeacon Parry], with others of like sentiment, seem to expect people in their senses to believe that all the Clergy must necessarily be men "fearing God and working Righteousness," which to me appears to be a species of popery. When these Schools and Institutions were formed, Who were the Clergy? The Rector of S$^{t}$. Paul's, the Rev$^{d}$. Arthur Freeman, was a notorious drunkard, who, it was generally understood, never retired to rest but in a state of intoxication, and was at last found dead in his bed. The Rev$^{d}$. George Collins, long contemptible in the eyes of every one, & lately made Chaplain of the Garrison on the [Sugar] Ridge, because he was deemed by the Bishop unfit to hold the Living any longer, was then Rector of S$^{t}$. Philip's (Willoughby Bay) [see Fig. 6.7]. M$^{r}$. Weston, Rector of S$^{t}$. Mary's, was a medical man, but too superficial in intellect to succeed, so went home, took Orders, came out, waited 'till his father died, and then succeeded him in the Living. Like all the others, he had empty Pews every Sunday (Dawes, 1825j).

Despite the practical working relations he had initially established with representatives of other denominations, Dawes remained "... unabated ..." in his attachment to the Church of England. He reaffirmed his commitment by referring to the (unspecified) time when his "... first aspirations for "true repentance" ascended to the throne of God in the sacred language of her liturgy ..." and the call he received "... was heard and answered ..." (Clarke, 2015: 94–95). Given his strong religious stance, it did not take long before he fell out with his former collaborators among the United Brethren and the Anglican church alike, citing moral objections to some of their activities (Clarke, 2015: 102):

**Fig. 6.7** Shipping sugar, Willoughby Bay. Ladies' Society for Promoting the Early Education of Negro Children (London, ca. 1833). (Courtesy Slavery Images: A Visual Record of the African Slave Trade and Slave Life in the Early African Diaspora; Creative Commons Attribution-Non Commercial 4.0 International license)

> … who, with more decency of conduct, formally adhering to the Church of England while strangers to the heart-warming influences of the Holy Spirit, diffuse a baneful influence around them. (Dawes, 1824d)

Pleased to be able to maintain the Church Missionary Society's moral standards, Dawes regularly reported that many of the mixed race and slaves' children went on to successful marriages in the established Christian tradition.

Yet, although in his mind he was a firm adherent to the Evangelical wing of the Church of England, in reality he tended somewhat to the strict(er) practices of the Methodists. His main aim was to provide a level of education that would allow one to read the scriptures, act according to the Church's moral compass and, hence, eventually find salvation, the ultimate aim. That attitude is reflected in a note he wrote following a visit to the Hope School on 14 July 1822:

> [Teaching is] so important a Work; which by the Blessing of God may be means of saving their Souls. I could not help lifting up my heart to God in prayer, that not one of them might be lost. (Dawes, 1822d)

Dawes' evangelical beliefs are underscored by a request for replacement books (Dawes, 1823a, 1823c, 1823d) he made to the Church Missionary Society in early 1823, following the destruction by fire of the victualling store and book depository at English Harbour. He asked for copies of Edward Bickersteth's *A Scripture Help*, Thomas Scott's *A Commentary on the Whole Bible*, Henry Venn's *The Compleat Duty of Man*, a complete series of the *Christian Observer*, English and Scottish manuscripts "… about the time of the Reformation …", as well as 'Hymn & Psalm tunes', but "… more plain and solemn", indeed a typical evangelical attitude.

Dawes' increased responsibilities, which came with the expansion of the Church Missionary Society's remit across the West Indies, took a toll on his health. As we already saw in relation to his involvement in Sierra Leone, his constitution was ill prepared for the type of hot and humid climate he would encounter in the Caribbean. He suffered from regular episodes of fever and also started to develop rheumatism. His first reference to this latter ailment, "Rheumatic complaints were worse than ever", is found in a report from October 1821 (Dawes, 1821a); these complaints were likely exacerbated by a fall from his horse on arrival home late one night (Dawes, 1821a, 1821b, 1821c). A family emergency occupied most of his mind in early 1822, when he stated that "… the death of my dear little Grand Daughter has so operated on my mind … that I cannot write more at present" (Dawes, 1822a).

By 1824, the Church Missionary Society's educational reach covered children from some 60 estates across Antigua:

> About Town, The Hope, Cedar Grove, Bethesda, Sion Hill, Grace Bay. Total 2002: 80 black men, 55 black women, 8 coloured men, 12 coloured women, 759 black boys, 850 black girls, 131 coloured boys, 107 coloured girls. 108 in the Bible, 217 in the Testament, 229 3 & 4 Syllables, 268 2 Syllables, 410 1 Syllable, 770 Alphabet. The Hope is the largest school with 574 total, 375 average attendance. (Church Missionary Society, 1824d)

Tobago, Dominica, St Vincent and St Kitts had also been embraced by the Society's expansion drift. That year, Governor D'Urban reported that the Moravians taught 620 children across their school system, compared with the almost 2000 pupils attending Church Missionary Society Schools. The emphasis was on religious education:

> The method of instruction which our Missionaries have pursued for many years, at all our Stations, has been to invite the negroes and people of colour, both children and adults, bond or free, to attend them, at such hours as could be spared from their ordinary duties. In this way, many have been taught to read; and from the number of those who have been thus instructed, both children and adults, a large proportion of the teachers have been derived, who are now employed in our various schools. These schools are held at the Missionary stations, on Sunday and Monday evenings, and likewise on the Estates at noon and night, ... Besides learning to read, our scholars commit to memory texts of scripture, &c. which they repeat twice a week, in a meeting held at the church, the Missionary adding some explanatory remarks. Many of the adults possess copies of the Bible, and the children are supplied with New Testaments. (Taylor, 1825: 7)

Dawes conceded that religious education was the prime objective of the Church Missionary Society, "… a body of good men in England who earnestly desired their spiritual welfare, and thought the best means of promoting it was teaching them to read the Bible …", meanwhile emphasising "… their awful responsibility to God, if they neglected to improve … their [the children's] minds" (Dawes, 1825f).

The most important lessons would be "… pasted on boards …" to give students "… opportunities of learning at home during the week" (Dawes, 1820b). This practice appears to have been commonplace, given that the Rev. Lewis Stobwasser (d. 1832), an outspoken Moravian missionary, explained that Sunday school pupils were given "… a lesson, pasted on a small board, which they put into a small bag or pocket they have for that purpose". The lesson was to be revised on a daily basis, under the supervision of an adult "Negro" who had been trained "… for the instruction of others"—apparently without compensation (Stobwasser, 1823: 4).

The Church Missionary Society's success was largely owing to the resources Dawes had access to, including funds for infrastructure development (school buildings), paid teachers and assistants, a local printing press and a well-oiled missionary system. Access to the printing press was particularly beneficial. Under Dawes' guidance, it produced lessons, tracts and a hymn book; Bibles, catechisms and tracts were imported from England. Dawes sent sample lessons to the Church Missionary Society in London. They contained

> … spelling and pronunciation exercises (for example, 'ab eb ib ob ub' and 'can cane car care'), word lists, word groups (Eatables, Apparel, Belong to a House, Parts of the Body, The World, Trees, Plants Fruit &c., Number Weights &c, Titles), lists showing separated syllables, and reading text prepared for the one-syllable class on religious and moral themes. (Dawes, 1820b)

By 1825, activity at the navy base at English Harbour was winding down, and the Female Refuge Society's focus gradually shifted to supporting the daughters of white male plantation employees (Dawes, 1825b). Dawes felt vindicated to some extent, since he had long been highly critical of what he called "… libidinous Planters …" and plantation employees—most commonly middle- and lower-ranking free men, such as overseers, bookkeepers, managers and estate attorneys (Lazarus-Black, 1994: 81). He cited the "… enormous wickedness of the inhabitants of this Island at this time and that of the Antediluvians [*sic*]," whom he accused of "… extreme depravity …" (Dawes, 1822b). In an 1824 letter to Mrs Luckock, the wife of a missionary for the Society for the Conversion of Negro Slaves, Anne Gilbert cites from *Hebrews* 11:25 to express her disgust at the practices of concubinage and prostitution: "… [M]y heart frequently yearns over them with affectionate sympathy … [for their] privations …" (cited by Dawes, 1824c). She calls out the local churches for collusion in sexual exploitation and writes with contempt about

> Ladies, who wish to be considered extremely decorous and delicate, promoting … [sexual 'seasoning'][9] by the sale of Slaves for such purposes, and by exactions of hire, which they know cannot be raised in any other way. (cited by Dawes, 1824c)

[9] Enslaved women in the Caribbean, often having arrived directly from Africa, were frequently subject to the dehumanising practice of 'seasoning', that is, an introduction to the slave colonies' routines through cross-racial rape and sexual abuse (Thomas, 2013b).

## 6.5 The Beginning of the End

The arrival, in late January 1825, of Bishop Coleridge and Archdeacon Thomas Parry (1794–1870) to take charge of the newly established Diocese of Barbados signalled the start of a rapid decline in the Church Missionary Society's influence in Antigua. It also put an effective stop to Dawes' long-term efforts to educate slaves and former slaves as a means of spreading Christianity across the West Indies.

Dawes' harshly critical report to Governor D'Urban about the ecclesiastical state of affairs in the West Indies came full circle to cause him serious trouble. At a meeting in November 1825 between the new Bishop and the Rev. Edward Bickersteth (1786–1850), Secretary of the Church Missionary Society, the Bishop confirmed that he had seen an uncensored copy, in which "... several improper things were stated" (Church Missionary Society, 1825b). However, Dawes held firm to his opinions, telling Bickersteth that the consequence of the Bishop's new ecclesiastical regime was "... not the inculcation of real vital Xtianity [Christianity] but almost exclusively, the accomplishment of what will produce a directly opposite effect" (Dawes, 1825h). Meanwhile, he also reported that Coleridge's "... grand object ..." consisted of "... supplanting the C.M.Soc.$^{y}$'s Schools, and abolishing the Wesleyan Methodist Mission altogether" (Dawes, 1825h). He was indeed greatly concerned that since the arrival of Coleridge and Parry, not a single vacant curacy had been filled by Evangelical clergy:

> ... but it is grievous to think, tho' the Bishop found two Livings vacant, the only evangelical Minister in the Church is one who was here when he [the Bishop] came. His Lordship appears to have a great dread of what he calls fanaticism, ... (Dawes, 1825f; emphasis in original)

The attitude of Coleridge and Parry was almost the polar opposite of Dawes' approach. The Anglican clergy's ultimate aim in pushing through reform of Antiguan Sunday schooling was the marginalisation of evangelicism while imposing patriarchal parochial authority (and thereby relegating women from public life), formalising links between Sunday school and church attendance, and suppressing cooperation of the Church Missionary Society with the island's Methodists (Dawes, 1825f; Thomas, 2011a). They prioritised the development of good working relationships with the plantation elites. Anne Gilbert described the senior Anglican clergy as "... ungodly Agents ..." (cited by Dawes, 1825g).

Dawes was always attuned to threats to morality, and so he vigorously defended the Sunday schools system that had been established on the island:

> The situation of a certain class of young persons in this Island, many of whom are educated in the Schools of the C[hurch]. M[issionary]. S[ociety]., often excites painful emotions in the minds of those whose duty it is to watch & labour for the benefit of the rising Generation. The E[nglish]. Harbour Sunday School was the mean [*sic*] of bringing forward to notice so many affecting cases of moral & physical wretchedness in the young people & children of colour in that neighbourhood, as ultimately led to the formation of "the Female Refuge Society." The Methodist Sunday School in S$^{t}$. John produced a similar effect in this quarter of the Island & brought into existence "The Distressed Females' Friend[s] Society." Both these societies, confined as their operations have necessarily been from the inadequacy of funds, have been instrumental thro' the Divine Blessing of much good. ... Many of the unhappy

> children now in question, owe their existence to illicit connexions formed by Managers or Overseers with Slaves under their controul [*sic*]. The standard of morality is so much higher than formerly, in the minds of this class of people, that generally speaking, both Parents have it in contemplation to bring up their female offspring virtuously, & as a preparatory step, send them to one or other of the Country Schools. ... The Sunday School is open to them, & that they gladly avail themselves of, but, destitute of the means of subsistence, & from their very modesty & propriety of behaviour, more desirable objects to licentious men; they are assailed by importunities, & who can say that they will be able to resist them? ... At the same time, another stumbling block in the way of their persevering in the path of chastity is the countenance afforded by the Moravian system of admitting into Church Membership, women living in a state of concubinage. (Dawes, 1824c)

He was particularly concerned about the threat to moral standards that unexpectedly came from the Established Church. Although their arrival had been "… long looked for and long hoped for …", it had led to "… disappointment & grief of heart …" owing to the relaxing of moral standards, which would likely "… end in rebellion against all authority …" (Dawes, 1825h). "'Serious Christians' …", Dawes wrote, "… needed to construct a social order in which the scenes of domestic or church life would always be preferred to 'the gaieties and follies of an ensnaring world' and this meant building a framework within which individuals could find help, care and support." The 'gaieties and follies' Dawes referred to was the apparent seal of approval given to dancing by the presence of Coleridge and Parry at a dance at the Antiguan Government House (see Fig. 6.8); enslaved people were subsequently told, "… the King's Parson had just said, it was no harm to dance" (Dawes, 1825i). In quick succession, Dawes wrote four letters to the Church Missionary Society to complain about this new tolerance of worldly follies, and he would continue to do so on every occasion he felt that the Established Church threatened the colony's moral standards (Thomas, 2011a).

Dawes was greatly concerned that dancing might have a deteriorating influence on "… the local negroes …", with Thwaites adding that he had been compelled to dismiss six of his teachers "… for adultery & fornication" (Dawes, 1825i). For Dawes, the "… primary importance …" within the schools under his supervision was the "… promotion of Chastity among the young" (Dawes, 1820b). Dawes' distress was exacerbated by the apparent revival of dances—Benna dances,[10] balls hosted by local plantation owners and slave balls[11]—which he believed were responsible for

[10] Anne Gilbert distinguished Benna dance, "… African dance …", from a "… grand negro dance …" (the slave ball; Thomas, 2011a). Dawes explains, in a letter to Bickersteth marked 'private', that "[t]he Benna is an obscene African dance, nearly obsolete in this Island, but now reviving." However, "… [t]he answer to everyone is 'The Bishop says it is no harm to dance.'" (Dawes, 1825i).

[11] At slave balls, "… the negroes personated the gentry … used their plate, carriages, watches, ladies jewels &c. &c. The dancing ladies are so extravagant as to change their dresses three times in one night, & so much as sixteen dollars has been paid to open the dance" (Dawes, 1825f).

**Fig. 6.8** Government House, Antigua. The Hall (The Original Dining Room). Leeward Islands. November 1934. (Courtesy UK National Archives; CO 1069/233; no known copyright restrictions)

seducing students away from his Sunday school network[12]—Dawes and his Evangelical circle considered dancing of similar moral turpitude as gambling and 'obeah', an offense against school discipline often leading to expulsion:

> Suppose the Schools were put under the Bishop's direction, what would be the result? All the pious Teachers & catechists would be dismissed unless they would engage never to go into a Methodist Chapel. Their places would be supplied with immoral or even profligate young men, ... The Schools would be countenanced by libidinous Planters, the female scholars debauched & the males driven back into profligacy of life & promiscuous intercourse; to say nothing of the dissipations of dancing, gambling, obeah & all their consequent horrors. (Dawes, 1825j)

The bigger problem for Dawes and his associates was, however, the 'independence' portrayed by children exposed to these 'loose' moral standards, and the consequent difficulties in maintaining discipline at school. Elizabeth Thwaites supports that latter concern by quoting from a song sung by "… young slaves …" in 1825:

> Me no care for Daddy, Me no care for Mammy,
> Me will break thro' the rule,

[12] He considered the women's extravagant dresses the "… price …" of the "… unchastity …" of the Sunday School students "… seduced out of the C[hurch]. M[issionary]. S[ociety]. Schools by their wicked masters …" and the resulting "… anguish of soul …" of their mothers (Dawes, 1825i).

Me will have um, Me will have um, Aye.

Dawes includes these lines from Elizabeth Thwaites' letter in one of his own letters to the Church Missionary Society in which he complains,

> I am sorry to say that the demoralizing principles & practices, which [have been] issued from Government House to the ruin of many of our young people, continue to spread. After the example of Capt$^{n}$. Lyons, who made a dance for his people, and spend [*sic*] it with a girl whom it is said he seduced, Mr. Salmon had a dance at the end of the crop, & carried down the dance with two characters of like description. This, with the countenance given by the Bishop, has had a most baneful influence on the morals of the slaves, especially the young. There are dances on several of the Estates this way, on Saturday nights, accompanied I fear with the usual consequences, adultery, fornication, stealing &c. (Dawes, 1825i)

> I think it not too much to say that he has compromised his Dignity, his Religion & every proper feeling, to procure favor with the Planters, & that, of some of the most licentious among them". (Dawes, 1825j)

We get a rare view into Dawes' personal feelings from a letter he sent in June 1825, lamenting the departure of his daughter, Judith Jones (née Dawes),[13] to England:

> The parting of my Daughter & her family without any prospect of ever meeting again in this world has so occupied my mind that I have not [been] able to add more. (Dawes, 1825c)

Not even a year later, as he was struggling to deal with Parry's incursions into his Sunday schools network, he simply stated that he had heard that Judith had passed away in Liverpool (Dawes, 1826b).

The effects of the new ecclesiastical regime were felt as soon as July 1825, prompting Dawes to write a lengthy letter of distress to the Church Missionary Society. Parry had already managed to force a number of Sunday school teachers to resign, replacing them with Church of England catechists. Coleridge had expressly prohibited the teaching of writing to enslaved children.

Dawes was particularly concerned about Parry's attitude regarding female teachers of male children. In his eight-page letter to the Church Missionary Society's Secretary, he quoted Parry's statements that "... women [are] very improper persons to be known on such occasions [taking charge of benevolent societies]; ...", "... want of delicacy in women of character [does] more injury to the morale of Society than any thing else ...", "... women should be made to know their inferiority ..." and "... should not be supposed to know so much as they really did know ..." (Dawes, 1825f). He reported that Anne Gilbert dismissed such comments by stating that Parry

> ... appears to have spent almost the whole of his intelligent existence in a College & knows no more of the moral perplexities & miseries of the colonies than if he had dropt [*sic*] from one of the other planets. Unless his eyes are opened by divine agency, he will reject & oppose every thing of a benevolent nature in which the Clergy are not the prominent instruments. (Dawes, 1825f)

---

[13] Judith Dawes had married John Jones, an officer stationed at the dockyard at English Harbour.

Dawes' response to Parry's misogynistic commentary was much stronger; he accused the Archdeacon of supporting the "... idea of clerical infallibility and female insignificance ..." (Dawes, 1825f; original emphasis). He was clearly distressed by the turn of events since the arrival of Coleridge and Parry:

> The Bishop has shewn himself to be adverse to the British and Foreign Bible Society and has expressed a displeasure at seeing [the Rev.] M$^{r}$. Horsford's name to a circular sent to the other Islands exhorting and inviting the Clergy and others to form auxiliary societies. Neither the Bishop [n]or Archdeacon is known to have expressed any satisfaction at the societies formed and the good done already in the Island. – M$^{r}$. Parry indeed goes so far as highly to disapprove of the female Societies, and M$^{rs}$. Parry has declined the office of Vice President of the F[emale]. R[efuge]. S[ociety]. (which was offer'd her before her husband's sentiments were known) on the plea of her inadequacy to so "public a situation". (Dawes, 1825e; emphasis in original)

In criticising Parry's attacks on the female leadership of the Female Refuge Society, Dawes writes,

> I trust the cry of the destitute ones, against whose woes their fellow-creatures stop their ears and shut up their bowels of compassion, will enter into the ears of the Lord of Sabaoth. (Dawes, 1825f)

Here, Dawes has adapted a Bible verse from *James* 5:4 that is openly critical of the wealthy and powerful, where Dawes equates slavery with an ungodly although commonly tolerated excess (Thomas, 2011a):

> Behold the hire of the labourers who have reaped down your fields, which is by you kept back by fraud, crieth: and the cries of them which have reaped are entered into the ears of the Lord of Sabaoth.

Although he had initially wanted to send Elizabeth and Charles Thwaites to England to expose the 'scandalous' policies of Coleridge and Parry to Parliament, Dawes abandoned that plan in September 1825 given the treatment received by William James Shrewsbury (1795–1866), a Methodist missionary in Barbados. Shrewsbury had been 'banished' from the islands as a consequence of his opposition to the prevailing doctrine. Dawes warned Bickersteth of the danger inherent to releasing his confidential letters to the Church Missionary Society, expressing fear of successful defamation prosecutions:

> I should in the first place have requested you to bear in mind that tho' my communications are such, when I speak with confidence, as I think I have undoubted authority for, yet, that authority resting in many instances on slave evidence, if I should in consequence of their transpiring [being made public], be prosecuted for defamation, I should be unable to prove the truth of my statements, & must bear the odium and pay the penalty. (Dawes, 1825h)

There was indeed a good reason why Dawes mooted the use of ciphers in his correspondence with the Society on a number of occasions (see Fig. 6.9).

The personal relationship of Dawes and Coleridge had started off poor. It deteriorated rapidly even more when, at their first face-to-face meeting, the Bishop stated that he did not consider the Church Missionary Society a church society (Dawes, 1825f, 1825j, 1826a).

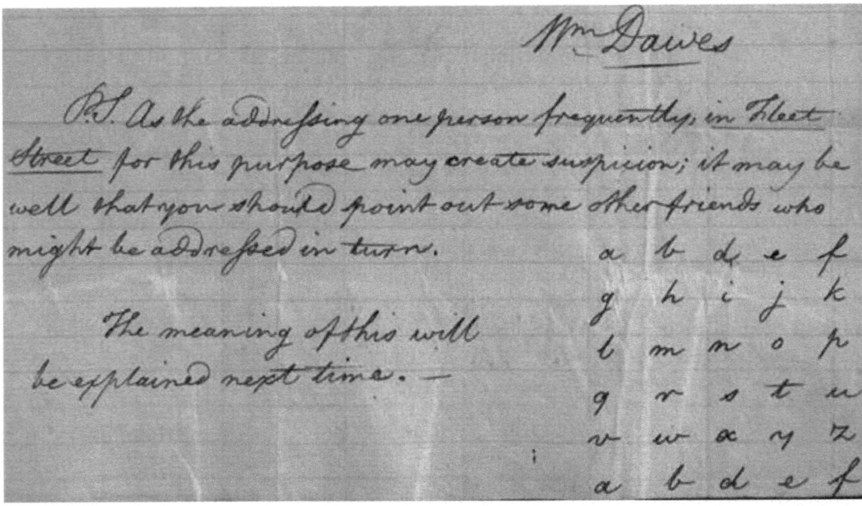

Wm Dawes

P.S. As the addressing one person frequently, in Fleet Street for this purpose may create suspicion; it may be well that you should point out some other friends who might be addressed in turn.

The meaning of this will be explained next time. —

| | | | | |
|---|---|---|---|---|
| a | b | d | e | f |
| g | h | i | j | k |
| l | m | n | o | p |
| q | r | s | t | u |
| v | w | x | y | z |
| a | b | d | e | f |

**Fig. 6.9** Proposed cipher included in a private letter from William Dawes to Edward Bickersteth, Secretary of the Church Missionary Society. (Dawes, W., 1825. Letter to the Church Missionary Society Secretary, dated 3 September 1825. In: *The papers of William Dawes, missionary in Antigua. I: Cultural Contacts, 1492–1969*. University of Birmingham Library. C W/O 31; out of copyright)

> The Rev.[d] N[athan]. G[ilbert]. replies, some of the best of our Bishops & Clergy were Members of it; & instanced the Bishop of Calcutta. The B[ishop]. of Barbados denied the B[ishop]. of Calcutta being so, but M[r]. G[ilbert]. begging pardon for contradicting His Lordship, assured him he asserted it from a knowledge of the fact which silenced the Bishop. (Dawes, 1825f)

In a follow-up meeting with Bickersteth in London in November 1825 (Church Missionary Society, 1825a), Coleridge repeated that as he had no direct links with the Church Missionary Society in England, he was unable to "… lend his name to them …", that is, to the Society's schools in Antigua. It should not come as a surprise that this prompted Bickersteth to protest vigorously and lobby on behalf of the Church Missionary Society, confirming their operation was based on the "… principles of the Established Church". Coleridge eventually relented, offering his approval to an arrangement in which Society school masters were nominated by the local Church of England minister, and the schools would fall under the jurisdiction of the Bishop of Barbados (Church Missionary Society, 1825a). The arrangement as proposed complied with the guidelines adopted by the Barbados Branch Association of the Society for the Conversion and Religious Instruction and Education of Negro Slaves in the British West India Islands founded in 1825. (An Antigua branch was also formed.)

Yet, despite the higher-level agreement that had been reached, Dawes continued to run into obstacles of Parry's making. The Archdeacon continued to close some of the Society's schools, and he was set on removing Charles Thwaites for his Methodist connections. Dawes expended significant effort in trying to save Thwaites' employment. As a case in point, in September 1827 Dawes emphasised that their cooperation with Methodists and Moravians brought "... many benighted souls within the pale of the Church Universal" (Dawes, 1827b). In addition, the Moravian Rev. Jens Olufsen (d. 1829) provided oversight of the Grace Bay Sunday School (see Fig. 6.10), which was "... supported ..." by Dawes and was touted as the "... most flourishing school ..." run by the United Brethren:

> The most flourishing school is Gracebay, under Brother Olufsen's inspection. About 100 children meet here every Sunday; and a number of them are already tolerable proficients in reading. This school has hitherto been supported by the agent of the Church Missionary Society, a favor which claims our sincere acknowledgments. (Richter, 1825: 7)

Nevertheless, Dawes eventually lost the fight to retain Thwaites' services as Parry dismissed him in 1828 over his refusal to renounce Methodism.

**Fig. 6.10** View of Grace Bay, showing a plantation house in the background. (Johann Heinrich Stobwasser, Ansichten von Missions-Niederlassungen der evangelischen Brüder-Gemeinde/Vues des establishments missionaries fondés par la Communauté évangélique des Fréres-Unis; Basel: Birmann, 1830). (Courtesy Slavery Images: A Visual Record of the African Slave Trade and Slave Life in the Early African Diaspora; Creative Commons Attribution-Non-commercial 4.0 International license)

## 6.6 Collapse of the Sunday Schools Network

The changes imposed by Parry on behalf of Bishop Coleridge eventually led to the collapse of the schools' network. Student numbers started to decline precipitously. At the end of 1824 the number of pupils in the schools was 2009 but by 1829 the Church Missionary Society taught just 39 students (Church Missionary Society, 1829b). In view of declining student numbers, Dawes' annual salary had already been reduced to £200 at a meeting of the Society's Committee on 18 April 1828 (Church Missionary Society, 1829a). A special expenditure committee concluded in August 1829 that the costs of running the Church Missionary Society's Sunday Schools network far surpassed their income, so that they recommended to discontinue the Antiguan operations, with "... deep regret ..." (Church Missionary Society, 1829b).

Like Dawes had earlier, they had come to the conclusion that the impact of the Bishop's policies had caused the death knell to the Society's Sunday Schools. In particular the requirement for catechists to be licensed by the Bishop of Barbados led to ongoing difficulties in recruiting potential teachers in England, whereas the Society could not find enough potential catechists already based in Antigua who would be able to meet the Bishop's requirements. It was clear to all involved in the colony that Bishop Coleridge had completely misread the situation in Antigua, where the success of the Church Missionary Society's Sunday Schools was owing to the absence of a requirement for teachers to be affiliated with the Established Church.

In addition to the professional impediments thrown up by the new ecclesiastical regime, Dawes' personal circumstances also caused him much hardship. His financial health continued to be troublesome, to the extent that in late 1826 he felt compelled to apply to the British Government for compensation of work done half a lifetime earlier in New South Wales (Tench, 1926 [1826]):

> That Governor Phillip deemed these extra Services, performed by Your Lordship's Memorialist [Dawes], deserving a compensation of five Shillings pr.diem, in addition to his Pay as an Officer of the Marines, and offered to recommend to His Majesty's Ministers that that Allowance should be made to him from the time of his appointment to those Offices [as Engineer and Officer of Artillery], and to be continued to him, if he would accept of an Ensigny in the newly formed New South Wales Corps; but he did not deem it expedient to exchange the rank of [second] Lieutenant, which at that time he had enjoyed eleven years for that of Ensign, and consequently, such recommendation he believes was never made [by Phillip], and he never received any compensation whatever for such extra Services; although he acted in the capacities of Engineer and Officer of Artillery to the colony nearly the whole time he was in New South Wales, which was almost four years.
>
> That Your Lordship's Memorialist is now in circumstances of great pecuniary embarrassment.

In his letter of support, attached to Dawes' appeal for compensation, his friend, now-Lieutenant-General Watkin Tench, clarified (Dawes, 1826d),

> [t]hat the Impression & Belief was general in the Colony, that Governor Phillip had in contemplation to recommend that a recompense of Five Shillings per Day from His Majesty's Government, should be paid to the Memorialist for the Zeal, ability, & exertions he had used

> during the period he filled the station of an Officer of Engineers & Artillery at Port Jackson ...

However, the Government deemed that too much time had elapsed and so the request was denied.

In addition to Dawes' ongoing financial difficulties, old age came with increasingly feeble physical health. Throughout the second half of the 1820s, we encounter a range of complaints about his deteriorating health. For instance, in August 1827 Dawes writes,

> For some weeks past, I have at times been much more than usually affected with Rheumatism, debility and languor from the frequent and great changes of weather, accompanied often, as now, with distressing headaches. (Dawes, 1827a)

Some time later, in March 1828, he fractured his thigh. His wife Grace wrote to the Church Missionary Society that her husband was to be bed-ridden for eight weeks (Dawes, 1828a, 1828b). In July 1828, Dawes complained that his mobility was restricted since he still needed to use crutches (Dawes, 1828a, 1828b). Summarising his health over "... at least ten years ...", in 1829 he described his life in the "... torrid zone ..." as "... gradually dying of heat, as if I should be now placed before a large fire roasting to death" (Dawes, 1829a). This prompted him to advise the Society that if they wanted to hire future European catechists for employment in the Caribbean, they ought to supplement their salaries to allow for the not inconsiderable expenditure associated with regular medical attention (Dawes, 1825d):

> From the best information I can procure & from my own experience, I think a man should not have less than £100 Stg. [Sterling] P[e]r.Ann[um].; his Wife £60 and for each child not less than £25. I also think it an indispensable condition, than an [*sic*] European, who may not have resided in a tropical climate before, so as to be in the same degree seasoned, should be allowed the attendance of a regular Medical Practitioner. In this case it is probable that the Doctor attending the Estate would attend the family at a certain sum a head P[e]r.Ann[um]. The usual charge to an Overseer or an Estate, or to a coloured Clerk in a Counting House is £5 Curr.y [currency] Pr.Ann[um].; but I don't think less than double that sum a head P[e]r.Ann[um]. would be deemed sufficient for such a family as above. It is really alarming to see the amount of a Doctor's bill for a very short attendance, without any operation or other extra matter. – If circumstances should require him to keep a horse, not less than £35 in the Country or £50 Stg. in or near a Town, should be added. This latter sum is what my horse costs me now, when corn is comparatively cheap, ... All the above (except the £50) is on the supposition that he will reside on an Estate in this Island, be provided with a house and nothing else.

In fact, following the reduction of his salary to £200 per annum, Dawes had attempted to find gainful employment in England, as he anticipated that his appointment with the Church Missionary Society would not be "... of long continuance ...". Leaving the tropics was a primary aim; a decade earlier, he had already written to a friend in search of "... a situation ten degrees at least further removed from the Equator". Nevertheless, if he would be able to find salaried work, he "... was willing to remain & die a somewhat premature death, rather than quit a post which I conceived had been assigned me by the Providence of God" (Dawes, 1829a).

With the Church Missionary Society Sunday Schools network in Antigua having been wound up, the Society offered Dawes, then 67 and clearly suffering from poor health, an annual pension of £75. Throughout the tenure of Coleridge and Parry, Dawes and his associates had increasingly felt slighted and misunderstood by the Church Missionary Society in London. To express his indignation at the interference in his management of the schools system, Dawes started to use the term 'molestation'. He felt abandoned and disempowered in the face of the Church Missionary Society's interpretation of his "... distrust ..." in the Bishop's intentions and his alleged "... jealousy ..." of the man's projects.

The Society, in turn, had become alarmed at the "... irregularities [that] crept in ..." to the Antiguan schools system through the appointment of non-conformist teachers and the lack of an established link between Sunday school and church attendance (Thomas, 2011a). Dawes was, indeed, outspoken about the reason for the absence of such a link:

> I consider the familiar & personal instruction in religion which they receive at School, as far better calculated to benefit them, than the best Sermons, of which they could understand very little; & therefore if the Church were next door, I should not promote their going there in School hours. (Dawes, 1825j)

However, the Church Missionary Society, at arm's length in London, relied on a written response from Parry flouting the "... essential ... well-ordered Parish in England ... under the direction of the Clergyman of the Parish; and the children invariably attend Church". Parry's criticism of Dawes' ideal of a school "... for all denominations ..." was that this would be, "... more properly speaking, for none". With the Society being unaware of the complexity of the conditions on the ground, time and again, Dawes' instructions remained to cooperate with the new ecclesiastical authorities.

Following his retirement, Dawes reopened his chemist's dispensary, now sharing the business premises with his nephew, William Dawes, a watchmaker. In an attempt to improve his financial circumstances, when he advertised the reopening of his business as 'Licensed Vender and Dispenser of Medicine', he offered for sale his astronomical and meteorological instruments, his astronomy and mathematics books, his horse and also his "... unfashionable gig" (Dawes, 1829b). We learn from Dawes' great-granddaughter Ella Jones that

> [d]uring the latter part of his life, his circumstances were very much strained on account of the strong feeling against Mr Dawes, as a staunch friend of Mr Wilberforce. Fortunately, his son in England, Mr Rutter Dawes, a well-known astronomer, was able to send means to his father, sufficient to keep him in happiness. (Wright, 1927: 64)

Upon Dawes' retirement, Charles Thwaites, now employed as schoolmaster for the Wesleyan Missionary Society, tried to persuade the Wesleyan Methodist Society in England to continue Dawes' practice of publishing local material from Antigua. A "... small book of hymns ... [was] selected and printed ..." in Antigua for the Wesleyan schools network (Thomas, 2011a).

Meanwhile, Dawes occupied himself with disposing of the Church Missionary Society's assets, while gradually disappearing from the Society's minutes. He must have been rather disappointed by the slipping of educational standards witnessed in the former schools network he had managed: in 1830s Antigua, schooling focused more on rote learning of scripture and religious texts than on actual skills acquisition (Thomas, 2011a).

Dawes spent most of his time with his in-laws in Antigua (Clarke, 2015: 106), but he kept in touch with his sister Elizabeth until at least March 1832. At that time, he notified the Church Missionary Society that he would take care of an outstanding account she owed the Society—despite being close to destitute himself; he relied on his son-in-law John Jones as regards Society matters (Dawes, 1832).

## 6.7 Reflections on a Troubled Life

A brief announcement of his death, aged 74, was published in the *Missionary Register* of 1836. In contrast, the Church Missionary Society minutes do not record Dawes' passing. The exact date of his death is unknown. In response to a request for information, on 27 September 1926 the Adjutant-General of the Royal Marines in London advised,

> He is shown in Official Lists on Half Pay up to 1836, but his name is omitted from the List of Royal Marine Officers dated 1st January, 1837. (Wright, 1927: 63)

His obituary, published on 11 October 1836, praises him as a 'man of science' with an "… earnest desire to aid in the amelioration of the Negro race, and to instruct the rising generation of that people …":

> [T]here was no one who had general intercourse or particular access to the deceased who was not impressed with the highest respect for his character, and admiration of his talents as a Man of Science. (*Antigua Weekly Register*, 1836)

Dawes was buried in the Gilbert family plot in English Harbour. He shares his grave with John Gilbert, John's wife Anne and Grace Dawes' sister, who is only identified as 'M. M. Gilbert' on the headstone (Clarke, 2015: 106). In the mid-1920s, the Colonial Secretary at Antigua responded to a request for a copy of Dawes' death certificate, stating "I regret being unable to furnish you with a copy of his death certificate, as there was no official registration of deaths in Antigua prior to 1856" (Wright, 1927: 64).

Following William's death, and in view of the dire financial circumstances his wife Grace found herself in, a letter in December 1836 from the 'Dowager' Lady Grey,[14] itself in response to a letter from Charles Thwaites, requested that the Church Missionary Society recognise Grace as a missionary's widow. Although the Society was unable to do so, it granted Grace Dawes a one-off payment of £100 in recognition of her services (Church Missionary Society, 1836). This represents the final record

[14] The honorific 'Dowager' appears incorrect, given that Lady Grey's husband, the former British Prime Minister (Earl) Charles Grey, was still alive; he died in 1845.

pertaining to Grace Dawes in the Church Missionary Society's archive. She "… survived her husband several years" (Wright, 1927: 64).

And so have we come to the end of an eventful life that looked so full of promise, on so many occasions, but which did not always live up to expectations. In the context of his times, Dawes was exceptional, perhaps far ahead of his time, possibly an unrecognised genius, a great mind of the early modern world. Yet he did not quite grasp his place in the rapidly changing world around the turn of the nineteenth century, or the politics at work around him, occasionally lacking a minimum of common sense or a diplomatic touch, and so he often missed his best chance to make a difference to those both near and far.

A genius or a polymath, a theorist par excellence yet lacking enough doses of savvy and pragmatism, his politically woefully inept worldview, combined with an abrasive personality, unmovable, alienating stances and religious convictions set in stone, rendered him a mere footnote in history, fading away from almost all opportunities to make a tangible real-world impact. He could have been so much more than the minor celebrity he almost did not become after most of his family papers, notes and correspondence perished with the ship they were on in that notorious Caribbean hurricane of 1871.

A full understanding of the man requires a careful unpeeling of the many layers that made up his personality, complete with deep flaws and sparks of brilliance. In the pages of this book, we have seen why William Dawes should have become famous, what his most important achievements were, and why he became an ardent supporter of Abolitionism. Yet, his legacy is mostly defined by what is absent from the historical record, by all of the exciting opportunities to make a real impact that he somehow managed to miss or which he simply did not recognise.

Instead, while witnessing the rise to prominence of many of his contemporaries, Dawes butchered his chances, lost his allies and supporters, and grew old and disgruntled in a remote corner of the British Empire. The woeful loss of many of his personal papers in that wretched hurricane has made it that much harder to gain insights into the person that was William Dawes. Yet, in recent decades we have seen evidence of a minor revival of studies into the man's achievements, with more emphasis on his complex personality and character traits, in turn resulting in a number of new controversies and the suggestion that his obsessive compulsion might have been a symptom of Asperger's syndrome, a high-functioning autistic personality (Gibson, 2012: 50–51).

In this book, we have attempted to remain impartial and focus on the facts as far as we known them from the historical record. We have opted to present these as completely as possible, so that the reader can make up their own mind. Nevertheless, it seems clear that William Dawes cannot simply be pigeon-holed as a 'good' or 'bad' historical character. Just like most other people, his personality contained many layers, some 'good', some flawed, ultimately revealing a rather vulnerable man underneath a tough exterior persona.

# References


*Antigua Weekly Register.* (11 October 1836).

Bhabha, H. (1994). *The location of culture.* Routledge.

Church Missionary Society. (1813a). *Minutes*, 26 April 1813. Church Missionary Society Archive. G/C/1/1, Reel 75.

Church Missionary Society. (1813b). *Minutes*, 10 May 1813. Church Missionary Society Archive. G/C1/2.

Church Missionary Society. (1816a). *Minutes*, dated 21 March 1816. Church Missionary Society Archive. G/C1/2.

Church Missionary Society. (1816b). *Missionary Register* (p. 191), dated 1 May 1816.

Church Missionary Society. (1816c). *Minutes*, dated 19 August 1816. Church Missionary Society Archive. G/C1/2.

Church Missionary Society. (1817). *Minutes*, dated 13 January 1817. Church Missionary Society Archive. G/C1/2.

Church Missionary Society. (1819a). *Minutes*, dated 1 January 1819. Church Missionary Society Archive. G/C1/3.

Church Missionary Society. (1819b). *Minutes*, dated 9 August 1819. Church Missionary Society Archive. G/C1/4.

Church Missionary Society. (1819c). *Minutes*, dated 17 September 1819. Church Missionary Society Archive. G/C1/4.

Church Missionary Society. (1820a). *Minutes*, dated 13 March 1820. Church Missionary Society Archive. G/C1/4.

Church Missionary Society. (1820b). *Minutes*, dated 11 September 1820. Church Missionary Society Archive. G/C1/5.

Church Missionary Society. (1821). *Minutes*, dated 8 January 1821. Church Missionary Society Archive. G/C1/5.

Church Missionary Society. (1822). *Minutes*, dated 15 July 1822. Church Missionary Society Archive. G/C1/6.

Church Missionary Society. (1823a). *Minutes*, dated 17 May 1823. Church Missionary Society Archive. G/C1/6.

Church Missionary Society. (1823b). *Minutes*, dated 8 September 1823. Church Missionary Society Archive. G/C1/6.

Church Missionary Society. (1824a). *Minutes*, dated 13 January 1824. Church Missionary Society Archive. G/C1/6.

Church Missionary Society. (1824b). *Minutes*, dated 27 January 1824. Church Missionary Society Archive. G/C1/6.

Church Missionary Society. (1824c). *Minutes*, dated 27 February 1824. Church Missionary Society Archive. G/C1/6.

Church Missionary Society. (1824d). *General View of Sunday School Returns, Antigua*, dated 31 December 1824. Section V, Church Missionary Society Archive. W/O/7, Reel 9.

Church Missionary Society. (1825a). *Minutes*, dated 7 October 1825. Church Missionary Society Archive. G/C1/8.

Church Missionary Society. (1825b). *Minutes*, dated 28 November 1825. Church Missionary Society Archive. G/C1/8.

Church Missionary Society. (1829a). *Minutes*, dated 18 April 1828. Church Missionary Society Archive. G/C1/9.

Church Missionary Society. (1829b). *Minutes*, dated 7 August 1829. Church Missionary Society Archive. G/C1/10.

Church Missionary Society. (1836). *Minutes*, dated 20 December 1836. Church Missionary Society Archive. G/C1/13.

Clarke, R. M. (2015). *Vanguards of Empire: The lives of William Dawes, Watkin Tench and George Worgan.* Ph.D. Thesis, The Australian National University.

Currer-Jones, A. (1930). *William Dawes, R. M., 1762–1836*. Torquay UK, Smith.

Dawes, G. (1821a). Letter to the Church Missionary Society Secretary, dated 15 August 1821. Church Missionary Society Archive. W/O 32/1.

Dawes, G. (1828a). Letter to the Church Missionary Society Secretary, dated 10 March 1828. Church Missionary Society Archive. W/O 32/11.

Dawes, W. (1816a). *Missionary Register* (p. 144), dated 1 April 1816.

Dawes, W. (1816b). Foreign intelligence. *Missionary Register*, dated 3 April 1816. Cited by Thomas (2013a: 200).

Dawes, W. (1820a). Letter to the Church Missionary Society Secretary, dated 1 April 1820. Church Missionary Society Archive. Section If: *Missions to the Americas. I: Mission 1819–1861* (1999). Marlborough, Adam Matthew.

Dawes, W. (1820b). Letter to the Church Missionary Society Secretary, dated 9 June 1820. Church Missionary Society Archive. Section V, C/W/0/4, Reel 8. Cited by Thomas (2011a).

Dawes, W. (1820c). Letter to the Church Missionary Society, n.d. (ca. October 1820). Church Missionary Society Archive. W/O 31/8b.

Dawes, W. (1821a). Letter to the Church Missionary Society Secretary, dated 15 October 1821. Church Missionary Society Archive. W/O 31/14.

Dawes, W. (1821b). Letter to the Church Missionary Society Secretary, dated 17 December 1821. Church Missionary Society Archive. W/O 31/15.

Dawes, W. (1822a). Letter to the Church Missionary Society Secretary, dated 13 January 1822. Church Missionary Society Archive. W/O 31/127.

Dawes, W. (1822b). *Work journal*, 18 February 1822. Cited by Thomas (2011a).

Dawes, W. (1822c). Letter to the Church Missionary Society Secretary, dated 1 June 1822. Church Missionary Society Archive. W/O 31/21.

Dawes, W. (1822d). *Rough journal*, ca. July 1822d; filed immediately before Church Missionary Society Archive. W/O 31/129. Cited by Clarke (2015: 101).

Dawes, W. (1823a). Letter to the Rev. Josiah Pratt, dated 7 March 1823. Church Missionary Society Archive. W/O 31/28.

Dawes, G. (1823b). Letter to the Church Missionary Society Secretary, dated 15 March 1823. Church Missionary Society Archive. W/O 32/3.

Dawes, W. (1823c). Letter to the Church Missionary Society Secretary, dated 15 April 1823. Church Missionary Society Archive. W/O 31/30.

Dawes, W. (1823d). Letter to the Church Missionary Society Secretary, dated 8 May 1823. Church Missionary Society Archive. W/O 31/31.

Dawes, W. (1824a). Letter to the Rev. Josiah Pratt, dated 26 March 1824. Cited by Thomas (2011a).

Dawes, W. (1824b). Letter to the Church Missionary Society Secretary, dated 12 May 1824. Church Missionary Society Archive. In *The papers of William Dawes, missionary in Antigua. I: Cultural Contacts, 1492–1969* (p. 15). University of Birmingham Library. CW O 31.

Dawes, W. (1824c). Letter to the Church Missionary Society Secretary, dated 24 August 1824. Church Missionary Society Archive. Section V: *Missions to the Americas. I: West Mission 1819–1861* (1999). Marlborough: Adam Matthew.

Dawes, W. (1824d). Letter to the Church Missionary Society Secretary, dated 20 September 1824. Church Missionary Society Archive. W/O 31/135. Section V, C/W/M/1, Reel 2.

Dawes, W. (1825a). Report, dated 31 March 1825. CW/O,6,8. Oxford, Rhodes House Archives. Cited by Pybus (2009).

Dawes, W. (1825b). Letter to the Church Missionary Society Secretary, dated 11 May 1825. Church Missionary Society Archive. W/O 31/52. Section V: *Missions to the Americas. I: West Indies Mission 1819–1861* (1999). Marlborough, Adam Matthew.

Dawes, W. (1825c). Letter to the Church Missionary Society Secretary, dated 18 June 1825. Church Missionary Society Archive. Church Missionary Society Archive. W/O 31/56. Section V: *Missions to the Americas. I: West Indies Mission 1819–1861* (1999). Marlborough, Adam Matthew.

Dawes, W. (1825d). Letter to the Church Missionary Society Secretary, dated 21 June 1825. Church Missionary Society Archive. W/O 31/57.

Dawes, W. (1825e). Letter to the Church Missionary Society Secretary, dated 2 July 1825. In *The papers of William Dawes, missionary in Antigua. I: Cultural Contacts, 1492–1969* (p. 50). University of Birmingham Library. C W/O 31.

Dawes, W. (1825f). Letter to the Rev. E. Bickersteth, dated 29 July 1825. Church Missionary Society Archive. C W/O 31/59b.

Dawes, W. (1825g). Letter to the Church Missionary Society Secretary, dated 2 September 1825. Cited by Thomas (2011a).

Dawes, W. (1825h). Letter to the Church Missionary Society Secretary, dated 19 September 1825. Church Missionary Society Archive. W/O 31/63.

Dawes, W. (1825i). Letter to the Church Missionary Society Secretary, dated 14 October 1825. Church Missionary Society Archive. W/O 31/65.

Dawes, W. (1825j). Letter to the Rev. Edward Bickersteth, dated 15 November 1825. Cited by Thomas (2011a).

Dawes, W. (1825k). Letter to to the Church Missionary Society Secretary, dated 28 November 1825. Church Missionary Society Archive. In *The papers of William Dawes, missionary in Antigua. I: Cultural Contacts, 1492–1969* (p. 89). University of Birmingham Library. C W/O 31.

Dawes, W. (1826a). Letter to the Church Missionary Society Secretary, dated 6 January 1826. In: *The papers of William Dawes, missionary in Antigua. I: Cultural Contacts, 1492–1969* (p. 92). University of Birmingham Library. C W/O 31.

Dawes, W. (1826b). Letter to the Church Missionary Society Secretary, dated 24 April 1826. Church Missionary Society Archive. W/O 31/71.

Dawes, W. (1826c). Letter to the Church Missionary Society Secretary, dated October 1826. Cited by Thomas (2011a).

Dawes, W. (1826d). Memorial of William Dawes, December 1826. Mitchell Library, State Library of New South Wales. Ad. 49, Microfilm CY1414, frames 31–37.

Dawes, W. (1827a). Letter to the Church Missionary Society Secretary, dated 1 August 1827. Church Missionary Society Archive. W/O 31/89a.

Dawes, W. (1827b). Letter to the Church Missionary Society Secretary, dated 4 September 1827. Cited by Thomas (2011a).

Dawes, W. (1828b). Letter to the Church Missionary Society Secretary, dated 4 July 1828. Church Missionary Society Archive. W/O 31/97.

Dawes, W. (1829a). Letter to D. Coates, dated 5 May 1829. Cited by Thomas (2011a).

Dawes, W. (1829b). *Advertisement.* Antigua Free Press, 26 June 1829.

Dawes, W. (1832). Letter to the Church Missionary Society Secretary, dated 20 March 1832. Church Missionary Society Archive. W/O 31/118.

D'Urban, B. (1824). Letter to Earl Bathurst, dated 12 March 1824. Kew, UK National Archives. Colonial Office 7/10/183.

English Harbour Sunday School Society (1822a). *Resolutions passed at the General Meeting of 13 May 1822.* Section V, Church Missionary Society Archive. W/O/6, Reel 9.

English Harbour Sunday School Society (1822b). *Annual Report.* Cited by Thompson (2011a).

Gibson, R. (2012). *26 Views of the Starburst World. William Dawes at Sydney Cove 1788–91.* Crawley WA, University of Western Australia Publishing.

Gilbert, J. (1835). *Memoir of John Gilbert, Esq., Late Naval Storekeeper at Antigua. To Which Are Appended, a Brief Sketch of His Relic Mrs. Anne Gilbert, by the Rev. William Box, Wesleyan Missionary. And a Few Additional Remarks, by a Christian Friend* (Vol. 3). Liverpool, D. Marples and Company.

Gilbert, M. (1822). *Release from Miss Martha Gilbert and Mr William Dawes and Grace his Wife of two annuities of £20 and two legacies of £500 charged contingent upon the Estates of The Reverend Nathaniel Gilbert in Antigua.* The Beinecke Lesser Antilles Collection. Hamilton College, Clinton NY. MSS M517; cited by Hough and Hough (1994: 294).

Glen, R. (2001). The history of early Methodism in Antigua: a critique of Sylvia R. Frey and Betty Wood's 'Come shouting to Zion: African American Protestantism in the American South and British Caribbean to 1830'. *Journal of Caribbean History, 35*, 253–VII.

Hole, R. C. (1896). *The early history of the Church Missionary Society for Africa and the East to the end of A.D. 1814.* London, Church Missionary Society.

Hough, S., & Hough, P. (eds.). (1994). *The Beinecke Lesser Antilles Collection at Hamilton College: A catalogue of books, manuscripts, prints, maps and drawings, 1521–1860.* Gainesville FL, University Press of Florida.

James, S. M. (1993). Mothering: A possible black feminist link to social transformation? In S. M. James, & A. P. A. Busia (eds.), *Theorizing black feminisms: The visionary pragmatism of black women.* London, Routledge.

Lambert, D. (2005). *White Creole culture, politics and identity during the age of Abolition.* Cambridge University Press.

Lazarus-Black, M. (1994). *Legitimate acts and illegal encounters: Law and society in Antigua and Barbuda.* Smithsonian Institution Press.

London Metropolitan Archives. (1811). *Marriage registers.* Old Church St. Pancras, London.

Midgely, C. (1992). *Women against Slavery: The British Campaigns 1780–1870.* Routledge.

Oliver, V. L. (1896). *The History of the Island of Antigua, one of the Leeward Caribbees in the West Indies, from the first settlement in 1635 to the present time* (Vol. 2). Mitchell and Hughes.

Parish Registers of St Pancras Old Church. (1660–1916). Church of England. St Pancras Old Church.

Porteus, B. (1808). In T. Cadell, T. Payne, F.C., & J. Rivington (eds.), *A letter to the governors, legislatures, and proprietors of plantations in the British West-India Islands.* London, T. Cadell, T. Payne, & F. C. and J. Rivington.

Pybus, C. (2007). 'A less favourable specimen': The Abolitionist response to self emancipated slaves in Sierra Leone, 1793–1808. *Journal of Parliamentary History, 26*, 97–112.

Pybus, C. (2009). 'Not fit for your protection or an honest man's company': A transnational perspective on the Saintly William Dawes. *History Australia, 6*(1), 12.1–12.7.

Richter, Br. C. F. (1825). Letter to the Committee of the London Association in Aid of the Moravian Missions, dated 12 February 1825. In *Particulars respecting the schools for Negro children, &c. under the direction of the Moravian Missionaries in the West Indies.* London, 1826.

Stobwasser, L. (1823). Letter to the Committee of the London Association in Aid of the Moravian Missions, dated 3 June 1823. In *Particulars respecting the schools for Negro children, &c. under the direction of the Moravian Missionaries in the West Indies.* London, 1826.

Taylor, Br. J. (1825). Report to the Committee of the London Association in Aid of the Moravian Missions, dated 14 November 1825. In *Particulars respecting the schools for Negro Children, &c. under the direction of the Moravian missionaries in the West Indies.* London, 1826.

Tench, W. 1926 (1826). The memorial of Lieutenant William Dawes (Antigua, December 1, 1826). *Journal and Proceedings of the Royal Australian Historical Society, XII*, 227–228.

*The Missionary Register.* (1836) P. 528.

Thomas, S. N. A. (2011a). William Dawes in Antigua. *Journal of Colonialism and Colonial History, 12*(1). https://doi.org/10.1353/cch.2011.0011

Thomas, S. (2011b). Anne Hart Gilbert, Creole benevolence and anti-slavery, 1815–1834. *Nineteenth-Century Contexts, 33*, 227–245.

Thomas, S. (2013a). A transnational perspective on William Dawes' treatment of women. *History Australia, 10*, 187–204.

Thomas, S. (2013b). Affective dynamics of colonial reform and modernisation in Antigua, 1816–1835. *Feminist Review, 104*, 24–41.

Thompson, E. W. (1960). *Nathaniel Gilbert, Lawyer and Evangelist.* London, Epworth.

Van Sant, A. J. (1993). *Eighteenth-century sensibility and the novel: The senses in social context*. Cambridge University Press.

Watts, I. (1715). *Divine songs attempted in easy language for the use of children with some additional composures*. Early English Books Online Text Creation Partnership, 2011. Retrieved August 23, 2022, from https://quod.lib.umich.edu/e/ecco/004819385.0001.000/1:5.8?rgn=div2;view=fulltext

Watts, I. (1825). *Dr. Watts's plain and easy catechisms for children: The shorter catechism of the assembly of divines at Westminster with explanatory notes, and a preservation from the sins and follies of childhood and youth*. Cambridge, Hilliard and Metcalf.

Wright, H. (1927). Further note on Lieutenant William Dawes. *Journal of the Royal Australian Historical Society, 13*, 63–64.

# Chapter 7
# Astronomical Tent Observatories in the Late Eighteenth Century

## 7.1 Ship or Shore?

By the end of the 'Age of Sail', roughly corresponding to the period from the mid-fifteenth to the mid-nineteenth centuries, it had become common for long-haul oceanic voyages of discovery and exploration to enlist professional astronomers among ships' crews. Their services were indispensible for accurate geographic position determinations once the expedition had strayed beyond familiar waters.

As we have discussed already, determination of one's latitude was straightforward. Longitude determination, on the other hand, required access to an accurate timepiece that reflected the time at the ship's home port or at a reference meridian. Alternatively, longitudes could be determined by means of extensive, cumbersome calculations based on lunar distance measurements (e.g., de Grijs, 2020), combined with up-to-date nautical almanacs that included tables of lunar distances and their timings for one's reference meridian.

Accurate, compact (box) chronometers eventually became more commonplace on voyages of discovery by the mid-nineteenth century. Until then, however, expeditions' time measurements relied on a combination of early marine chronometers and lunar distance observations. Early marine timepieces often gained or lost time over the course of a voyage, and so lunar distance observations were essential for calibration purposes, that is, lunar distances were used to check the 'going rates' of shipboard clocks.

When in port or close to shore, such checks were often preferentially done on land and based on the most accurate time measurements available, those obtained from the type of long-case regulators carried on most long-haul voyages. Such 'grandfather' clocks often took several days to set up and settle, however (e.g., Davies, 2022: Part 2, Chap. 10), and they could only be used appropriately on land. In addition to those rather awkward clocks, the astronomer's arsenal of instruments required for accurate time measurements included one or more sextants, a universal theodolite,

---

This chapter is an abbreviated version of de Grijs (2023).



R. de Grijs and A. Jacob, *William Dawes*, Springer Biographies,
https://doi.org/10.1007/978-3-031-38774-6_7

chronometers and nautical almanacs, and often also a smaller 'journeyman' clock. To protect these instruments from the vagaries of the weather—and sometimes from the local population—from the mid-eighteenth century onward shipboard astronomers would often set up canvas tent observatories on shore.

Tent observatories were most often set up on a wooden frame. The instruments were generally placed on wooden casks, originally used to transport food, water, ale, wine or rum, which were filled with heavy, wet sand as ballast for stability (e.g., Orchiston, 2004, 2017). Having established stable shore-based observatories allowed the astronomers to obtain more accurate measurements than could be obtained on rocking and pitching ships.

At sea, they would obtain observations to determine their geographic coordinates, as well as magnetic and meteorological observations. On land, they would additionally record tidal observations, whereas the stable operation of a pendulum clock would also allow them to obtain gravity measurements (e.g., Morrison & Barko, 2009; Bosloper, 2010, 2017; see also Chap. 3.1).

Tent observatories were not in use until James Cook's first voyage to the Pacific, New Zealand and Australia of 1768–1771. They were, in fact, only used for a relatively short time period, until the development of reliable, compact maritime timepieces had made the use of tent observatories on maritime voyages of exploration redundant by the end of the eighteenth century. One of the last major sea-bound explorations carrying a tent observatory was Matthew Flinders' 1801–1802 circumnavigation of Australia on H.M.S. *Investigator*. Matthew Flinders (1774–1814), and increasingly his brother Samuel Ward Flinders (1782–1834), used a Ramsden universal theodolite, a Hadley sextant and an Earnshaw astronomical regulator to determine their longitudes using the lunar distance method. Pitching a tent observatory on shore, they then used their lunar distances to check the 'going rate' of their clocks (Kenneth, 2017).

## 7.2 Captain Cook's Tent Observatories

In the late 1760s, the Royal Society of London actively encouraged observations of the 1769 transit of Venus across the face of the Sun, funding several expeditions to do so. Venus transits, particularly when combining observations from multiple, geographically distinct vantage points, offer a unique opportunity to determine the distance from the Earth to the Sun, and hence to determine the size of the solar system as a whole. Cook's first voyage to the Pacific on H.M.B. *Endeavour* was one such expedition. It was reportedly the first scientific expedition equipped with an experimental prototype tent observatory. Constructed under the supervision of Nevil Maskelyne and designed by John Smeaton (1724–1792), a civil engineer, it was a heavy and rather cumbersome canvas structure (see Fig. 1.5).

The *Endeavour* arrived in Matavai Bay, on Tahiti's northern coast, on 13 April 1769, well in time to prepare for the transit observations of 3 June of that year. Cook selected Point Venus, a narrow peninsula known locally as Te Auroa, as the expedition's base. The protruding landmass represented an ideal observatory site; it could be accessed easily from the ocean and it was readily fortified as well. The *Endeavour*'s crew proceeded to build a banked, ditched and pallisaded enclosure containing the expedition's tents and equipment to "... protect the observers and the instruments from the natives ..." (Badger, 1969: 37),[1] which became known as Fort Venus. Its defences were bolstered by some of the cannons and swivel guns from the *Endeavour*.

Cook set up a tent containing a clock with a grid-iron pendulum, established on a wooden frame firmly fixed in the ground:

> The astronomical clock, made by Shelton and furnished with a gridiron pendulum, was set up in the middle of one end of a large tent, in a frame of wood made for the purpose at Greenwich, fixed firm and as low in the ground as the door of the clock-case would admit, and to prevent it being disturbed by any accident, another framing of wood was made round this, at the distance of one foot [30.5 cm] from it. (Green & Cook, 1771: 397)

The pendulum's length was adjusted to the same length as that at Greenwich to allow for gravity observations (e.g., Bosloper, 2010, 2017; Chap. 3.1). Green and Cook (1771: 398) later described the camp's design, pointing out that facing the tent containing the long-case regulator "... stood the observatory, in which were set up the journeyman clock and astronomical quadrant: this last, made by Mr. [John] Bird [1709–1776], of one foot radius, stood upon the head of a large cask fixed firm[ly] in the ground, and well filled with wet heavy sand". The observatory also contained three pivoted, reflecting telescopes made by James Short (1710–1768; Turnbull, 2004).

Cook was apparently lulled into a false sense of security: "I now thought myself perfectly secure from anything that these [native] people could attempt ..." (cited by Captain Cook Birthplace Museum, n.d.). However,

> ... when Mr. [Charles] Green and I went to set up the Quad.[t] it was not to be found, it had never been taken out of the Packing case ... and the whole was pretty heavy, so that it was a matter of astonishment to us all how it could be taken away, as a Centinal [*sic*; sentinel] stood the whole night within 5 yards [4.6 m] of the door of the Tent where it was put ... (*Ibid.*)

The quadrant was eventually recovered—and repaired—well before it was required for the Venus transit observations.

On Cook's second and third voyages to the Pacific of 1772–1775 and 1776–1779, the heavy and cumbersome Smeaton observatory design was discarded. Instead, Cook's astronomer William Bayly designed "... undoubtedly one of the most convenient portable Observatories that had yet been made", equipped with hinges that allowed "... the roof to open and close like an umbrella." (Wales & Bayly, 1777: vii). Bayly's and Smeaton's observatory designs replaced the less flexible, rigid wooden huts commonly used on earlier voyages.

[1] Another example involves a voyage back from Sydney, via China, to England in 1802, where James Inman (1776–1859) paid £8 as "A Present to [a] Convict Servant on leaving Port Jackson [Sydney Harbour], for his great Care, & Attention in Watching at the Tent Observatory, when Indisposition prevented me from sleeping at it myself". (Inman, 1802).

Cook undertook his second voyage to the Pacific and down to Antarctic waters with two ships, H.M.S. *Adventure* (commanded by Captain Tobias Furneaux, 1735–1781) and H.M.S. *Resolution*, under Cook's command. Bayly was assigned to astronomy duties on the *Adventure*, whereas William Wales served as the *Resolution*'s astronomer. Both were equipped with identical tent observatories, which "... will be of infinite use to them whenever they may have occasion to make observations on shore ..." (cited by Beaglehole, 1969: 722). Wales and Bayly (1777) provided a detailed description of their observatory's construction:

> The upright sides consist of eight staves, *AB*, *CD*, &c. (see Plate II) [Fig. 2.2] about two inches [5.1 cm] diameter, and five feet and an [*sic*] half [1.68 m] long, which supported a circular ring, 1, 2, 3, 4, &c. to 21. of eight feet [2.44 m] diameter, and the covering, *r*, *q*, 9, 10, &c. to 21, *o*, *p*, of oiled canvass. The staves are of beach-wood [*sic*; beech-wood], armed at the bottom with spikes, to stick into the ground, and at the top with small iron pins, fitted to holes which are made to receive them in the ring. The ring is composed of eight circular arches, of about three feet [91 cm] long, two inches broad, and an inch [2.5 cm] thick, made of beach-wood, and are readily put together, or taken asunder by means of strong iron plates, screwed fast with wood screws to the end of one arch, and by screws and nuts to the end of another, for the purpose of frequent screwing and unscrewing without danger of wearing out the holes, as would be the case with wood screws entering the wood itself. Into the outer edge of this ring are drove small staples, 1, 2, 3, 4, &c. and to the upper edge of the canvass, answerable thereto, are sewed several small hooks, which being hooked into these staples, serve to support the upper edge of the canvass, while its lower edge just reaches to the ground: The two parts of the canvass, 2, 1, *o*, *p*; 9, *q*, *r*, are supposed to be unhooked from the staples 1, 2, 3, 4, and 5, 6, 7, 8 respectively, and thrown back to shew inside the Observatory, and the manner of fixing up the Clock, ...: *BE* is a brace of the same sort of wood, screwed fast to the top of the staff *AB*, by a screw at *B*, and to the bottom of the staff *CD* at *E*. These braces, from the top of one staff to the bottom of the next, kept the whole upright circular frame very steady. *FGHIKLMN* is another circular ring, exactly of the same dimensions and construction with the former, on which it rests: To this the roof of the Observatory is screwed by means of ten long screws, which pass through the end of the rafters at *FGHIK*, &c. into iron nuts fixed in this upper ring for that purpose. The rafters *MP*, *RP*, *IU*, *KE*, &c. are attached to the crown-piece *PTU* by hinges, as represented at *T*, and *U*; and the two short rafters *FQ*, *NO*, are attached to the two *RP*, *MP*, also by hinges at *O*, and *Q*: By means of these hinges the roof is made to open or close like an umbrella, and of course, if disengaged from the circular ring *FRH*, &c. will fold together, and may be packed up in a very small compass.
>
> The covering of the roof is of very thick canvas, many times painted, and comes down so far as to hang over at the eves about four inches [10.2 cm]. The crown-piece, *PTU*, is about eight inches [20.3 cm] in diameter, and covered with a circular piece of canvas like that the roof is covered with. An eye-bolt *n* ⊙ passes through its center [*sic*], and is fastened on the inside by the nut ¤. This eye-bolt is intended for the reception of the hook *n*, which is fastened to the cord *mbgcd*, passing over a pulley at *W*, fixed in the top of the pole *VZ*. Towards the bottom of this pole there is fixed a lever *gh*, by means of the clamp *ef*, and its fellow on the opposite side, and the lever, turns on the iron bolt *f*. The cord *mbcd* passes through a hole *c* in the lever, and is drawn right when the end *b* of the lever is turned upwards, and then made fast: Now if the end *b* of the lever be brought down towards *z*, and there fastened by means of the becket [loop of rope], or endless cord *ik*, the roof of the Observatory will be drawn up from off the ring 1, 2, 3, &c. and may be turned round by twisting or unwinding of the cord, until the opening *NOPQE* is towards the sun, or any other object, of which an observation is wanted to be made. When the observation is completed, the lever may be released, and the roof is let down again to rest with its whole weight on the lower ring, as it will then be less

> liable to be disturbed by the wind: There are also eight small staples on the inside edge of the lower ring 1, 2, 3, &c. and as many small hooks, corresponding to them, on the upper, or that to which the rafters of the roof are fastened. These hooks, when the roof is lowered down, are to be hooked into the staples, and the cord then drawn tight, to prevent, yet farther, the effect of the wind. The opening *N*, *OP*, *QF*, is covered when not in use, by the flap, or roll of spare canvas *QRGS*, which is of the same sort, and painted in the same manner, as that which covers the roof. The whole of this Observatory, except the three poles *WZ*, *WX*, and *WY*, when taken down and packed up properly, is contained in a chest six feet and nine inches [2.06 m] long, and about 20 inches [50.8 cm] square: The poles, which form the tripod, are of about fifteen feet [4.57 m] long, and four inches diameter, may be laid amongst the spare booms of the ship, or if they should be thought too cumbersome there, may be cut out of the woods, or purchased for a trifle at any place where they are wanted. (Wales & Bayly, 1777: viii–x)

Wales and Bayly were both equipped with a similar complement of instrumentation as Cook and Green had carried on Cook's first voyage. They carried Gregorian reflecting telescopes (manufactured by Bird), astronomical quadrants, sextants, clocks and watches. In addition, the astronomers now had access to two Dollond refracting telescopes, a transit telescope constructed by Bird and four chronometers, which were modeled after John Harrison's prize-winning H4 'longitude clock' (Orchiston & Howse, 1998). Wales was given a chronometer made by John Arnold (1736–1799) and also Larcum Kendall's 'K1', an exact copy of Harrison's H4. Bayly was equipped with two Arnold chronometers (Betts, 1993).

The *Resolution* and *Adventure* left Plymouth Sound on 13 July 1772. Bayly's 'observations book' offers an interesting account of the difficulties shipboard astronomers would often encounter in attempting to carry out their duties on foreign shores. We get the distinct impression that setting up their tent observatories and the larger astronomical instruments and pendulum clocks was anything but straightforward (e.g., Baker, 2012). Nevertheless, the scientists exhibited patience and creativity in overcoming most obstacles.

For instance, in July 1772 Bayly screwed his astronomical regulator to a heavy bookcase in the Consul's House at Funchal (Madeira), because "… the book-case stood on a floor that was paved with Bricks and it was full of Books which rendered it very steady" (Bayly, 1772a). Upon arrival at Table Bay by October 1772, he had to overcome additional problems. His initial attempt at weighing down the stand of his astronomical quadrant by filling it with water failed when the water began leaking, and so he resorted to using sand instead. Moreover, on 4 November 1772, they "… Had a strong wind blowing at the SSE which brought great Quantities of sand from the Table mountains which greatly Affected the Instruments by Covering them with sand and shaking them and It was with difficulty we secured the tents from oversetting" (Bayly, 1772b). Both ships left the Cape of Good Hope on 22 November 1772.

On 7 April 1773, the *Adventure* reached Queen Charlotte Sound in New Zealand, where they anchored in Ship Cove near Hippah Island, initially referred to as 'Observatory Island'. Once again, Bayly was forced to creatively solve a number of practical problems before he could proceed with his astronomical observations. Since Bayly did not get along with Furneaux and the other officers on the *Adventure*, the astronomer could only rely on his own servant and an unofficial assistant, "… a good natured Welshman, who would always work if I gave him Brandy …" (Bayly, 1773a: 207). On 12 April, Bayly recorded in his observations book that

> On the top of this Island saw the only favourable spot I could find any where near the Ship to Observe Equal Altitudes with any propriety, but twas with the greatest difficulty I got the Instruments up, being first obliged to make steps in the Rocks, but as I had only two men sent with me, [they] were only able to cut away the Weeds in order to set up my Tent & make the Aforesaid steps that day. (Bayly, 1773b: 27r)

On 16 April, he "[c]arried [his] Instruments and the Tent Observatory on Shore, together with the iron blocks and [illegible] belonging to the Astronomical Clocks, and it was with the utmost difficulty we got them to the top of the Island" (*ibid.*). They pitched their tent observatory and built a temporary hut for the transit telescope.

On 19 April Bayly set up his astronomical regulator on Hippah Island:

> In one corner of the House (I had built) I sank a hole in the Rock about 12 Inches [30 cm] from Each Side of the House, & about 16 Inches [41 cm] deep; in the bottom of this hole on the Solid Rock, the Block of Iron [to support the clock] was laid as nearly Horizontal as possible, so that the bottom of the Clock-case was about 14 Inches [36 cm] below the floor of the House, & the Clock stood quite independent of every part of it. (*ibid.*)

Eventually, on 21 April he set up the transit telescope, with further adjustments made on 24 April:

> In the morning I moved the Transit Instrument very near the true Meridian by means of the Adjusting Screws, where it Accidentally cut two good marks on the tops of two Hills one to North distant about 1½ Mile [2.4 km]; & the other to the South about 4½ Miles [7.2 km]; the Instrument was constantly kept to these marks during the whole time it remained up, by examining it three or more times a day, & frequently before each observation when in the Day: as was likewise the Level, & the Posts was so well rammed that I never found it [out] more than a bredth [*sic*] of the Wire ... (*ibid.*)

Having become separated from the *Adventure* in heavy fog on 8 February 1773, the *Resolution* eventually arrived at their pre-arranged rendezvous location in Queen Charlotte Sound on 17 May 1773, following a short sojourn at the safe and deep anchorage in Pickersgill Harbour at Dusky Bay, today known as Dusky Sound.[2] The *Resolution*'s qualified astronomers (Cook and Wales) had not been idle, however. They used their reprieve from the high seas to acquire astronomical observations in Dusky Bay, specifically to determine the site's longitude and latitude.

During Cook's second voyage to the Pacific, the *Resolution* returned twice more to Queen Charlotte Sound, in 1773 and 1774. On all three occasions, Wales diligently recorded their latitude and longitude (see Orchiston, 1997; Orchiston & Howse, 1998). The *Adventure* also returned to Queen Charlotte Sound, but only once, on 1 December 1773, and they missed a rendezvous with the *Resolution* by six days on that occasion. Position determination was also Bayly's principal concern, particularly given the Sound's great potential as a replenishment destination for future Pacific voyages. Bayly's diligent record keeping provides us with useful insights into his use of the tent observatory during their second stop-over in Queen Charlotte Sound:

---

[2] Richard Pickersgill (1749–1779) was the *Resolution*'s third lieutenant.

> [On 3 December 1773, I] Carried my Tent Observatory & Instruments on Shore & set all up by the side of the Garden a little distance from the watering place [in Ship Cove] ... The place where my Observatory stood when we were here before bearing E 5½° W per Compass distant ... (Bayly, 1773c)

Bayly set out to obtain a series of gravity observations with the cumbersome astronomical regulator. Unfortunately, however, during this stop-over, security became a serious concern, as we learn from his notes of 14 December (see also Henry & Marra, 1775; Bosloper, 2017):

> I was up late observing & having taken some altitudes of Stars to the East & having set my Alarem [*sic*] to call me, to take them to the West, I went to bed, having nailed my old great coat at the entrance of my tent, at the inside of which I always placed the outside case of my Astronomical Quadrant ... In this box I kept my Lumber, such as tools, nails, &c. ... After I had been in bed & had slept some time, I was awaked [*sic*] by the rattling or noise of the lid of the box. I jumped up in the bed & took my gun ... calling at the same time, who was there, but could neither hear nor see anything ... Soon after I searched for my hat but could not find it ... on going out of the Tent Obs.$^{y}$I found it open & half the lid of the box at some dist. from the tent, & by feeling I found my hatchet & saw & hammer were gone out of the box. (Bayly, 1773a: 215–216)

Bayly's observatory design had proved is efficacy in storing and protecting the large astronomical regulators commonly carried on long-haul oceanic voyages. And so, on Cook's third voyage to the Pacific, tent observatories of the same type were deployed once again in Ship Cove, Queen Charlotte Sound, in February 1777 (e.g., Greenhill, 1970: 27; Orchiston, 2016). Cook, acting as the shipboard astronomer on H.M.S. *Resolution* once again, and Bayly, on H.M.S. *Discovery*, pitched their tent observatories close to one another in both Ship Cove and also in Nootka Sound at what is now Vancouver Island, Canada (Orchiston & Wells, 2020), between 29 March and 26 April 1778. Cook initially called their Nootka Sound mooring site Ship Cove, after his preferred site in New Zealand waters. It is now called Resolution Cove and is located at the southern end of Bligh Island.

## 7.3 Standard Equipment

Cook's three voyages to the Pacific served as practical trials of the latest tent observatory design. Satisfied that Bayly's design served the needs of long-haul expeditions well, tent observatories soon became standard equipment on voyages of exploration supported by the British Admiralty and the British Board of Longitude. By the late 1780s, Maskelyne—in his dual roles as Britain's Astronomer Royal and Commissioner of the Board of Longitude—proceeded to define a list of standard equipment for expeditionary voyages (Howse, 1979).

Standard observations Maskelyne expected his astronomers to undertake included determinations of latitudes and longitudes, magnetic variation and dip, air temperature, positions of headlands, islands and harbours, and the heights and times of tides (e.g., Beaglehole, 1999: 1500–1503). Maskelyne's instructions explicitly directed his astronomers to use the tent observatory they would be provided with:

> Observations to be made on shore:-
>
> 1. Whenever you land, if time permits, you are to set up the tent Observatory, and astronomical clock; either setting the pendulum exactly to the same length as it was of at Greenwich, before the voyage or noting the difference by the revolutions of the screw and divisions of the nut at bottom; ... (Maskelyne, 1801)

As a case in point, William Gooch (1770–1792), the astronomer on board the Vancouver expedition of 1791–1795, wrote to his parents (Gooch, 1791; our emphasis),

> Perhaps you'll like to know what Instruments I'm to take abroad, ... most of them the same that went with Cap.$^{n}$ Cook:
>
> 1. An Astronomical Clock
>
> 2. A Journeyman Clock
>
> 3. An Alarum Clock
>
> [4.] A Good Watch w$^{th}$ Second Hand
>
> [5.] An Achromatic Telescope of 46 Ins. Focus w$^{th}$ a divided Object-Glass Micrometer
>
> 6. A Reflecting Telescope
>
> 7. A Vertical Circle with an Azimuth-circle for taking altitudes and azimuths
>
> 8. A Transit Instrument of 4 Feet with a Level & upright wooden Posts
>
> 9. A Marine dipping needle [a freely suspended magnetic needle used to determine the local magnetic inclination]
>
> 10. A small Pocket compass
>
> 11. A Set of Magnetic Bars to change the Poles of the dipping Needle
>
> 12. A Burton's Theodolite with Stand
>
> 13. A Hadley's Sextant by Dollond
>
> 14. Another by Troughton
>
> 15. Two large Thermometers
>
> 16. Two Thermometers with wooden Scales, by Ramsden
>
> 17. A Portable Barometer by Burton
>
> 18. A Bason [basin] to hold Quicksilver [mercury] with Glass Roof
>
> 19. Quicksilver in a Bottle
>
> 20. A Night Telescope
>
> 21. A Steel Gunter Chain
>
> 22. A Knight Azimuth Compass
>
> **23. A Portable Tent Observatory**
>
> besides Books

Maskelyne had explicitly ordered that a portable observatory tent was made for Gooch to take on the Vancouver expedition (Maskelyne, 1791). By this time, however, it appears that tent observatories could now be supplied directly by the Navy Board, without the need to involve the Astronomer Royal (e.g., MacDonald & Withers, 2016: 31).

Gooch's standardised set of equipment broadly resembles that taken by William Dawes on the First Fleet in 1787–1788 (see Chap. 1).

## References


Badger, G. M. (1969). Cook the scientist. In G. M. Badger (Ed.), *Captain Cook: Navigator and Scientist. Papers Presented at the Cook Bicentenary Symposium, Australian Academy of Science, Canberra* (pp. 30–49).

Baker, A. (2012). "Precision," "Perfection," and the reality of British scientific instruments on the move during the 18th century. *Material Culture Review, 74*. Retrieved December 21, 2022, from https://journals.lib.unb.ca/index.php/MCR/article/view/20446.

Bayly, W. (1772a). *Observations taken at Funchal Town on the Island of Madiera* [*sic*], 30 July 1772. RGO 14/57, folio 3r. Cambridge University Library.

Bayly, W. (1772b). *Observations book*. RGO 14/57, folio 12r. Cambridge University Library.

Bayly, W. (1773a). Extract from the journal kept by William Bayly, astronomer, on HMS *Adventure*, Captain Furneaux, during Capt. Cook's second voyage. In R. McNab (Ed.), 1914. *Historical Records of New Zealand*. Wellington, Government Printer (pp. 201–218).

Bayly, W. (1773b). *Observations book*. RGO 14/57, folio 27r. Cambridge University Library.

Bayly, W. (1773c). *Observations book*. RGO 14/57, folio 55r. Cambridge University Library.

Beaglehole, J. C. (ed.) (1969). *The Journals of Captain James Cook on his voyages of discovery. II. The voyages of the* Resolution *and* Adventure, *1772–1775*. Cambridge, Cambridge University Press.

Beaglehole, J. C. (ed.) (1999). *The Journals of Captain James Cook on his voyages of discovery. III. The voyage of the* Resolution *and* Discovery, *1776–1780*. Part II. Woodbridge, Boydell Press.

Betts, J. (1993). The eighteenth-century transits of Venus, the voyages of Captain James Cook and the early development of the marine chronometer. *Antiquarian Horology, 21*, 660–669.

Bosloper, C. (2010). William Dawes' gravity measurement in Sydney Cove, 1788. In *XXIV FIG International Congress 2010: Facing the Challenges—Building the Capacity. TS 6M History of Surveying* (14pp.). Retrieved September 3, 2022, from https://www.icsm.gov.au/sites/default/files/CongressPaper4534.pdf

Bosloper, C. (2017). The first absolute gravity observations in Australia and New Zealand. In V. Janssen (ed.). *Proceedings of the 22nd Association of Public Authority Surveyors Conference* (pp. 247–257). Retrieved September 3, 2022, from https://www.apas.org.au/files/conferences/2017/APAS2017-proceedings.pdf

Captain Cook Birthplace Museum. (n.d.). *Themes: Transit of Venus*. Retrieved December 20, 2022, from https://www.captcook-ne.co.uk/ccne/themes/venus.htm

Davies, A. C. (2022). *The rise and decline of England's watchmaking industry, 1550–1930*. Routledge.

de Grijs, R. (2020). A (not so) brief history of lunar distances: Longitude determination at sea before the chronometer. *Journal of Astronomical History and Heritage, 23*, 495–522.

de Grijs, R. (2023). Astronomical tent observatories: Relics of a bygone era. *Journal of Astronomical History and Heritage, 26*, 159–178.

Gooch, W. (1791). Letter to his parents, undated. Ms. Mm. 6.48, folios 18v–19r. Cambridge University Library.

Green, C., & Cook, J. (1771). Observations made, by appointment of the Royal Society, at King George's Island in the South Seas. *Philosophical Transactions of the Royal Society, 61*, 397–421.

Greenhill, B. (1970). *James Cook, the opening of the Pacific*. London, Her Majesty's Stationary Office.

Henry, D., & Marra, J. (Eds.) (1775) *Journal of the* Resolution'*s voyage, in 1772, 1773, 1774, and 1775, on discovery to the southern hemisphere, by which the non-existence of an undiscovered continent, between the equator and the 50th degree of southern latitude, is demonstratively proved. Also a journal of the* Adventure'*s voyage, in the years 1772, 1773, and 1774. With an account of the separation of the two ships, and the most remarkable incidents that befel [sic] each. Interspersed with historical and geographical descriptions of the islands and countries discovered in the course of their respective voyages.* London, F. Newbery. Oxford Text Archive. Retrieved from http://hdl.handle.net/20.500.12024/K111410.000.

Howse, D. (1979). The principal scientific instruments taken on Captain Cook's voyages of exploration, 1768–80. *Mariner's Mirror, 65*, 119–135.

Inman, J. (1802). *An Account of Expences* [sic] *during a Voyage from England to New South Wales and back by the Way of China in 1802, 1803, & 1804.* Acts of Parliament and Awards, Cambridge University Library. RGO 14/1, folio 191.

Kenneth, M. (2017). Finding longitude: The *Investigator* expedition, 1801–1803. *International Journal of Maritime History, 29*, 771–787.

Maskelyne, N., (1791). Letter to Joseph Banks, dated 24 February 1791. *Papers of Nevil Maskelyne* (REG09/000037). P. 5:153. Cambridge University Library.

MacDonald, F., & Withers, C. W. J. (2016). *Geography, technology and instruments of exploration.* London, Routledge.

Maskelyne, N. (1801). *Instructions for John Crosley to go on a voyage to New Holland on board HMS* 'Investigator' *on a scientific expedition*, 7 March 1801. RGO 4/185, folios 4v–4a(r). Cambridge University Library.

Morrison, D., & Barko, I. (2009). Dagelet and Dawes: Their meeting, their instruments and the first scientific experiments on Australian soil. *Historical Records of Australian Science, 20*, 1–40.

Orchiston, W. (1997). Early astronomy in New Zealand: The South Sea Voyages of James Cook. Wellington, Carter Observatory. *Information Sheet, 10.*

Orchiston, W. (2004). James Cook's 1769 transit of Venus expedition to Tahiti. In D. W. Kurtz (Ed.). Transits of Venus: New Views of the Solar System and Galaxy. *Proceedings of the International Astronomical Union Colloquium* (Vol. 196, pp. 52–66).

Orchiston, W. (2016). *Exploring the history of New Zealand astronomy. Trials, tribulations, telescopes and transits.* Cham, Springer.

Orchiston, W. (2017). Cook, Green, Maskelyne and the 1769 transit of Venus: The legacy of the Tahitian observations. *Journal of Astronomical History and Heritage, 20*, 35–68.

Orchiston, W., & Howse, D. (1998). From transit of Venus to teaching navigation: the work of William Wales. *Astronomy & Geophysics, 39*, 6.21–6.24.

Orchiston, W., & Wells, W. (2020). Cook's Third Voyage to the Pacific and scientific astronomy on the northwestern coast of America: The sojourn at Nootka Sound in April 1778. *Journal of Astronomical History and Heritage, 23*, 174–208.

Turnbull, D. (2004). (En)-Countering knowledge traditions. The story of Cook and Tupaia. In T. Ballantyne (Ed.), *Science, empire and the European exploration of the Pacific* (pp. 55–76). Routledge.

Wales, W., & Bayly, W. (1777). *The original astronomical observations, made in the course of a voyage towards the South Pole, and round the world, in his Majesty's ships the* Resolution *and* Adventure, *in the years MDCCLXXII, –III, IV, and V*. London, W. and A. Strahan.

# Chapter 8
# The Unexpected Appearance of Dawes' Observatory on the '1808 Sydney Cove Map'

## 8.1 Preamble: Sydney's First Observatory

At the end of his three-year term, in December 1791, Dawes left the fledgling colony, returning to England on H.M.S. *Gorgon*. Meticulous to a fault, Dawes had carefully planned for his observatory's future. As early as 10 November 1788, he advised his patron, Dr. Nevil Maskelyne, that he hoped to find a suitable successor before his term in the colony came to an end. He even planned for the eventuality that he would fall seriously ill or die, in which case his close friend and the colony's chronicler, Watkin Tench, was on standby.

However, Dawes' back-up plan was never activated. The observatory building fell into disrepair, and it eventually collapsed. Although detailed European maps and charts of Sydney Cove continued to include Dawes' observatory until at least 1798 (Chap. 2), the last contemporary record of the observatory's physical appearance is found in the 28 August 1795 diary entry of John Crosley, astronomer on board H.M.S. *Providence*:

> I went on shore and examined the place where Mr. Dawes' observatory was built but found nothing standing but the uprights which supported the roof and the pillar on which he placed his quadrants. (cited by James, 2012)

## 8.2 The 1808 'Sydney Cove Map'

In this context, one would expect that by 1808—thirteen years after Crosley's diary entry cited above—nothing of any note would have been left to remind one of Dawes' Observatory. Yet, the '1808 Sydney Cove Map' includes a cluster of buildings that are clearly identifiable as Dawes' Observatory. The map, spanning a diameter of

This chapter is based on de Grijs (2022).

 
R. de Grijs and A. Jacob, *William Dawes*, Springer Biographies,
https://doi.org/10.1007/978-3-031-38774-6_8

3.6 m, is located in present-day First Fleet Park to the west of Circular Quay (see Fig. 8.1), the bustling ferry terminal adjacent to Sydney's central business district on the shore of Sydney Cove. The Sydney Cove Map is explicitly meant to represent the topography of the growing settlement at the time of Governor William Bligh's (1754–1817) administration in 1808. Governor Bligh was the fourth governor of New South Wales (1806–1808).

This apparent anachronism piqued our interest, and so we decided to explore the map's provenance and the reasons as to why Dawes' Observatory might have been included. The design and construction of the Sydney Cove Map were commissioned by the New South Wales Department of Public Works, as part of its West Circular Quay special project (Atkinson, 1994), and by Caltex Oil (Australia) Pty. Ltd. (presently trading as Ampol Ltd.). It was meant as a gift from Caltex to the people of Sydney in commemoration of Australia's Bicentenary commemorations of 1988. The artwork was meant to mark the original entry, at high tide, of the Tank Stream into Sydney Cove (Atkinson, 2004). The Tank Stream, Sydney Cove's fresh water supply, occupies a highly symbolic place in the context of the growing town

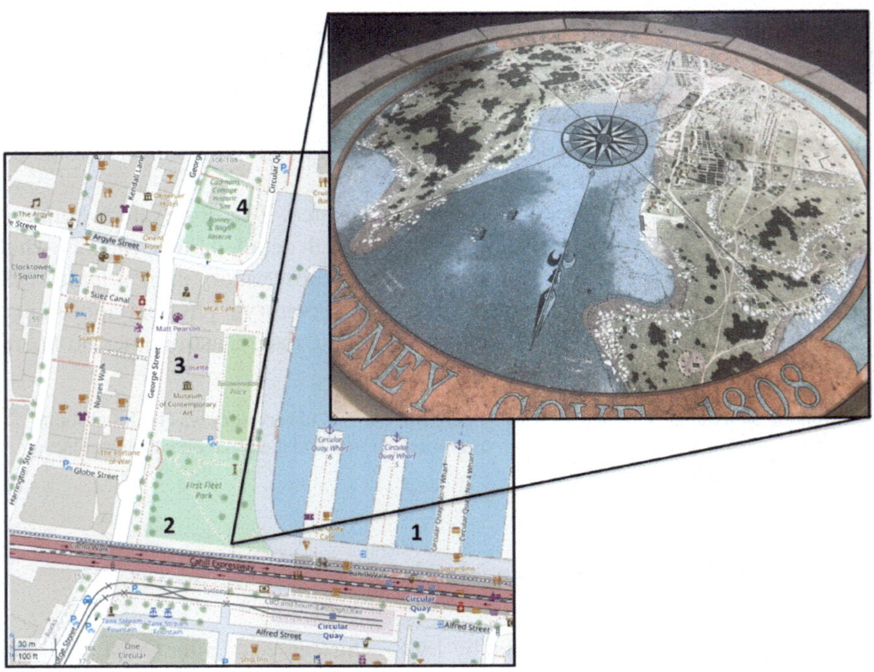


**Fig. 8.1** Location of the Sydney Cove Map (inset top right; viewed from the South) in First Fleet Park (2) to the West of the ferry wharfs of Circular Quay (1), adjacent to Sydney's central business district. (3) Previously the Maritime Services Building (presently the Museum of Contemporary Art). (4) Cadman's Cottage. The red highway at the bottom of the map segment is the Cahill Expressway, which is located one level above the railway line passing through Circular Quay railway station. North is at the top of the map. (Map © OpenStreetMaps, CC BY-SA 2.0. Photo: Author supplied)

(for an accessible history and context of the Tank Stream, see Sydney Water, 1994). It witnessed the construction of the first British buildings and, as the settlement grew, it gradually turned from a source of drinking water into an open sewer. It has long been buried deep underground, in a brick tunnel, although the water continues to flow below the city's foundations.

At the time of the Sydney Cove Map's construction, the Tank Stream's outlet into Sydney Cove was deemed a most appropriate location, given that First Fleet Park connects Circular Quay with the former Maritime Services Board building (the present-day Museum of Contemporary Art) and Cadman's Cottage, the site of Sydney's oldest house (see Fig. 8.1). At the time of the Bicentenary commemorations, a bust of Governor Arthur Phillip overlooked the park. That statue was relocated in 2014 to First Government House Place, outside the Museum of Sydney, and unveiled on 28 August 2014 to commemorate the bicentenary of Phillip's death. In addition, in 1988 a life-sized statue of Governor Bligh had also just gone on display near Cadman's Cottage (Irving, 1988b). Bligh's statue remains on the same site today.

Historical research (see Sect. 8.3) was done by architecture students at the University of New South Wales (Sydney) under the supervision of the architect and historian Robert (Bob) C. L. Irving (b. 1926; for his background, see Whitaker, 2001; updated by the Royal Australian Historical Society, 2017). In 1987, while undertaking research for the map's design, Irving—an authority on Sydney's development prior to 1810 (City of Sydney, 1988; *Southern Courier*, 1988)—was President of the Royal Australian Historical Society, which acted as consultant for the project (Irving, 1988b) and whose name appears at the base of the map.

Conversion from the initial pencil sketch provided by Irving's team to the realisation of the actual map (for a detailed view of the area around Dawes' Observatory, see Fig. 8.2) was coordinated by project architect and construction manager Beverley Atkinson of Conybeare Morrison & Partners (Conybeare Morrison & Partners, 1988). Construction of the round 'terrazzo'[1] and brass map, surrounded by Kent Street sandstone seating, took approximately three months (Atkinson, 1994, 2004). The terrazzo surface was prepared by craftsmen at Terazzo[2] and Co. Pty. Ltd. under the direction of Remo Raffin (*Southern Courier*, 1988).

The Sydney Cove Map was officially unveiled by Air Marshal Sir James Rowland, Governor of New South Wales and Patron of the Royal Australian Historical Society, on Sunday 15 May 1988 at 11:15 a.m. (Caltex, 1989; *Sydney Morning Herald*, 1988). Its management passed from the City of Sydney to the Sydney Harbour Foreshore Authority. In 2015, the Authority's role was consolidated with other State Government property and precinct management tasks into Property NSW, presently the Department of Planning, Industry and Environment.

Following its public unveiling, the map quickly became a favourite destination among tourists, educators and the general population of Sydney alike (Atkinson,

[1] Terrazzo is a special form of fine concrete made with marble aggregate.

[2] Note that the company's name only includes a single -r- (it was misspelt by *Southern Courier*, 1988; Artlab Australia, 1994), whereas the special surface is referred to using -rr- (see, e.g., Terazzo and Co. Pty. Ltd., n.d.).

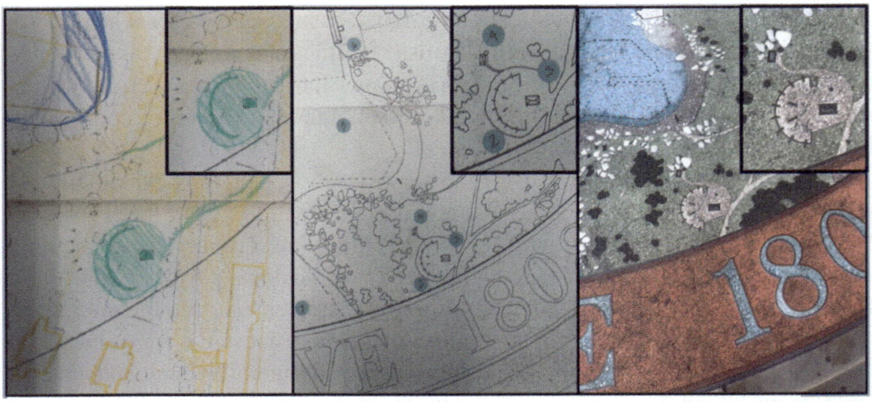

**Fig. 8.2** Pencil sketch (left), detailed architectural drawing (middle, including building labels; see the text) and final rendition (right) of a section of the Sydney Cove Map showing the locations of Dawes' Battery and Dawes' Observatory (see the insets for zoomed-in regions). (Photos: Author supplied. Left and middle panels © Conybeare Morrison International Pty. Ltd.; reproduced with permission)

1994, 2004; Irving, 1988b; *Sydney Morning Herald*, 1988). It was voted as 'Best Building of the Year' in 1990, along with other structures (Atkinson, 1996), published in the *Sydney Morning Herald* (Atkinson, 1994). However, not everyone appeared impressed by the A$140,000 price tag (although sponsored in full by Caltex): "For that sort of money, I'd expect a bit more. I suppose it's reasonable, but the colours don't really catch the eye" (Mr. Colin Griffin, quoted by the *Sydney Morning Herald*, 1988).

Although plans were afoot more than a decade ago to relocate the map (Sydney Harbour Foreshore Authority, 2011: 5), the Sydney Cove Map remains a centrepiece in First Fleet Park. Given its location almost directly under a railway line and the Cahill Expressway, noise and air pollution, as well as skateboard antics, have affected the map's appearance and longevity (Atkinson, 1994). An appraisal of the map's condition in early 1993 (see, e.g., International Conservation Services, 1993; City of Sydney Archives & History Resources, 1996) prompted the urgent planning of major maintenance and repairs (Artlab Australia, 1994, 1996; Atkinson, 1994).

## 8.3 Detailed Map Analysis

### *8.3.1 Map Inscriptions*

The Sydney Cove Map contains two main inscriptions on its top surface: "SYDNEY COVE 1808" and "CIRCULAR QUAY 1988". Along the upper ridge, two quotes from those who sailed on the First Fleet have been engraved:

> Sydney Cove is one of the smallest in the Harbour but the most convenient as ships with the greatest burden can with ease go into it and heave out close to shore. John White, Surgeon.

> We got into Port Jackson early in the afternoon, and had the satisfaction of finding the finest harbour in the world, in which a thousand sail of the line[3] may ride in the most perfect security. Captain Arthur Phillip.

On the lower ridge, information about the provenance of the historical sources is engraved. Most importantly for the present narrative, under '1808' we read,

> Sydney Cove is shown here as it was seen by the artists Thomas Watlin[g] in 1799 and 1800 and [George] W[illiam] Evans in 1803, and the Surveyors Augustus Alt in 1788, Charles Grimes in 1800 and James Meehan in 1807.

This inscription suggests that the eye-catching label '1808' should rather be interpreted such that the map's topographical features encompass a time period of some 20 years up to 1808. The latter year was selected by the Sydney Cove Map's designers as benchmark for Sydney's early development, given that in 1807 Governor Bligh commissioned James (Jamie) Meehan (1774–1826) to undertake a major survey of the settlement—indeed, the survey referred to on the lower ridge of the Sydney Cove Map. In 1808, Bligh was deposed by the New South Wales Corps in a *coup d'état* now commonly referred to as the Rum Rebellion.

## *8.3.2 Source Material*

Irving and his team relied on a variety of historical sources, including official surveys and plans (whose land grants and leases he called "... accurate but lifeless ..."; Irving, 1988a: 1), amateur maps and paintings. They verified all sources using contemporary documents, and they analysed the views represented by more than one hundred artists, which gave

> ... the most rewarding impression of daily life in the small settlement, ships and boats, tracks and paths, the stocks and pillory, backyards and washing on the line, people gathering firewood, soldiers parading, animals working and grazing in the town, aboriginal inhabitants and their activities. (Irving, 1988b)

> All of the amateur maps were also studied: because they are less formal thay [*sic*] are often more 'human'. For instance, some of them show not only sites and buildings but also streets, fences and gardens, and even outbuildings such as kitchens and, occasionally, privies. Later maps, for instance those compiled during [Governor Lachlan] Macquarie's time [1810–1821], were also studied because they show things that had survived from earlier periods. Modern surveys were used to check the parts of the old town that were still there, though not much of the 1808 town can still be seen ... (Irving, 1988a)

"You look on that map and you'll see a huge great fig tree that was left in the corner of Government Garden," said Berverley Atkinson (quoted by the *Sydney Morning*

[3] A sail (or ship) of the line was a type of naval warship during the Age of Sail, designed for the naval tactic known as the 'line of battle'.

*Herald*, 1988). Irving (1988b) explains that the map shows "… streets and allotments, buildings, trees and fences, all represented in symbolic colours". The map's design used three different colours for different varieties of trees, five for various types of gardens, as well as a range of light and shade tones (*Southern Courier*, 1988).

> Early buildings, houses (including convict houses and Bennelong's[4] house), planted fields and ships being built at that time in the docks, are clearly marked out on this new map of Sydney Cove. The tanks cut into the sandstone after which the Tank Stream was named are also to be seen on the edge of the map.
>
> In the map, the texture of the west is very different from the texture of the east: in the east the government and establishment built their substantial residences, whilst in the west the military and convicts existed in their wattle and daub houses. If you look closely you will see how the indigenous bushland at this time was being invaded by European species: this map shows the imposition of European settlement with its bright green exotic vegetation, into the grey green of the virgin eucalypt stands. A prominent feature of the map are the four windmills, … (Connybeare Morrison & Partners, 1988)

### *8.3.3 A Confusing Timeline*

Nevertheless, despite the extensive historical research undertaken, various sources assign different dates to the Sydney Cove Map's representation. Even among the map's designers, dates and date ranges differ between publications. As a case in point, Atkinson's publications include the following phrases (our emphasis):

1. Sydney Cove Map: **1808**. (Coneybeare Morrison & Partners, 1988).
2. …, it collects features of Sydney evident **between 1788 and 1808** according to charts, paintings and engravings, … (Artlab Australia, 1994; Atkinson, 1994).
3. The aim was to reconstruct the ground plan of the Sydney Cove convict settlement as it was in **1800**, using paintings and surveys of the time. (Atkinson, 2004).

Irving invariably refers to the map's date or date range as follows (our emphasis):

1. The map is a type of topographical portrait of Sydney **at the time of the Rum Rebellion**. It shows the Sydney that was familiar to Governor Bligh. (Irving, 1988a).
2. The map shows Sydney Cove as it was in **1808**. …, the map is like a bird's-eye view of the town **at the time of the Rum Rebellion**, … (Irving, 1988b).

Other publications, both official leaflets and newspapers, also differ in their interpretation of the map's temporal representation (our emphasis):

1. … shows the settlement at Sydney Cove exactly as it was **between 1799 and 1808** – every tree, every roof-top, every pathway and pub. (City of Sydney, 1988).

[4] Woollarawarre Bennelong (ca. 1764–1813) was an Aboriginal elder of the Wangal people in the Port Jackson area at the time of the arrival of the First Fleet. He served as interlocutor between the newly arrived colonists and the Indigenous population, including during a visit to the UK (1792–1795).

2. … and shows the settlement of Sydney Cove **between 1799 and 1808**. (*Sydney Morning Herald*, 1988).
3. The main information came from an official survey of the colony, commissioned by Governor Bligh in **1807**. (*Southern Courier*, 1988).
4. In intricate detail, it depicts Sydney Cove as it was in **1808** (the year our first Government House was built),[5] … (Caltex, 1989).
5. …, the Map … illustrates the settlement at Sydney Cove precisely as it developed **between 1799 and 1808** – every tree, every rooftop, every pathway and pub as drawn by foremost historical authorities. (Terazzo and Co. Pty. Ltd., n.d.).

Perhaps Jackson (2018: 25) summarised the map's convoluted representation best: "This map seems to reinterpret several pre-1810 maps of Sydney Cove." However, since her conclusion is hidden in a figure caption at the end of a lengthy document, it has not seen much attention or wide adoption.

### *8.3.4 Dawes' Observatory*

Interestingly, the same figure caption (Jackson, 2018: 25) also refers to the inclusion of Dawes' Observatory on the Sydney Cove Map: "The first observatory site is marked (perhaps inaccurately) just west of what is now the Park Hyatt Hotel." As we will see below, the location indicated is actually as accurate as possible based on extant historical sources. Irving himself spares few words about Dawes' Observatory. The only reference to the building is found in his informal notes (Irving, 1988a: 2; his emphasis): "Sydney's first **observatory** was built by Lieutenant William Dawes on the point which now bears his name." Meanwhile, however, the relevant suburbs have been renamed. At the present time, the headland to the west of Sydney Cove includes the Sydney suburbs of The Rocks on the eastern side and Dawes Point on the western side of the promontory's ridge. The area of historical interest is no longer contained by the current boundary of the Dawes Point suburb, which was formally established in 1993 (Jackson, 2018).

Yet, despite this cursory reference to Dawes' Observatory, Irving and his team appear to have been aware of the fact that the location of Dawes' Observatory pertained to 1788. The cache of documents deposited by Beverley Atkinson at the State Library of New South Wales includes large-scale architectural maps and drawings, including a coloured sketch map of Sydney Cove (see Fig. 8.3) and a final blueprint of the eventual map that was to be realised (see Fig. 8.4). The latter also includes a version with features of interest labelled, which is accompanied by a "*LIST OF LABELS on the segments of the SYDNEY COVE MAP*". This latter list includes, for the "NW—N" segment under label No. 4, "Dawes' Observatory, 1788"—indeed, a specific reference to the year in which the Observatory was established.

[5] This is an odd statement, given that Governor Arthur Phillip already laid the foundations of the first Government House in 1788; for an official account, see https://www.governor.nsw.gov.au/government-house/.

**Fig. 8.3** Pencil sketch map providing the main historical context for the eventual realisation of the Sydney Cove Map of 1808. (Photo: Author supplied. © Conybeare Morrison International Pty. Ltd.; reproduced with permission)

**Fig. 8.4** Detailed architectural drawing (left) and final rendition of the Sydney Cove Map (right). (Photos: Author supplied. Left panel © Conybeare Morrison International Pty. Ltd.; reproduced with permission)

We note that whereas the coloured sketch map includes a building on the approximate site of Dawes' Observatory, it does not include road access. However, on the final blueprint, a curved road or track has appeared from Dawes' Battery to the observatory. Yet, none of the contemporary maps we investigated to identify the most likely location of Dawes' Observatory (Chap. 2) traced a road or pathway between Dawes' Battery and Dawes' Observatory; on those maps where roads or tracks were indicated, they led directly to Dawes' Observatory from elsewhere along the length the headland. As such, we can only conclude that some artistic license has been applied.

The inscription on the lower ridge of the Sydney Cove Map implies that the design team based part of their map design on a survey undertaken in 1788 by Augustus Theodore Alt, the fledgling colony's official Surveyor-General. However, as we implied in Chaps. 2 and 3, while that survey may have carried Alt's name, it is possible that Dawes undertook the actual work.

In any case, none of Alt's survey maps are known to have included Dawes' Observatory. The only accurate contemporary maps featuring the observatory were produced by Dawes, Captain John Hunter of H.M.S. *Sirius*, or both (Chap. 2).

In Chap. 2 (see Fig. 2.17) we compared the locations of Dawes' Observatory implied by the contemporary maps of the Sydney Cove area we had uncovered. A comparison of the architectural blueprint of the 1808 Sydney Cove Map strongly suggests that the location adopted here was taken from the map labelled "Sydney Cove, Port Jackson, surveyed by Captain Hunter, 1 March 1788" (Fig. 2.5). That map does not include any roads, but it clearly indicates the location and appearance of the planned observatory buildings. All other contemporary maps suggest locations that are slightly further east with respect to the location implied by the 1808 Sydney Cove Map.

We thus conclude that the '1808 Sydney Cove Map' is likely a representation of the topography of the early settlement at Sydney Cove spanning an extended period up to 1808. Although most of the topographical features were taken from the 1807 survey by James Meehan, the reliance of the Map's designers on a range of historical sources spanning a period of 20 years has inadvertently introduced an anachronism by including Dawes' Observatory. The latter building had been established in mid-1788 (Chap. 2), but it fell into disrepair in the mid-1790s, as we know from contemporary documents (James, 2012; Chap. 2).

It is possible that Irving and his team were unaware of the eventual fate of Dawes' Observatory, given that detailed European maps and charts of Sydney Cove continued to include Dawes' observatory until at least 1798, whereas easy access to contemporary resources has only become possible thanks to the widespread digitisation efforts undertaken in recent years. Attempts to contact the historian have remained unanswered, which is probably not a surprise given his advanced age.

## References

Atkinson, B. (1986–2004). *Photographs and papers mainly relating to the design, construction and restoration of the Sydney Cove Map, First Fleet Park, Sydney, 1986–2004*. Call number: PXE 928; reference code: 440376. State Library of New South Wales. https://search.sl.nsw.gov.au/permalink/f/1ocrdrt/ADLIB110324669.

Atkinson, B. (1994). The Sydney Cove Map 1808–1988. *Magazine of the Royal Australian Historical Society, 3*.

Atkinson, B. (1996). *The Sydney Cove Map restoration 1996*. Available as part of Atkinson (1986–2004).

Atkinson, B. (2004). The Sydney Cove Map. *Australiana*, 23–24.

Artlab Australia. (1994). *Sydney Cove Map report for Sydney City Council*, p 67. Unique ID: LIB-00013882. [Accessed at the City of Sydney Archives, 9 Feb 2022].

Caltex. (1989). The most impressive arrival at Circular Quay since the Q.E. II. [Complete with its own brass fittings.] (Advertisement). *Australian Natural History, 22*, 537.

City of Sydney Archives and History Resources. (1996). *Sydney Cove Map*. Source system ID: 091\091865. Box: CSA114208. https://archives.cityofsydney.nsw.gov.au/nodes/view/599254 [Accessed 11 Feb 2022].

City of Sydney. (1988). *Sydney Cove 1808 Circular Quay 1988*. Official programme on the occasion of the unveiling of the Sydney Cove Map (2 versions). Available as part of Atkinson (1986–2004).

Conybeare Morrison & Partners. (1988). *Sydney Cove Map: 1808. Architects' Statement*. Available as part of Atkinson (1986–2004).

Dawes, W. (1788). Letter to Nevil Maskelyne, dated 17 November 1788. RGO 14/48, folios 290–292.

de Grijs, R. (2022). The unexpected appearance of Dawes' Observatory on the "1808 Sydney Cove Map." *Journal of Astronomical History and Heritage, 25*, 83–90.

International Conservation Services. (1993). *Conservation survey of the Sydney City Council Open Museum*. Condition Report 27.

Irving, R. (1988a). *Sydney Cove Map. Some information derived from historical material, which might be of use in compiling speeches*. p. 4. Available as part of Atkinson (1986–2004).

Irving, R. (1988b). Old Sydney in a new park. *Newsletter of the Royal Australian Historical Society, 5*.

Jackson, D. (2018). *Dawes Point: Geographical review*. http://walshbayhistory.net/wp-content/uploads/2018/09/dawes-point-geography-review-19-feb-2018.pdf [Accessed 11 Feb 2022].

James, A. (2012). The William Dawes Observatory. *Southern astronomers and Australian astronomy*. http://www.southastrodel.com/Page031b.htm [Accessed 11 Feb 2022].

Royal Australian Historical Society. (2017). Annual report 2016. Fellows. *History*, 23.

*Southern Courier*. (1988). *Sydney Cove Map etched in bronze*. Available as part of Atkinson (1986–2004).

Sydney Harbour Foreshore Authority. (2011). *First Fleet Park revitalization*. Community consultation. http://www.shfa.nsw.gov.au/content/wcq/CommunityConsultationBoards.pdf [Accessed 11 Feb 2022].

*Sydney Morning Herald*. (1988). *Map monument shows way we were*. p. 7. Available as part of Atkinson (1986–2004).

Sydney Water. (1994). *The Tank Stream*. SW94 09/10. https://www.sydneywater.com.au/content/dam/sydneywater/documents/tank-streamheritage-fact-sheet.pdf [Accessed 20 Sep 2022].

Terazzo and Co. Pty. Ltd. (n.d.) *A guide to the magic of terrazzo*. Available as part of Atkinson (1986–2004).

Whitaker, A.-M. (2001). Biographical notes on the Fellows of the RAHS. Irving, Robert Charles Lewis. *Journal of the Royal Australian Historical Society*, 87, 61–87. https://www.thefreelibrary.com/Biographical+notes+on+the+Fellows+of+the+RAHS.-a075247073 [Accessed 11 Feb 2022].

# Glossary

**Alt, August(us) Theodore Harman (1731–1815)** First New South Wales Surveyor-General; sailed on board the *Prince of Wales* as part of the First Fleet.

**Banks, Joseph (1743–1820)** Botanist, later President of the Royal Society of London (1778–1820); patron of **James Cook**'s first voyage to the Pacific on H.M.B. *Endeavour* (1768–1771).

**Bayly, William (1737–1810)** Astronomer on **James Cook**'s second and third voyages to the Pacific (1772–1775 and 1776–1779, respectively); headmaster of the Royal Naval Academy at Portsmouth (1785–1807).

**Bickersteth, Edward (1787–1850) (Reverend)** English evangelical deacon and priest; Secretary of the Church Missionary Society (1824–1831).

**Bligh, William (1754–1817)** British naval officer infamous for his role in the mutiny on H.M.S. *Bounty* (1789); fourth Governor of New South Wales (1806–1808).

**Bowes-Smyth, Arthur (1750–1790)** Surgeon-General on the First Fleet (1787–1790); sailed on the *Lady Penrhyn.*

**Bradfield, John Job Crew (1867–1943)** Chief Engineer of the Sydney Harbour Bridge.

**Bradley, James (1692–1762)** Britain's third Astronomer Royal (1742–1762); Savillian Professor of Astronomy at Oxford University (1721–1762).

**Bradley, William (1758–1833)** British naval officer and cartographer; sailed on board H.M.S. *Sirius* as part of the First Fleet.

**Campbell, John (ca. 1758–1795)** Commanding officer of the New South Wales detachment of Marines (March 1790–December 1791).

**Cavendish, Henry (1731–1810)** English natural philosopher and scientist renowned for his discovery of hydrogen and for his determination of the gravitational constant, *G*, in 1798.

**Clarkson, John (1764–1828)** Abolitionist; Governor of Sierra Leone (August–December 1792); younger brother of **Thomas Clarkson**.

**Clarkson, Thomas (1760–1846)** Abolitionist; founding Director of the Sierra Leone Company; older brother of **John Clarkson**.



R. de Grijs and A. Jacob, *William Dawes*, Springer Biographies,
https://doi.org/10.1007/978-3-031-38774-6

**Coleridge, William Hart (Reverend) (1789–1849)** Bishop of Barbados (1824–1842).

**Collins, David (1756–1810)** British Marine officer, Judge Advocate and chronicler of the early colony of New South Wales; sailed on board H.M.S. *Sirius* as part of the First Fleet.

**Columbine, Edward Henry (1763–1811)** Governor of Sierra Leone (February 1810–May 1811).

**Cook, James (1728–1779)** British naval officer, explorer and cartographer; famous for his three voyages to the Pacific, New Zealand and Australia between 1768 and 1779.

**Crosley, John (1762–1817)** English astronomer appointed by the British Board of Longitude to serve on H.M.S. *Providence* during part of the 'Vancouver expedition' (1793–1795).

**D'Urban née Wilcocks, Anna (d. 1843)** Wife of Sir **Benjamin D'Urban**.

**D'Urban, Benjamin (1777–1849)** British general and colonial administrator; Governor of Antigua (1820–1824).

**Dagelet, Joseph Lepaute (1751–1788)** French astronomer who accompanied the ill-fated Lapérouse expedition (1785–1788).

**Dawes, Benjamin (1737–1813)** Father of William Dawes; clerk-of-works at the Ordnance Office of the Portsmouth Naval dockyard.

**Dawes née Sinnatt, Elizabeth (b. before 1740)** Mother of William Dawes.

**Dawes, William Rutter (1799–1868)** Third son of William Dawes and **Judith Rutter**; astronomer, medical doctor and pastor.

**de Lalande, Joseph Jérôme Lefrançois (1732–1807)** French astronomer; renowned for this tables of planetary positions.

**DuBois, Isaac (b. ca. 1764)** Second husband of **Anna Maria Falconbridge**; employed by the Sierra Leone Company.

**Dundas, Henry (Lord Melville) (1742–1811)** British Secretary of War (1794–1801).

**Falconbridge, Alexander (ca. 1760–1792)** Former slave ship surgeon and Abolitionist; seconded to Sierra Leone (1791–1792).

**Falconbridge, Anna Maria (Horwood) (1769–1835)** First English woman to compose a narrative account of her experiences in Africa; wife of Alexander Falconbridge.

**Flinders, Matthew (1774–1814)** English navigator and cartographer in charge of the first comprehensive circumnavigation of present-day Australia (1801–1803).

**Fowkes, Francis (fl. ca. 1788–ca. 1800)** Convict artist, sometimes tentatively identified with the 'Port Jackson Painter'.

**Furzer, James (Lieutenant) (d. 1799)** Quarter Master of H.M.S. *Sirius* during the voyage of the First Fleet (1787–1788).

**Gilbert née Hart, Anne (1768–1834)** Sister-in-law of William Dawes; co-founder (1815), with Elizabeth Thwaites née Hart (1771–1833), of the Female Refuge Society in English Harbour, Antigua.

**Gilbert, Grace (ca. 1776–1844)** Second wife of William Dawes, of mixed European and Caribbean descent; cousin of the Rev. **Nathaniel Gilbert**.

**Gilbert, Nathaniel (Reverend) (1761–1807)** First clergyman appointed in Sierra Leone, originally from Antigua; vicar of Bledlow, Buckinghamshire, England (1798–1807).

**Gordon, Robert Jacob (1743–1795)** Commander of the Dutch troops at Cape Town (1780–1795) at the time the First Fleet called into Table Bay; of Scottish descent.

**Graham, George (1673/5–1751)** English clockmaker who determined the isochronal pendulum length at 39.126 'English inches' (99.380 cm) in 1722.

**Grenville, William (Lord Grenville) (1759–1834)** British politician who rose to eventually become Prime Minister of the United Kingdom (February 1806–March 1807); relevant to William Dawes' career, Lord Grenville was Secretary of State for the Home Department, in charge of the Colonial Office (June 1789–June 1791)

**Grey, Charles (Earl) (1764–1845)** British politician who served as Prime Minister of the United Kingdom from 1830 to 1834.

**Halley, Edmond (1656–1742)** Britain's second Astronomer Royal (1720–1742); famous for having correctly predicted periodic return dates of a comet now known as 'Halley's Comet'.

**Harrison, John (1693–1776)** Clockmaker and carpenter from northern England; constructed a range of increasingly accurate maritime timepieces, of which 'H4' eventually won him a significant share of the 1714 Longitude Prize awarded by the British Parliament.

**Horne, Melville (Reverend) (ca. 1761–1841)** Cousin of William Dawes' second wife, **Grace Gilbert**; Dawes and Horne travelled together from England to Sierra Leone on the *York* in 1792.

**Howe, Richard (Lord Howe) (1726–1799)** British Admiral of the Fleet (1783–1788).

**Hunter, John (1737–1821)** Second captain of H.M.S. *Sirius* on its voyage as flagship of the First Fleet; second Governor of New South Wales (1795–1800).

**Jenkinson, Robert (Lord Liverpool) (1770–1828)** British politician; Secretary of State for War and the Colonies (1809–1812).

**Johnson, Richard (Reverend) (ca. 1753–1827)** First clergyman seconded to the New South Wales colony; sailed on the *Golden Grove* as part of the First Fleet.

**Jones, Judith née Dawes (1795–1826)** Oldest daughter of William Dawes and **Judith Rutter**.

**Kendall, Larcum (1719–1790)** English watchmaker; constructed the first exact copy of **John Harrison**'s award-winning longitude clock H4, a maritime timepiece now known as K1 (1769).

**King, John (1759–1830)** English official; Permanent Under-Secretary of State at the Home Office (1791–1806).

**King, Philip Gidley (1758–1808)** British naval officer and colonial administrator; sailed on board H.M.S. *Sirius* as part of the First Fleet; third Governor of New South Wales (1800–1806).

**Lapérouse (Jean-François de Galaup, comte de Lapérouse) (1741–1788)** French naval officer and explorer; best known as leader of the ill-fated circumnavigation of the world named after him (1785–1788).

**Lowes, John (?) (b. between 1752 and 1761?)** Surgeon's mate on H.M.S. *Sirius* during the voyage of the First Fleet; later associated with the Sierra Leone Company.

**Ludlam, Thomas (1775–1810)** Colonial administrator; three-time Governor of Sierra Leone (May 1799–1800; August 1803–January 1805; 1806–July 1808).

**Macarthur, Elizabeth (1766–1850)** Wife of John Macarthur (1767–1834), Lieutenant on the Second Fleet.

**Macaulay, Zachary (1768–1838)** Abolitionist; intermittently Governor of Sierra Leone between 1794 and 1799.

**Maskelyne, Nevil (1732–1811)** Britain's fifth Astronomer Royal (1765–1811); patron of William Dawes during his time on the First Fleet and in New South Wales.

**McIntire (McEntire, McIntyre), John (ca. 1754–1788)** Governor **Arthur Phillip**'s convict 'gamekeeper', speared to death on 9 December 1790 by an Aboriginal known as Pemulwuy (ca. 1750–1802).

**Meehan, James (1774–1826)** Surveyor-General commissioned in 1807 by Governor **William Bligh** to undertake a comprehensive survey of the settlement of Sydney.

**Miller, Andrew (ca. 1759–1790)** Commissary of Stores in the newly established colony of New South Wales; sailed on board H.M.S. *Sirius* as part of the First Fleet.

**Mills, Selena (Selina) Anne (1767–1831)** Wife of **Zachary Macaulay**.

**Newton, Isaac (1642–1726/7)** English natural philosopher; key figure during the Enlightenment and founder of the field of classical mechanics.

**Newton, John (Reverend) (1725–1807)** Anglican clergyman and Abolitionist; former slave ship captain and investor in the slave trade.

**Parry, Thomas (1794–1870)** Archdeacon of Antigua (from 1825); subsequently second Bishop of Barbados (1842–1869).

**Phillip, Arthur (1738–1814)** Commander of the First Fleet (1787–1788); first Governor of New South Wales (1788–1792).

**Porteus, Beilby (1731–1809)** Anglican clergyman; Bishop of London (1787–1809).

**Richer, Jean (ca. 1630–1696)** French astronomer; assistant at the French Academy of Sciences.

**Ross, Robert (Major) (ca. 1740–1794)** Commanding officer of the Marines in New South Wales on detachment from H.M.S. *Sirius* upon arrival of the First Fleet in Sydney Cove.

**Rutter, Judith (before 1744–ca. 1800)** First wife of William Dawes.

**Scott, Thomas (Reverend) (1747–1821)** Anglican clergyman.

**Sharp, Granville (1735–1813)** Abolitionist; founder of the St. George's Bay Company (forerunner of the Sierra Leone Company); considered one of the founders of Sierra Leone.

**Shelton, John (1712–1777)** English clockmaker.

**Smith, Robert (Lord Carrington) (1752–1838)** British politician; slave owner and early Director of the Sierra Leone Company.

**Southwell, Daniel (ca. 1764–1797)** Master's mate on H.M.S. *Sirius* during the voyage of the First Fleet.

**Stewart, Robert (Lord Castlereagh) (1769–1822)** Anglo–Irish politician; Secretary of State for Foreign Affairs (United Kingdom) (1812–1822).

**Tench, Watkin (1758–1833)** British Marine officer detached to the First Fleet on the *Charlotte.*

**Thompson, Thomas Perronet (1783–1869)** British politician; first Governor of Sierra Leone appointed by the British Crown (1808–1810).

**Thornton, Henry (1760–1815)** English economist, banker and politician; Chairman of the Sierra Leone Company (1791–1808).

**Thorpe, Robert (1773–1836)** Dublin-born Chief Justice of Prince Edward Island (Canada; 1801–1804), politician in Upper Canada (1806–1807) and Chief Justice of Sierra Leone (1808–1815).

**Thwaites, Charles (1780–ca. 1860)** Philanthropist and co-founder (1809) of the English Harbour Sunday School, Antigua, jointly with his wife Elizabeth and **Anne Gilbert**.

**Townshend, Thomas (Lord Sydney) (1733–1800)** British politician; Home Secretary (1783–1789).

**Twiss, William (Captain) (1745–1827)** Royal Engineer in the British Army, based at the Portsmouth naval dockyard.

**Wales, William (1734?–1798)** English astronomer who sailed on **James Cook**'s second voyage to the Pacific (1772–1775); headmaster of the Royal Mathematical School of Christ's Hospital (1776–1798); Secretary of the British Board of Longitude (1795–1798).

**Watson, Brook (Sir) (1735–1807)** British politician and Lord Mayor of London (1796); Deputy Governor of the Bank of England (1798–1806)

**Watson, James Henry (1841–1934)** President of the Royal Australian Historical Society (1926–1927).

**White, John (1756–1832)** Surgeon on the *Charlotte* as part of the First Fleet; Surgeon-General of New South Wales (1788–1796).

**Wilberforce, William (1759–1833)** English politician and Abolitionist.

# Index





R. de Grijs and A. Jacob, *William Dawes*, Springer Biographies,
https://doi.org/10.1007/978-3-031-38774-6

The manufacturer's authorised representative in the EU is Springer Nature Customer Service Centre GmbH, Europaplatz 3, 69115 Heidelberg, Germany. If you have any concerns regarding our products, please contact ProductSafety@springernature.com

Printed and bound by CPI Group (UK) Ltd, Croydon, CR0 4YY
05/10/2026
02228122-0007